# Lecture Notes in Computer Science 4400

Commenced Publication in 1973
Founding and Former Series Editors:
Gerhard Goos, Juris Hartmanis, and Jan van Leeuwen

T0223161

James F. Peters   Andrzej Skowron
Victor W. Marek   Ewa Orłowska
Roman Słowiński   Wojciech Ziarko (Eds.)

# Transactions on Rough Sets VII

Commemorating the Life and Work
of Zdzisław Pawlak, Part II

 Springer

Editors-in-Chief

James F. Peters
University of Manitoba, Winnipeg, Canada
E-mail: jfpeters@ee.umanitoba.ca

Andrzej Skowron
Warsaw University, Poland
E-mail: skowron@mimuw.edu.pl


Volume Editors

Victor W. Marek
University of Kentucky, Lexington, USA
E-mail: marek@cs.uky.edu

Ewa Orłowska
National Institute of Telecommunications, Warsaw, Poland
E-mail: E.Orlowska@itl.waw.pl

Roman Słowiński
Poznan University of Technology, Poznan, Poland
E-mail: Roman.Slowinski@cs.put.poznan.pl

Wojciech Ziarko
University of Regina, Canada
E-mail: ziarko@cs.uregina.ca

Library of Congress Control Number: 2007922187

CR Subject Classification (1998): F.4.1, F.1, I.2, H.2.8, I.5.1, I.4

LNCS Sublibrary: SL 1 – Theoretical Computer Science and General Issues

ISSN        0302-9743
ISBN-10     3-540-71662-9 Springer Berlin Heidelberg New York
ISBN-13     978-3-540-71662-4 Springer Berlin Heidelberg New York

Springer is a part of Springer Science+Business Media

springer.com

© Springer-Verlag Berlin Heidelberg 2007
Printed in Germany

Typesetting: Camera-ready by author, data conversion by Scientific Publishing Services, Chennai, India
Printed on acid-free paper     SPIN: 12042814     06/3180     5 4 3 2 1 0

# Preface

Volume VII of the *Transactions on Rough Sets (TRS)* is a sequel to volume VI of the TRS. Both volumes commemorate the life and work of Zdzisław Pawlak (1926-2006)[1]. It is evident from the wide spectrum of contributions to these volumes that Zdzisław Pawlak's legacy is rich and varied. Prof. Pawlak's research contributions have had far-reaching implications inasmuch as his works have served as cornerstones in establishing new frontiers for scientific research in a number of fields.

From an early age, Zdzisław Pawlak devoted his life to scientific research. His pioneering work included research on modeling industrial processes, the design of computers, information retrieval, modeling conflict analysis and negotiation, genetic grammars and molecular computing. His research led to the introduction of knowledge representation systems during the early 1970s and the discovery of rough sets during the early 1980s. Added to that was Prof. Pawlak's lifelong interest in painting, photography and poetry. During his lifetime, he nurtured worldwide interest in approximation, approximate reasoning and rough set theory and its applications[2]. Evidence of the influence of Prof. Pawlak's work can be seen in the growth of rough-set literature that now includes over 4000 publications by more than 1600 authors in the rough set database[3] as well as the growth and maturity of the International Rough Set Society[4]. Moreover, numerous biographies of Zdzisław Pawlak have been published[5].

This volume of the TRS presents papers that reflect the profound influence of a number of research initiatives by Zdzisław Pawlak. In particular, it introduces a number of new advances in the foundations and applications of artificial intelligence, engineering, logic, mathematics, and science. These advances have significant implications in a number of research areas. In addition, it is evident from the papers included in this volume that rough set theory and its application form a very active research area worldwide. A total of 42 researchers from 13 countries are represented in this volume, namely, Australia, Canada, Germany, India, Italy, Japan, Poland, P.R. China, Sweden, Thailand, Taiwan, UK (Wales)

---

[1] Prof. Pawlak passed away on 7 April 2006.

[2] See, *e.g.*, Pawlak, Z., Skowron, A.: Rudiments of rough sets, *Information Sciences* 177 (2007) 3–27; Pawlak, Z., Skowron, A.: Rough sets: Some extensions, *Information Sciences* 177 (2007) 28–40; Pawlak, Z., Skowron, A.: Rough sets and Boolean reasoning, *Information Sciences* 177 (2007) 41–73.

[3] http://rsds.wsiz.rzeszow.pl/rsds.php

[4] http://roughsets.home.pl/www/

[5] See, *e.g.*, Peters, J.F. and Skowron, A., Zdzisław Pawlak: Life and Work. *Transactions on Rough Sets* V, LNCS 4100 (2006) 1-24. See, also, R. Słowiński, Obituary, Prof. Zdzisław Pawlak (1926-2006), *Fuzzy Sets and Systems* 157 (2006) 2419-2422.

and the USA. Evidence of the vigor, breadth and depth of research in the theory and applications of rough sets can be found in the articles in this volume.

Most of the contributions of this commemorative volume of the TRS are on an invitational basis and every paper has been refereed in the usual way. This special issue of the TRS contains 19 papers that explore a number of research streams that are either directly or indirectly related to research initiatives by Zdzisław Pawlak. These research streams are represented by papers on intelligent signal processing techniques (Andrzej Czyżewski), belief networks (Jerzy W. Grzymała-Busse, Zdzisław S. Hippe, Teresa Mroczek), relational attribute systems (Ivo Düntsch, Günther Gediga, Ewa Orłowska), dominance-based rough set approach (Salvatore Greco, Benedetto Matarazzo, Roman Słowiński), rough sets in bioinformatics (Torgeir R. Hvidsten, Jan Komorowski), selection of important attributes for medical diagnosis systems (Grzegorz Ilczuk, Alicja Wakulicz-Deja), rough clustering (Pawan Lingras), case-based reasoning classifiers (Yan Li, Simon Chi-Keung Shiu, Sankar Kumar Pal, James Nga-Kwok Liu), Web information gathering (Yuefeng Li, Ning Zhong), rough sets in pattern recognition (Sushmita Mitra, Haider Banka), possibilistic information (Michinori Nakata, Hiroshi Sakai), hybrid rough sets-population-based system (Puntip Pattaraintakorn, Nick Cercone), intelligent system for survival analysis based on hybrid rough sets (Puntip Pattaraintakorn, Nick Cercone, Kanlaya Naruedomkul), classifying remotely sensed images (B. Uma Shankar), rough feature selection (Qiang Shen), granulation in information security (Da-Wei Wang, Churn-Jung Liau, Tsan-sheng Hsu), definability and approximation (Yiyu Yao), audiovisual emotion recognition (Yong Yang, Guoyin Wang, Peijung Chen, Jian Zhou, Kun He).

The editors of this volume extend their hearty thanks to the following reviewers: Jan Bazan, Maciej Borkowski, Beata Konikowska, Bożena Kostek, Pawan Lingras, Son Nguyen, Władysław Skarbek, Marcin Szczuka, Sheela Ramanna, Dominik Ślęzak, Jerzy Stefanowski, Piotr Synak, Dimiter Vakarelov, Hui Wang, Piotr Wasilewski, Marcin Wojnarski, Jakub Wróblewski, and Yiyu Yao.

This issue of the TRS has been made possible thanks to the laudable efforts of a great many generous persons and organizations. The editors and authors of this volume also extend an expression of gratitude to Alfred Hofmann, Ursula Barth, Christine Günther and the LNCS staff at Springer for their support in making this volume of the TRS possible. In addition, the editors extend their thanks to Marcin Szczuka for his consummate skill and care in the compilation of this volume.

December 2006                                          Victor Marek
                                                      Ewa Orłowska
                                                     James F. Peters
                                                    Roman Słowiński
                                                   Andrzej Skowron
                                                   Wojciech Ziarko

# LNCS Transactions on Rough Sets

This journal subline has as its principal aim the fostering of professional exchanges between scientists and practitioners who are interested in the foundations and applications of rough sets. Topics include foundations and applications of rough sets as well as foundations and applications of hybrid methods combining rough sets with other approaches important for the development of intelligent systems.

The journal includes high-quality research articles accepted for publication on the basis of thorough peer reviews. Dissertations and monographs up to 250 pages that include new research results can also be considered as regular papers. Extended and revised versions of selected papers from conferences can also be included in regular or special issues of the journal.

# Table of Contents

## Contributed Papers

## Monographs

# Speech Coding Employing Intelligent Signal Processing Techniques

Andrzej Czyzewski

Multimedia Systems Department
Gdansk University of Technology
ul. Narutowicza 11/12, 80-952 Gdansk, Poland
ac@pg.gda.pl

**Abstract.** The concepts and experiments presented are focused on modifications of an existing parametric speech coding algorithm (CELP)
introduced in order to improve subjective speech quality in telephone
connections. The perceptual coding to bit rate limiting was added and
algorithms qualifying speech components to the categories of "voiced",
"unvoiced", "transients" using rough sets were studied. The speech signal quality achieved with the proposed hybrid codec was compared to
the quality offered by some standard speech codecs.

**Keywords:** CELP residual coding, hybrid codec architecture, perceptual speech coding, rough set decision algorithm.

## 1 Introduction

The majority of speech telecommunication systems in today's use offer a narrow-
band transmission, limited to about 200– 4000 Hz. The principal effect of band
limiting is the degradation of intelligibility of speech occurring mostly due to
the influence of upper band limiting to the perception of plosives and fricatives.
Moreover, recognizing co-talkers is impeded because of the meaning of vocal tone
band located in the range of low frequencies.

Typical applications of computer technologies in digital signal processing only
rarely consider the opportunities of data processing with the use of methods
which stem from artificial intelligence or soft computing. In the meantime the
area of DSP (Digital Signal Processing) has an extensive demand for applications of intelligent signal processing because of unrepeatability and uncertainty
of real-life signals and the lack of adequate mathematical models of signal production processes. That is why learning algorithms and data mining techniques
are important to this kind of applications.

In most of the applications related to transmission of speech signal, parametric
coding algorithms are used (CELP, ACELP, LD-CELP, etc.). These algorithms
reduce bit-rate of the signal significantly, sacrificing quality of the signal to some
degree. For many years, bit-rate and delay were the main criteria in speech codec
assessment, while subjective signal quality, expressed using the mean opinion
score (MOS) scale, was considered less important. Most of parametric speech

J.F. Peters et al. (Eds.): Transactions on Rough Sets VII, LNCS 4400, pp. 1–15, 2007.

codecs used in current applications provides signal quality from 3.2 to 4.0 in the MOS scale (where 5.0 means the best possible quality) [26].

The meaning of wideband speech is now recognized in some newer ITU-T standards. Two of them, called AMRWB and VMRWB, can be viewed as pure speech coding algorithms based on the ACELP technology. They do not provide, however, at least satisfactory quality of non-speech signals representation [19,24]. Therefore, an extended AMRWB+ codec was introduced to overcome this limitation. Unfortunately, as the AMRWB+ codec takes an advantage of hybrid ACELP/transform coding techniques, it introduces a coding delay up to 90 ms, thus in general it is not suitable for the real-time two-way communication. Accordingly, one can notice that there is still a need for A wideband, highquality, mid-delay speech codec with improved ability to encode non-speech signals.

Contrarily to coding techniques based on the speech production model, that is insufficient for more complex signals, the codec proposed in this paper employs the analysis technique for extracting sines, noise and transient parts of the signal. The analysis is supported by a soft computing algorithm. In the next step, each part of the entire signal is encoded using an adequate technique, including the perceptual criteria. It has to be mentioned that sines+residual model is widely used as a powerful tool for signal modification (e.g. pitch, time-scale) [20]. The sines+residual signal representation was also employed for efficient narrowband speech coding at about 8 kbps rate. Additionally, it was found that it is a robust method for coding both speech signals and mixed audio content [18,21,23]. Concerning this, it was also expected that extending the sines+residual model with transient selection module will further improve the signal representation accuracy. As the aim is to present the super-wideband signal to the listener, the spectrum components exceeding 7 kHz are reconstructed artificially in the proposed approach. It has to be mentioned that during some stages of encoding process the perceptual criterion was applied, allowing a reduction of the bit-rate requirements for the codec bit stream [5,22].

The main problem in the parametric approach to speech coding is how to encode transients, voiced and unvoiced signal components, efficiently. Encoding of transient states is especially important here, because an inappropriate encoding of transients may result in significantly decreased signal quality, Various parametric codecs use different approach to this problem, yet none of these approaches provide sufficiently accurate transient encoding, which is reflected in quality values (MOS). One of the concepts of the hybrid codec presented in this paper is extraction of transient, voiced and unvoiced components from the signal and using an appropriate approach for each of these groups. In the synthesis of musical instruments, the introduction of transient analysis and synthesis to the "sine and noise" model resulted in improved signal quality. Hence, it may be expected that using a similar "voiced-transient-unvoiced" approach to speech signal will provide an improvement of signal quality, as well. However, no research on this topic has been done so far by the author and his team of researchers.

The aim of the perceptual speech codec module is to improve further signal quality by incorporating the perceptual coding algorithm into the parametric codec. The calculation of the masking offset, playing a significant role in the masking threshold calculation based on the uncertainty measure can be also interpreted in terms of rough set theory [1]. That is because of a dependency occurring between the rough measure and the unpredictability measure. The noisy data processing is an evident example of making uncertain decisions, because Unpredictability Measure represents the margin of uncertainty while interpreting sound spectrum shape in terms of useful or useless components representation [14].

The novel approach to speech coding using the hybrid architecture is presented in the consecutive paragraphs of this paper. Advantages of parametric and perceptual coding methods are utilized together in order to create a speech coding algorithm assuring a better signal quality than in the traditional CELP parametric codec.

## 2  Voiced/Unvoiced Speech Selection Algorithms

Since the LP coding relies on a simple two-state model of speech production, each frame of the input signal is classified either as voiced or unvoiced. Usually, the classification is based on the observation that frames of voiced parts are strongly correlated with each other and have relatively higher energy than unvoiced parts [6]. This approach is also utilized in the engineered algorithm. However, an additional intelligent decision module is employed in order to ensure that frames classified as unvoiced will not contain transients.

The detector relies on three parameters which are calculated for every block of segmented signal $s[n_0,...,n_{N-1}]$ according to the following formulas [6,7,16]:

$$x_o = \frac{1}{2N} \sum_{n=1}^{N-1} |sgn\left(s\left[n\right]\right) - sgn\left(s\left[n-1\right]\right)| \tag{1}$$

$$x_1 = \frac{1}{N} \sum_{n=0}^{N-1} |s\left[n\right]| \tag{2}$$

$$x_2 = \max \left( \sum_{n=0}^{\frac{N}{2}-1} s\left[n\right] \cdot s\left[n + \frac{N}{2}\right] \right) \tag{3}$$

where: $s[n]$ – block of the signal, $N$ – frame length.

The frame is classified as voiced if the following expression is true:

$$w_0 + \sum_{k=1}^{M} w_k \cdot x_{k-1} > 0 \tag{4}$$

where: $w_k$ – elements of weighting vector, $M$ – number of parameters.

The $w_k$ elements of weighting vector were chosen in order to allow a proper frame classification for different speech samples. Since the detector does not take into account any information about the previous classification results, the voiced/unvoiced decision may change instantly from one frame to another when some special conditions occur. Thus, an appropriate hysteresis function was utilized previously [16] in order to prevent undesirable state changes of the detector. It has to be mentioned that not only pure-voiced frames but also frames containing transients are classified into the voiced part of the speech signal. Instead of the decision term (4) a soft computing decision algorithm was also employed to define the current frame as voiced or unvoiced. This algorithm based on the rough set approach uses an automatically reduced set of attribute values and learned rough rules. The conditional attributes are represented by: $x_0=\{0,1\}$; $x_1=\{low, medium, high\}$; $x_2=\{low, medium, high\}$. The decision attribute is binary d $=\{voiced, unvoiced\}$.

The rough set algorithm implemented earlier by R. Krolikowski was utilized [17]. The elaborated rule induction algorithm is based on the rough set methodology. Since the basic rough operators (the partition of a universe into classes of equivalence, $C$-lower approximation of $X$ and calculation of a positive region) can be performed more efficiently when objects are ordered, the algorithm often executes sorting of all objects with respect to a set of attributes. Reducing of values of attributes requires that all combinations of the conditional attributes must be analyzed. In general case, the decision table should be sorted as many times as is the number of all these combinations. Therefore, in every set of attributes $A$ ($A \subseteq C$), a subset of the conditional attributes $C$ should be exploited as many times as possible. The algorithm splits the decision table $T$ into two tables: consisting of only certain rules ($T_{CR}$) and of only uncertain rules ($T_{UR}$). For them both, there is additional information associated with every object in them. The information concerns the minimal set of indispensable attributes and the rough measure $\mu_{RS}$. The latter case is applied only for uncertain rules. The elaborated algorithm consists of the following main group of modules:

a) Master procedure of all procedures related to the rough set-based induction algorithm

– **procedure** RS_algorithm

It is assumed that a decision table is fed to the procedure either at a call or supplied during its execution. By analogy, the procedure returns a table with generated rules.

b) Initial procedures

– **procedure** preprocessing

The procedure prepares 2 tables of certain $T_{CR}$ and uncertain rules $T_{UR}$. In addition, it computes the set of concepts $V$ with respect to the set of attributes $D$.

– **procedure** generate_rules($C$ : set_of_attributes)

<u>Input:</u> $C$ - the set of the conditional attributes

The procedure is a master procedure of all those procedures and functions which task is to generate rules by removing superfluous values of attributes. At the end, these values are replaced by 'do not care' value. However, the procedure affects the tables $T_{CR}$ and $T_{UR}$ in an implicit way.

– **procedure** postprocessing

The procedure prepares the output table of generated rules $T$ in such a way that a rule could be accessed in at most $|C| \cdot \log_2 N$ comparisons, where $N$ is the number of objects in $T$.

c) Procedures preserving the proper depth of the analysis of the conditional attributes

– **procedure** P $(C, A$ : set_of_attributes)

<u>Input:</u>   $C$ - the set of the conditional attributes,
        $A$ - an arbitrary set of attributes, where $A \subseteq C$
<u>Output:</u> potentially modified auxiliary data associated with
        the tables $T_{CR}$ , $T_{UR}$
The procedure provides the proper depth of the analysis of all combinations of the conditional attribute set $C$ and is executed recursively.

– **procedure** _P $(A$ : set_of_attributes; *depth* : integer)

<u>Input:</u>   $A$ - an arbitrary set of conditional attributes,
        *depth* - depth of recursions,
<u>Output:</u> potentially modified auxiliary data associated with
        the tables $T_{CR}$ , $T_{UR}$
Similarly to the procedure P(), this procedure provides the proper order of the analysis of all combinations of the conditional attribute set $C$. It contributes to a further recurrent processing of the conditional attributes and is similar to P().

– **procedure** left $(A$ : set_of_attributes; *depth* : integer)

<u>Input:</u>   $A$ - an arbitrary set of conditional attributes,
        *depth* - depth of recursions,
<u>Output:</u> potentially modified auxiliary data associated with
        the tables $T_{CR}$ , $T_{UR}$
The procedure is strictly related to the proper depth of the analysis of the conditions and concerns recurrent processing. The execution of the procedure enables the analysis of some subsets of the attributes without sorting the table.

d) Procedures eliminating superfluous values of attributes in certain and uncertain rules

– **function** process_certain_rules ($A$ : set_of_attributes): set_of_sets_of_objects

Input:    $A$ - an arbitrary set of conditional attributes,
Output: $Z$ - set of sets of objects belonging to the same equivalence class,
          potentially modified auxiliary data associated with the table $T_{CR}$

The function performs all necessary operations in order to reduce values of attributes. In consequence, it may cause a removal of an attribute. The procedure affects only certain rules. The output value $Z = U_{CR}/IND(A)$ is returned so that the procedure **process_uncertain_rules** doesn't need to compute the partition $U_{CR}/IND(A)$ .

– **procedure** process_uncertain_rules
  ($Z$ : set_of_sets_of_objects;$A$ : set_of_attributes)

Input:    $Z$ - set equivalence classes for the table$T_{CR}$ ($U_{CR}/IND(A)$) ,
          $A$ - an arbitrary set of conditional attributes
Output: potentially modified auxiliary data associated with the table $T_{UR}$

The procedure calculates the rough measure for each object of the decision table $T_{UR}$ (uncertain rules) and for each combination of attributes. The largest values of the measure are stored.

e) Functions which are related to the basic rough set operators

– **function** U_IND ($T$ : table; $A$ : set_of_attributes) :
  set_of_sets_of_objects

Input:    $T$ - a decision table,
          $A$ - an arbitrary set of attributes,
Output: a set of objects belonging to the same equivalence class

The function parts the universe $U$ (table $T$) into classes of equivalence $U/IND(A)$ according to indiscernibility relation with respect to a set of attributes $A$.

– **function** _CX ($T$ : table; $C, X$ : set_of_attributes) :
  set_of_objects

Input:    $T$ - a decision table,
          $C, X$ - arbitrary sets of attributes,
Output: a set of objects in $T$ which belong to the $C$-lower approximation
          of the set $X$

The function computes the $C$-lower approximation of $X$.

– **function** POS_REG (T : table; A1, A2 : set_of_attributes) :
  set_of_objects

Input:    $T$ - a decision table,
          $A1$, $A2$ - arbitrary sets of attributes,
Output: a set of objects in $T$ which constitute the positive region

The function calculates the positive region of classification $U/IND(A2)$ for the set of attributes $A1$.

f) Auxiliary procedure and function

– **procedure** checkU_IND ($Y$ : set_of_sets_of_objects; $A$ : set_of_attributes)

Input:    $Y$ - a set of equivalence classes after the partitioning $U_{CR}/IND(A)$,
          $A$ - an arbitrary set of conditional attributes
Output: potentially modified auxiliary data associated with the table $T_{CR}$

The procedure checks whether all objects belonging to the partition $U_{CR}/IND(A)$
have the same decision values. If so and if the number of current dispensable attributes is less than the number of such attributes for these objects, stored so far, it stores the information of the dispensable attributes.

– **function** intersection ($X$, $Y$ : set_of_objects) : set_of_objects

Input:    $X$, $Y$ - ascending sorted sets of objects
Output: $Z$ - the intersection of $X$ and $Y$

The function calculates the intersection of the input sets $X$ and $Y$. The product is returned by the output set $Z = X \cap Y$ . The assumption is that the input sets $X$ and $Y$ are ascending sorted sets and therefore in the worst case, there are $|X| + |Y|$ comparisons necessary.

The rule discovery procedure was executed in order to acquire the knowledge from the set of 100 speech utterances produced by 10 speakers. Each utterance was edited from the speech signal using 10 ms rectangular window cut. The edited utterances were labelled manually as voiced or unvoiced by a human operator basing on his knowledge of the utterance context. The values of attributes $x_1$ and $x_2$ were normalized in order to keep them in the range $\{0,1\}$. A uniform quantization of conditional attribute values $x_1$ and $x_2$ was exploited in the experiments for the sake of simplicity. Hence the values of $x_1$ and $x_2$ were considered low when $x_k$ < $0.333...$, medium in the range $0.333... < x_k$ <$0.666..$ and high for $x_k$ > $0.666...$ . The rules learned automatically from the data set were pruned in such a way that only those with the strength higher than 0.5 were retained. The pruning operation resulted in reducing the rule base to 68 most certain rules (39 % of all rules) applied subsequently to decision making with regard to different speech utterances produced by the same speakers. The result of classification for a particular speech sample is presented in Fig. 1b revealing supremacy of the rough set decision algorithm over the previously applied decision based on term (4), the result for the second case being plotted in Fig. 1a.

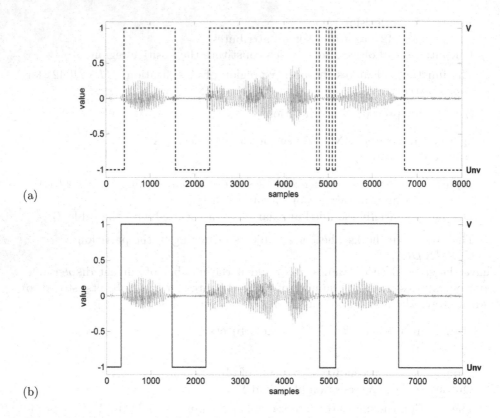

(a)

(b)

**Fig. 1.** Results of voiced (V)/unvoiced (Unv) classification: (a) criterion (4) - based decisions; (b) rough set algorithm based decisions

## 3  Detecting Transient States in Speech Signal

The proposed algorithm consists of two main stages [16]. In the first one, the rough decision is made about the voiced/unvoiced character of speech segments, and in the second one, both pure-voiced part and transients of the signal are selected. The traditional approach to the transient detection is based on the assumption that the energy of the signal increases rapidly when the transient occurs [8]. Although the energy tracking of the signal is useful for detecting the time-domain transients, it fails in the case of frequency-domain transients occurrence. If the frequency-domain transient occurs, the energy distribution changes over frequency, while total energy of the signal remains nearly constant. In order to detect that group of transients, it is necessary to analyze the energy variations in some subbands instead of tracking the energy of the entire signal [9]. The $s_{tr}[n]$ and $s_{pv}[n]$ signals represent the transient and the pure-voiced part of the input signal, respectively.

The transient selector operates as follows. In the first stage, the input signal is divided into some short segments, and for each segment the Fourier spectrum is calculated (with the use of the FFT algorithm). The part of the spectrum representing frequencies above 100 Hz is then divided into N uniform subbands, and for each subband the energy en(b) is obtained. It has been found empirically that the analysis of energy variations in eight subbands is sufficient for a transient detection in speech signals. In the next step, the value of f(b) for each subband is calculated. The f(b) function is formulated in the following way:

$$f_n(b) = \alpha \cdot en(b) + (1 - \alpha) \cdot f_{n-1}(b) \tag{5}$$

where: $\alpha$ – constant.

Further, the parameters related to transient measures are obtained according to the formula:

$$G(b) = \frac{en(b)}{f_n(b)} \tag{6}$$

Next, the value of the G(b) parameter is compared with the selected empirically $T$ threshold for each subband and the total transient measure parameter F is calculated:

$$F = \sum_{b=0}^{N-1} d(b) \tag{7}$$

where: $d(b) = \begin{cases} 1 & G(b) > T \\ 0 & G(b) \leq T \end{cases}$

In the last step, the value of F parameter is compared to the global threshold Tg:

$$F > T_g \tag{8}$$

If the term (8) is fulfilled then a particular frame is classified as a one containing transient.

It has to be mentioned that a similar algorithm is incorporated in the MPEG-4 AAC general audio encoder [10]. Since in this approach the band-limited speech signal is analyzed, an additional parameterization is necessary in order to allow a robust transients detection. An important observation is that the time-domain transients of speech signals are associated with transitions between their unvoiced and voiced parts. Thus, a one way to make the detection algorithm more efficient is to measure the zero crossing rate of the speech in parallel [9]. Because that operation can be viewed as a simple voicing detection, similar results may be yield when only the voiced part of the speech signal is fed into the transient detector [2]. Therefore, in practical experiments first the voiced/unvoiced decision has to be made, and then the transients should be detected within the voiced part of the speech signal. In order to allow rule-based decisions in the process of qualifying speech fragments as transients/non-transients a similar approach to the one described in the previous paragraph was made. The values of the function $f_n(b)$ for b={1,2,...,8} were considered as a set of conditional attribute values. The same as previously 100-element set of training examples was utilized which was labelled manually for the transient presence in all 10 ms speech utterance fragments. The rule pruning operation employing $\mu_{RS} > 0.5$

criterion resulted in reducing the rule base to 26 most certain rules applied subsequently to the decision making. The result of classification for a particular speech sample is presented in Fig. 2 allowing a comparison of results obtained using the decision term (8) and the rough rule-based classification.

It can be noticed from the Fig. 2 that the proposed algorithm is able to assign signal segments containing both: time-domain and frequency-domain transients. However, due to limited resolution of analysis, some fragments of the transients may be placed in the neighborhood of the assigned frames. Therefore, similarly as in the preceding paragraph, an additional hysteresis module has been employed in order to prevent transient segmentation during the selection process.

## 4    Psychoacoustic Coding of Residual Signal

The psychoacoustic models included in standards such as MPEG 1, AAC are based on the excitation pattern model in which the amount of masking depends on the excitation. In the sines+residual model the psychoacoustic model can also benefit from the tonal vs. non-tonal component determination. In the CELP speech coders the residual signal is obtained in the process of coding as a result of analyze-by-synthesis procedure. Contrarily to the CELP codec algorithmic solutions [3], in the proposed approach only the voiced part of the residual signal is perceptually encoded. The architecture of the proposed encoder is presented in Fig. 3. The input signal s[n] is encoded first by the CELP encoder operating at fixed bit-rate. The $s_{CELP}[n]$ signal represents the main bit stream, which is also decoded locally in order to obtain the residual signal $\Delta[n]$. In the next stage, the voiced and unvoiced parts of the residual signal are detected. It has to be noted that only the voiced part $\Delta_v[n]$ is further processed. As a sequel, the selected voiced part of residual signal is perceptually encoded and sent to the decoder in parallel to the CELP bit stream. The decoding process consists of two stages. In the first one, CELP and residual bit streams are decoded using the CELP and the perceptual decoder, respectively. Next, the resulting signals are added together in order to compose the entire decoded speech signal.

The G728 LD-CELP codec operating at 12.8 kbps rate was utilized during the experiments in a base layer of the codec. The residual signal was encoded employing the perceptual module [4,25] operating at 14 kbps average rate. Notwithstanding the codec was designed for compression of music and general audio, and is not optimized for perceptual speech coding, it seems to be useful in evaluating of the proposed codec concept [11].

The experiments concerned the application of unpredictability measure and modified unpredictability measure were carried out in order to verify the influence of these measures to accurate locations of tonal components. A voiced frame (containing noticeable tonal components) of the speech signal was processed during the experiments. Fig. 4 presents the values of unpredictability measure (upper plot) and modified unpredictability measure (lower plot). The issues related to the calculation of the masking offset, playing a significant role in the masking threshold calculation, based on the uncertainty measure was also

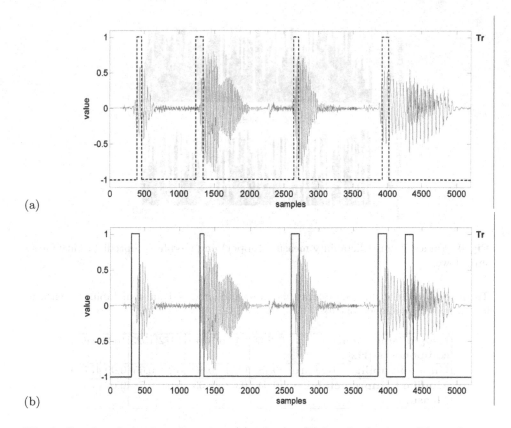

(a)

(b)

**Fig. 2.** Results of transient detection: (a) criterion (8) based selections; (b) rough set algorithm based selections

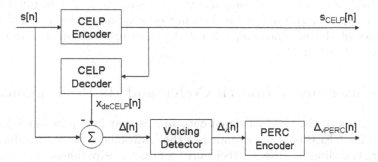

**Fig. 3.** Encoder architecture

interpreted in terms of the rough set theory [14]. That is because of a dependency occurring between the rough measure and the unpredictability measure. The noisy data processing is an evident example of making uncertain decisions, because Unpredictability Measure represents the margin of uncertainty while

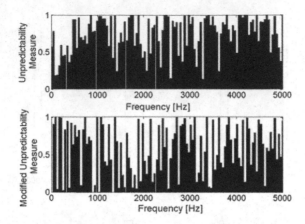

**Fig. 4.** Values of unpredictability measure (upper) and modified unpredictability measure (lower)

**Table 1.** Values of unpredictability measure and modified unpredictability measure for evaluated tonal components

| Freq. of sinusoidal components [Hz] | 123.8 | 247.6 | 376.8 | 1372 | 1523 | 1636 | 3407 |
|---|---|---|---|---|---|---|---|
| Unpredictability Measure | 0.18 | 0.44 | 0.36 | 0.75 | 0.65 | 0.99 | 0.76 |
| Modified Unpredictability Measure | 0.04 | 0.02 | 0.05 | 0.65 | 0.78 | 0.06 | 0.48 |

interpreting sound spectrum shape in terms of useful or useless components representation [14]. It can be noticed in Fig. 4 that the values of unpredictability measure indicate more noisy components (values above 0.5) than tonal components in the analyzed frame, what does not correspond to the actual situation. The values of the modified unpredictability measure indicate the right tonal components (see Tab. 1).

## 5    Architecture of Speech Codec and Its Performance

The drawback of the residual signal encoding is that it requires first local decoding of the CELP bit stream, and then the additional delay is introduced into the entire encoding process. Furthermore, conducted experiments have revealed that perceptual encoding of the speech signal transients is inefficient when the module operates in the low bit-rate mode. Therefore, the architecture of speech codec employing unvoiced/voiced/transient segmentation was investigated. The diagram illustrating the proposed encoder architecture is shown in Fig. 5. In the first step, the input signal is split into the unvoiced $s_{unv}[n]$, the pure-voiced $s_{pv}[n]$ and the transient $s_{tr}[n]$ parts. Further, each part is encoded employing an appropriate method. As the purpose of the proposed experiment was to check if hybrid

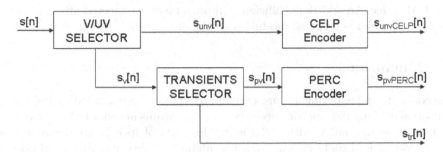

**Fig. 5.** Hybrid encoder architecture

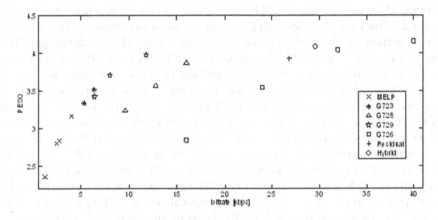

**Fig. 6.** Quality comparison of various speech codec architectures

speech encoding would result in a quality improvement, the transients were not encoded with a use of any lossy method. It was found that about 40% of input signal frames were classified as unvoiced ones and about 7% as transient segments [2]. Thus, the overall estimated bitrate for unvoiced bit stream is about 6.4 kbps (0.4*16000) and 9 kbps (0.07*16*8000) in case of 16-bits PCM transients representation. Consequently, the examined encoder operates at the average 29.6 kbps bitrate. Obviously, an appropriate method for transients coding would increase efficiency of the proposed codec architecture, and thus will be devised within a future work. It is expected that the hybrid encoder would then operate at the 24 kbps rate or even lower. In the decoder, the unvoiced and pure-voiced bit streams are decoded. Finally, these two parts are added together with the $s_{tr}[n]$ signal and the entire speech signal was obtained in this way.

Fig. 6 presents the mean PESQ scores [15] for various speech codec architectures and also the results obtained for the two codecs proposed in this paper. All of quality evaluation tests were performed with the use of the OPERA application [13]. When comparing the PESQ scores for various codec architectures shown in Fig. 6, it is visible that in order to achieve a similar quality of speech coding the hybrid codec requires a lower average bitrate than the

ADPCM codec. An additional efficiency improvement is expected after replacing the general audio perceptual module with the one dedicated to speech coding.

## 6 Conclusions

The concepts and experiments presented in this paper were focused on the modification of existing parametric speech coding algorithms in order to improve their subjective signal quality. Although the intelligibility of speech signal encoded by CELP codecs is usually satisfactory, the highest quality possible to obtain in this kind of a codec architecture is still limited. Instead of various techniques utilized during analyze-by-synthesis procedure, one straightforward method of improving the codec quality should rely on the additional encoding of the residual signal. This kind of a codec architecture has been previously proposed as an efficient method for the lossless wideband speech coding. Hence, the focus of this research was to investigate the lossy speech codec by providing the enhancement layer for an existing standardized CELP codecs. In order to improve the overall quality of the speech signal the residual signal was encoded with the use of the perceptual module. While proposed architecture is similar to the one utilized in MPEG-4 CELP scalable speech codec, the method of the residual signal coding employed in the described experiments is different and also different methods for discerning voiced speech components from unvoiced ones and for detecting transients in the continuous speech signal were applied and studied. The rough set algorithm implemented earlier at the Multimedia Systems Department TU Gdansk proved to be applicable to above experiments revealing one more time its robustness in making decisions on the basis of uncertain and noisy data.

**Acknowledgments.** Research funded by the Ministry of Science and Higher Education within the Grant No. 3 T11D 004 28. The author would like to thank his Ph. D. student, Mr. Maciej Kulesza for his valuable input to this paper.

## References

1. Pawlak Z., A Treatise on Rough Sets. Transactions on Rough Sets **IV** J. F. Peters, A. Skowron (Eds.), pp. 1 - 17, Springer, Berlin (2005).
2. Kulesza M., Szwoch G., Czyzewski A., Improving signal quality in speech codec using a hybrid perceptual-parametric algorithm, Multimedia and Network Information Systems' (MISSI), Wroclaw, Poland 21-22. 9. (2006) 181 - 192.
3. Ritz C. H., Lossless wideband speech coding, $10^{th}$ International Conference on Speech Science and Technology, Sydney, Australia, Dec. (2004).
4. Czyzewski A., Applications of Neural Networks and Perceptual Masking to Audio Restoration. *Journal of New Music Research*, pp.339-349, vol. 22, No. 5 (2001).
5. Verma T.S., Levine S.N., Meng T.H., Transient Modeling Synthesis: a flexible analysis/synthesis tool for transient signals. International Computer Music Conference, Greece (1997).

6. Chu W.C., Speech Coding Algorithms. Foundation and Evolution of Standardized Coders, John Wiley & Sons, Hoboken (2003).
7. Goldberg R., Riek L., A Practical Handbook of Speech Coders, CRC Press, Boca Raton, (2000).
8. Kliewer J., Mertins A., Audio subband coding with improved representation of transient signal segments, Proc 9th European Signal Processing Conference (EUSICPO-98), Rhodes, Greece, September (1998) 1245-1248.
9. Babu V. S., Malot A. K., V. M. Vijayachandran V.M., Vinay M. K., Transient Detection for Transform Domain Coders, AES $116^{th}$ Convention, Berlin (2004).
10. ISO/IEC 14496-3:2001 Information technology - Generic coding of moving pictures and associated audio information: Part 3: Advanced Audio Coding (AAC), (2001).
11. OGG Vorbis Specification:
    http://xiph.org/vorbis/
12. Painter T., Spanias A., Perceptual Coding of Digital Audio, Proceedings of IEEE, vol. 88, (2000) 451-513.
13. Opticom, Opera your digital ear, User manual, version 3.5 (2002).
14. Czyzewski A., Dziubinski M., Litwic L, Maziewski P., Intelligent Algorithms for Movie Sound Tracks Restoration. Transactions on Rough Sets V, J. F. Peters, A. Skowron (Eds.) (2006).
15. ITU-T Recommendation P.862, Perceptual evaluation of speech quality (PESQ): An objective method for end-to-end speech quality assessment of narrow-band telephone networks and speech codecs, (2003).
16. Kulesza M., Szwoch G., Czyzewski A., High quality speech coding using combined parametric and perceptual modules, 13th World Enformatika Conference Proc. pp. 244 - 249, Budapest, Hungary, 26-28.5 (2006).
17. Czyzewski A., Królikowski R., Neuro-Rough Control of Masking Thresholds for Audio Signal Enhancement. *Journal of Neurocomputing*, vol. 36, (2001) 5-27.
18. Annadana R., Ferreira A., Sinha D., A new low bit rate speech coding scheme for mixed content, $120^{th}$ AES Convention, Paris, France, May (2006).
19. Ahmadi S., Jelinek M., On the architecture, operation, and applications of VMR-WB: The new cdma2000 wideband speech coding standard, *IEEE Communication Magazine*, vol. 44, no. 5, May (2006) 74-81.
20. Chazan D., Hoory R., Sagi A., Shechtman S., Sorin A., Shuang Z., Bakis R., High quality sinusoidal modeling of wideband speech for the purposes of speech synthesis and modification, IEEE International Conference on Acoustic, Speech, and Signal Processing - ICASSP, Toulouse, May (2006).
21. Fuemmeler J., Hardie R., Gardner W., Techniques for the regeneration of wideband speech form narrow band speech, *EURASIP Journal on Applied Signal Processing*, vol. 2001, no. 4, Dec. (2001) 266-274.
22. Levine, S., Smith Iii J., Improvements to the Switched Parametric & Transform Audio Coder, Proc. 1999 IEEE Workshop on Application of Signal Processing to Audio and Acoustics, New York, Oct. (1999).
23. Najafzadeh-Azghandi H., Kabal P, Perceptual coding of narrowband audio signals at 8 kbit/s, Proc. IEEE Workshop Speech Coding, Pocono Manor, (1997).
24. Ojala P., Lakaniemi A., Lephanaho H., Jokimies M., The adaptive multirate wideband speech codec: system characteristics, quality advances, and deployment strategies, *IEEE Communication Magazine*, vol. 44, no. 5, May (2006) 59-65.
25. Kulesza, M. Litwic L., Szwoch G., Czyzewski A., High Quality Speech Codec Employing Sines+Noise+Transients Model. $53^{rd}$ Open Seminar on Acoustics, Zakopane, Poland, 11-15.09. (2006).
26. Yang M., Low bit rate speech coding, *IEEE Potentials*, vol. 23, No. 4,(2004) 32-36.

# Relational Attribute Systems II:
# Reasoning with Relations in Information Structures

Ivo Düntsch[1,*], Günther Gediga[2], and Ewa Orłowska[3]

[1] Brock University, St. Catharines, Ontario, Canada
[2] University of Osnabrück, Germany
[3] National Institute of Telecommunications, Warsaw, Poland

**Abstract.** We describe deduction mechanisms for various types of data bases with incomplete information, in particular, relational attribute systems, which we have introduced earlier in [8].

**Keywords:** Attribute, information relation, information system, relational attribute system, fuzzy information system, relational deduction, semantical framework.

## 1   Introduction

Rough sets were introduced by [29]; they served well as a vehicle for expressing dependencies in datatables, as well as for attribute reduction that depends on the equivalence classes induced by the attribute mappings. The information systems of the rough set model were single valued and could only express deterministic information. Already [15,16] had considered information systems where an object under an attribute function was allowed to take a set of values, which could also be empty; a similar road was taken by [26]. Common to both approaches is the replacement of an attribute function between objects and a single value of an attribute domain by an attribute relation where an object can be related to any set of attribute values. In [8] we have supplemented the notion of an indeterministic information system to a model of data called *relational attribute system* (RAS) in the spirit of non–invasive data analysis [7]. Its distinguishing feature is the provision of a semantical framework for the data table: Given an attribute $a$, an object $x$, and a set $a(x)$ of values which are associated with $x$, there are various ways in which $a(x)$ can be interpreted; for instance, as exemplified in [8],

1.  $a(x)$ is interpreted conjunctively and exhaustively. If $a$ is the attribute "speaking a language", then,

$$a(x) = \{\text{German, Polish, French}\}$$

    can be interpreted as

    $x$ speaks German, Polish, and French and no other languages.

* The author gratefully acknowledges support by the Natural Science and Engineering Council of Canada.

J.F. Peters et al. (Eds.): Transactions on Rough Sets VII, LNCS 4400, pp. 16–35, 2007.

2. $a(x)$ can also be interpreted conjunctively and non-exhaustively as in

   $a$ speaks German, Polish, and French and possibly other languages.

3. $a(x)$ is interpreted disjunctively and exclusively. For example, a witness states that

   The car that went too fast was either a Mercedes or a Ford.

   Here, exactly one of the statements
   - The car that went too fast was a Mercedes.
   - The car that went too fast was a Ford.
   is true, but it is not known which one.

4. $a(x)$ is interpreted disjunctively and non-exclusively. If $a$ is "cooperates with", then

   $$a(\text{Ivo}) = \{\text{Günther, Ewa}\}$$

   means that Ivo cooperates with Günther, or Ewa, or both.

The desired semantics can be given in the form of relational constraints, using machinery from the theory of relation algebras [33]. We have indicated the usefulness of the approach by two examples, one pertaining to interrater reliability, the other one to software usability. In a subsequent paper we will use the RAS model and the inference techniques described below to address in detail the practical aspects of our approach.

In this paper, our task is to develop a reasoning mechanism for the RAS model. We follow the general methodology for developing inference tools for information structures based on the object-property assignments, as surveyed in [5]. The specific feature of the methods presented in this paper is that, firstly, we define a class of algebras of relations suitable for the information structure under consideration, and, secondly, we develop deduction rules for this class of algebras. In applying this methodology we observe that in fact several other information structures besides our RAS model can be dealt with in a similar way. Thus, we present deduction systems for various information structures:

- *Information systems with incomplete information and no semantics,*
- *Relational attribute systems,*
- *Fuzzy information systems,*
- *Temporal information systems.*

Once an object–property assignment is given, with each object from the information structure under consideration there is associated a finite set which, in particular, may be empty or contain more than one value . Consequently, each information structure determines a family of sets specific for the structure, resulting from the assignment of the properties to the objects. The relationships among the objects can be articulated by comparing their sets of properties. The comparison is usually expressed in terms of binary set relations. This leads to the concept of information relations.

There are three fundamental ingredients of a definition of any information relation:

- A specification of a family of sets of properties of objects,
- A specification of set relations meaningful for this family,
- A specification of the information relation itself in terms of these set relations.

For the information structures listed above, we present deduction mechanisms for verification of constraints holding for those information relations. The deduction systems presented here belong to the family of *Rasiowa–Sikorski* (RS or dual tableau) style relational proof systems [31]; systems of such type were developed for a number of theories, for example, [11,17,9,25]. There are various implementations of RS systems: Relational attribute systems have been implemented in [3]; an implementation of more general relational proof systems can be found at [4]. The system presented there contains rules for deduction with binary relations as well as with typed relations, and thus, it is suitable for reasoning in relational databases [19]. The system is modular – a general feature of the RS style –, and the user can include specific deduction rules if needed. In particular, the deduction rules presented in the present paper can be incorporated. Some other implementations of relational deduction in nonclassical logics are presented in [10].

## 2  Deduction System for Standard Algebras of Binary Relations

In this Section we recall basic principles of relational proof systems and the deduction rules for standard algebras of binary relations [23]. The operations of Tarski's algebra of binary relations [33] are Boolean set operations of union ($\cup$), intersection ($\cap$) and complement ($-$), and relational operations of product ( ; ), converse ($^{-1}$), and the constants $1'$ of the identity and $1$ of universal relation. For binary relations $R$ and $S$ on a set U, $R \; S = \{(x,y) \in U \times U : \exists z \in U(x,z) \in R \wedge (z,y) \in S\}^1$ and $R^{-1} = \{(x,y) \in U \times U : (y,x) \in R\}$ A *relational term* is any expression built from relation variables and constants with these operations. If $x, y$ are object variables and $P$ is a relational term, then any expression of the form $xPy$ is a relational formula.

The semantics of relational formulas is determined in terms of the notion of model and satisfiability of formulas. A *model* is a system $\mathcal{M} = (U, m)$, where $U$ is a nonempty set (of objects) and $m$ is a meaning function that provides an interpretation of relational terms, i.e. $m(P) \subseteq U \times U$ for any relation variable $P$, $m(1')$ is the identity relation on $U$, $m(1) = U \times U$, and $m$ extends homomorphically to all terms. By a *valuation* in a model $\mathcal{M}$ we understand a function $v$ that assigns objects from $U$ to object variables, that is, $v(x) \in U$ for any object variable $x$. The *satisfiability relation* is defined by $\mathcal{M}, v \models xPy$ iff $(v(x), v(y)) \in m(P)$. A formula $xPy$ is *true* in a model $\mathcal{M}$ whenever $\mathcal{M}, v \models xPy$ for every valuation $v$ in $\mathcal{M}$, nd it is *valid* whenever is true in all models. Hence, validity of $xPy$ amounts to saying that $P = 1$ holds in every algebra of binary relations. A finite sequence of relational formulas is said to be valid whenever universally quantifird disjunction of its members is valid in the classical first order logic.

The proof system consists of two groups of rules, namely, *decomposition rules* and *specific rules*. Decomposition rules enable us to decompose formulas into a finite sequence of (usually syntactically simpler) formulas, or a pair of finite sequences of formulas (and then we separate the sequences with — in a definition of the rule) while the specific rules enable us to modify a sequence to which they are applied; they have

---

[1] Using Tarski's existential quantifier $\mathbf{E}$ (*i.e.*, some) [34], the assertion $\exists z \in U(x,z) \in R$ can also be written $\underset{z}{\mathbf{E}}(z \in U(x,z) \in R)$.

**Table 1.** Decomposition rules

$$(\cup) \quad \frac{K,x(P\cup Q)y,H}{K,xPy,xQy,H} \qquad\qquad (-\cup) \quad \frac{K,x-(P\cup Q)y,H}{K,x(-P)y,H \mid K,x(-Q)y,H}$$

$$(\cap) \quad \frac{K,x(P\cap Q),H}{K,xPy,H \mid K,xQy,H} \qquad\qquad (-\cap) \quad \frac{K,x-(P\cap Q)y}{K,x(-P)y,x(-Q)y,H}$$

$$(^{-1}) \quad \frac{K,xP^{-1}y,H}{K,yPx,H} \qquad\qquad (-^{-1}) \quad \frac{K,x(-P^{-1})y,H}{K,y(-P)x,H}$$

$$(--) \quad \frac{K,x(--P)y,H}{K,xPy,H}$$

$$(;) \quad \frac{K,x(P;Q)y,H}{K,xPz,H,x(P;Q)y \mid K,zQy,H,x(P;Q)y} \quad z \text{ is an object variable}$$

$$(-;) \quad \frac{K,x-(P;Q)y,H}{K,x(-P)z,z(-Q)y,H} \quad z \text{ is restricted}$$

**Table 2.** Specific rules

$$1'_1 \quad \frac{K,xPy,H}{K,x1'z,H,xPy \mid K,zPy,H,xPy} \quad z \text{ is an object variable}$$

$$1'_2 \quad \frac{K,xPy,H}{K,xPz,H,xPy \mid K,z1'y,H,xPy}$$

$$\text{sym } 1' \quad \frac{K,x1'y,H}{K,y1'x,H}$$

the status of structural rules. The role of axioms is played by what is called *axiomatic sequences*.

A proof system for Tarski's algebras of binary relations consists of the decomposition rules given in Table 1, where $K$ and $H$ denote finite, possibly empty, sequences of relational formulas; the specific rules are presented in Table 2. There, a variable is said to be *restricted in a rule* whenever it does not appear in any formula of the upper sequence in that rule. This system has been developed in [21].

The specific rules characterize the identity relation $1'$. Namely, $(1'_1)$ corresponds to the property that $1';R \subseteq R$ for any relation $R$. Similarly, $(1'_1)$ says that $R;1' \subseteq R$. Observe that the reverse inclusions also hold, since $1'$ is reflexive; thus, no more rules are needed for guaranteeing that $1'$ is a unit element of relational composition. (sym $1'$) expresses the symmetry of $1'$, and transitivity of $1'$ is an instance of $(1'_1)$.

A sequence of relational formulas is said to be *axiomatic* if it contains formulas of the following forms; here, $P$ is a relational term, and $x,y$ are object variables.

a1. $xPy,x(-P)y$.
a2. $x1y$.
a3. $x1'x$

(a2) reflects the fact that 1 is the universal relation, and (a3) says that $1'$ is reflexive. The rules listed in Table 1 and Table 2 are correct i.e., they preserve and reflect validity of the sequences of formulas: the upper sequence of the rule is valid if and only if all the lower sequences of this rule are valid. Axiomatic sequences are valid.

Although these rules and axiomatic sequences enable us to prove only that $1'$ is an equivalence relation, the given deduction system is complete with respect to the class of standard algebras of relations, where $1'$ is the identity. The proof uses the usual argument well known from first order logic. Namely, it can be shown that for every model of the relational language with $1'$ interpreted as an equivalence relation there is a model where $1'$ is an identity and both models verify the same formulas.

To check the validity of a relational formula, we successively apply decomposition and/or specific rules to it, thus obtaining a tree whose nodes consist of finite sequences of formulas. Such a tree is referred to as a decomposition tree. We stop applying the rules to the formulas of a node whenever the node contains an axiomatic sequence of formulas. A branch with such a node is declared closed. A decomposition tree is said to be closed whenever all of its branches are closed. The following soundness and completeness Theorem is well known (see e.g. [23,13]).

**Theorem 1.** *A relational formula is valid iff it possesses a closed decomposition tree.*

Hence, possession of a closed decomposition tree may be understood as provability. The proof of this theorem is based on the three lemmas. First of all, we assume that a decomposition tree of a formula is complete: if a rule is applicable to a node of the tree, then it has been applied. Then we prove a closed branch theorem which says that if a branch of the complete decomposition tree includes a node with the formula $xRy$ and a node with the formula $x(-R)y$, where $R$ is a relational term, and $x, y$ are object variables, then this branch has also a node with and axiomatic sequence. This follows from the fact that the rules appropriately transfer the formulas from the upper sequence to the lower sequences. Next, for an open branch, say $b$, of a complete decomposition tree we construct what is called a branch model, $M^b$. It is constructed from the syntactic resources of the relational language. Its universe is the set of object variables. The meaning of a relation variable or a relation constant, say $R$, is a binary relation defined as $(x, y) \in m^b(R)$ iff formula $xRy$ does not appear in any node of branch $b$. The second important lemma, referred to as a branch model theorem, says that a branch model constructed as above is a model of the relational language i.e., the relational constants admitted in the language are appropriately interpreted: $m^b(1)$ is the universal relation and $m^b(1')$ is an equivalence relation. The third lemma, referred to as a satisfaction in branch model theorem, says that if a formula is satisfied in a branch model $M^b$ by an identity valuation $v^b$ such that $v^b(x) = x$ for any object variable x, then it does not appear in any node of branch $b$. With these lemmas the completeness (validity implies provability) can be proved. The soundness (provability implies validity) follows from the correctness of the rules and from validity of axiomatic sequences.

If we extend the set of relational formulas to the first order language with binary predicates, then the appropriate deduction system can be obtained by adding the deduction rules of first order logic developed in [31].

The above relational logic with its system of rules is complete both for the class RRA of representable relation algebras and the class RA of relation algebras. The system can also be applied to solve the three major logical tasks for a number of logics and classes of algebras, namely checking *validity, entailment, satisfiability,* and *truth in a model* (often referred to as *model checking*). The details can be found in [13]. Once a representation of formulas of a logic or the terms over a class of algebras is provided in the form of relational terms over a class, say $C$, of appropriate algebras of relations [22,24], the relational representation of these logical tasks is as follows: Checking validity amounts to verifying whether $R = 1$ holds in every algebra of relations from $C$, for some relation term $R$. Entailment is the problem of checking whether from a finite number of identities of the form $R_1 = 1, \ldots, R_n = 1$ we can infer that $R = 1$. According to the Tarski rule, this problem can be reduced to checking the identity $1 ; -(R_1 \cap \cdots \cap R_n) ; 1 \cup R) = 1$. The satisfaction problem of checking whether $\langle a, b \rangle \in R$ for some relation $R$ and some objects $a, b$ amounts to verifying whether $A ; B^{-1} \subseteq R$, where $A$ and $B$ are the point relations representing the objects $a$ and $b$, respectively, and they satisfy the usual point axioms $P ; 1 = 1$, $P ; P^{-1} \subseteq 1'$, and $P \neq \emptyset$ [32].

## 3 Relations Derived from Information Systems

In this Section we recall the notion of an information system [16,28], and relations derived from such a system; an exhaustive list of those relations can be found in [5].

By an *information system* we understand a structure $S = (\text{OB}, \Omega, \{V_a : a \in \Omega\})$ such that $U$ is a nonempty set of objects, $\Omega$ is a finite nonempty set of attributes, each $V_a$ is a nonempty set of values of attribute $a$. An attribute is a function $a : U \mapsto \mathscr{P}(V_a)$ that assigns subsets of values of attributes to the objects. If for every $a \in \Omega$, $a(x)$ is a singleton set, then system $S$ is said to be *deterministic*, otherwise $S$ is *nondeterministic*.

Any set $a(x)$ can be viewed as a set of properties of an object $x$ determined by attribute $a$. For example, if attribute $a$ is 'color' and $a(x) = \{\text{green}\}$, then $x$ possesses property of 'being green'. If $a$ is 'age' and $x$ is 25 years old, then $a(x) = \{25\}$ and this means that $x$ possesses property of 'being 25 years old'. If $a$ is 'languages spoken' and if a person $x$ speaks, say, Polish (Pl), German (D), and French (F), then $a(x) = \{\text{Pl, D, F}\}$, and $x$ possesses properties of 'speaking Polish', 'speaking German', and 'speaking French'. In this setting any set $a(x)$ is referred to as the set of $a$-properties of object $x$ and its complement $V_a - a(x)$ is said to be the set of negative $a$-properties of $x$.

Let $S = (U, \Omega, \{V_a : a \in \Omega\})$ and $A \subseteq \Omega$. The following families of set information relations on set $U$ are the subject of investigation in a number of papers:

**Strong (weak) indiscernibility** $(x, y) \in \text{ind}_A$ iff $a(x) = a(y)$ for all (some) $a \in A$,
**Strong (weak) similarity** $(x, y) \in \text{sim}_A$ iff $a(x) \cap a(y) \neq \emptyset$ for all (some) $a \in A$,
**Strong (weak) forward inclusion** $(x, y) \in \text{fin}_A$ iff $a(x) \subseteq a(y)$ for all (some) $a \in A$,
**Strong (weak) backward inclusion** $\text{bin}_A$ iff $a(y) \subseteq a(x)$ for all (some) $a \in A$,
**Strong (weak) negative similarity** $(x, y) \in \text{nim}_A$ iff $-a(x) \cap -a(y) \neq \emptyset$ for all (some) $a \in A$,
**Strong (weak) incomplementarity** $(x, y) \in \text{icom}_A$ iff $a(x) \neq -a(y)$ for all (some) $a \in A$,
**Strong (weak) diversity** $(x, y) \in \text{div}_A$ iff $a(x) \neq a(y)$ for all (some) $a \in A$,
**Strong (weak) disjointness** $(x, y) \in \text{dis}_A$ iff $a(x) \subseteq -a(y)$ for all (some) $a \in A$,
**Strong (weak) exhaustiveness** $(x, y) \in \text{exh}_A$ iff $-a(x) \subseteq a(y)$ for all (some) $a \in A$,

**Strong (weak) right negative similarity** $(x,y) \in \text{rnim}_A$ iff $a(x) \cap -a(y) \neq \emptyset$ for all (some) $a \in A$,

**Strong (weak) left negative similarity** $(x,y) \in \text{lnim}_A$ iff $-a(x) \cap a(y) \neq \emptyset$ for all (some) $a \in A$,

**Strong (weak) complementarity** $(x,y) \in \text{com}_A$ iff $a(x) = -a(y)$ for all (some) $a \in A$.

In all the above definitions, complement is taken with respect to set $\text{VAL}_a$. If $A = \{a\}$ is a singleton set, then we write $R_a$ instead of $R_{\{a\}}$ for any information relation $R$. Observe that if $(x,y) \in \text{dis}_a$, then $a(x) \cap a(y) = \emptyset$, and if $(x,y) \in \text{exh}_a$, then $a(x) \cup a(y) = V_a$ which explains the names of the relations. In the earlier literature (e.g., [5]) the relations were referred to as right (resp. left) orthogonality.

The strong relations satisfy the following conditions for all $P, Q \subseteq \Omega$:

S1. $R_{P \cup Q} = R_P \cap R_Q$,
S2. $R_\emptyset = U \times U$.

The weak relations satisfy:

W1. $R_{P \cup Q} = R_P \cup R_Q$,
W2. $R_\emptyset = \emptyset$.

A specific family of sets associated to an information system $S = (U, \Omega, \{V_a : a \in \Omega\})$ is the family $\{a(x) : a \in \Omega, x \in U.\}$. It is easy to see that all the information relations defined above can be specified in terms of three families of set relations, namely, $\subseteq_a$, $\Sigma_a$, $N_a$, where $a \in \Omega$ and

$$x \subseteq_a y \Longleftrightarrow a(x) \subseteq a(y),$$
$$x \Sigma_a y \Longleftrightarrow a(x) \cap a(y) \neq \emptyset,$$
$$x N_a y \Longleftrightarrow a(x) \cup a(y) \neq V_a.$$

These relations can be extended in the usual way to the relations indexed with subsets of set $\Omega$. Now, an information relation derived from an information system is any relation generated from $\subseteq_a, \Sigma_a, N_a$, for $a \in \Omega$, with the standard relational operations.

For example, the information relations relations determined by an attribute $a$ are defined as follows:

$\text{ind}_a = \subseteq_a \cap \subseteq_a^{-1}, \text{sim}_a = \Sigma_a, \text{fin}_a = \subseteq_a, \text{bin}_a = \subseteq_a^{-1}, \text{nim}_a = N_a, \text{icom}_a = \Sigma_a \cup N_a,$
$\text{div}_a = - \subseteq_a \cup - \subseteq_a^{-1}, \text{dis}_a = -\Sigma_a, \text{exh}_a = -N_a, \text{rnim}_a = - \subseteq, \text{lnim}_a = - \subseteq_a^{-1}, \text{com}_a = -\Sigma_a \cup -N_a.$

In an abstract setting, by an IS-frame (information system frame) we mean a system $(U, \{\leq_P : P \subseteq A\}, \{\sigma_P : P \subseteq A\}, \{v_P : P \subseteq A\})$, where $U$ and $A$ are nonempty sets, $A$ is finite, and the following conditions are satisfied for all $x, y, z \in U$ and for every $p \in A$. For the sake of simplicity we write $\leq, \sigma, v$ instead of $\leq_p, \sigma_p, v_p$:

IS1. $\leq$ is reflexive, transitive , and antisymmetric.
IS2. $\sigma$ is symmetric, and $1' \cap (\sigma \,;\, 1) \subseteq \sigma$ (weakly reflexive, i.e., $x\sigma y$ implies $x\sigma x$).
IS3. $\sigma \,;\, \leq \,\subseteq \sigma$, i.e., $x\sigma y$ and $y \leq z$ imply $x\sigma z$.
IS4. $x\sigma x$ or $x \leq y$.

IS5. $v$ is symmetric and weakly reflexive.

IS6. $\leq^{-1}$ ; $v \subseteq v$, i.e., $x \leq^{-1} y$ and $yvz$ imply $xvz$.

IS7. $xvx$ or $x \leq^{-1} y$.

IS8. $-\sigma$ ; $-v \subseteq \leq$ i.e., $x \leq y$ or $x\sigma z$ or $yvz$.

IS9. $x\sigma x$ or $xvx$.

IS10. $-v$ ; $\leq \subseteq -v$

The above list of axioms is based on the axioms presented in [36]. Furthermore, we have to declare whether the relations are strong or weak by postulating the axioms (S1) and (S2) or (W1) and (W2).

By an IS-relation algebra we understand an algebra of relations generated by $\{\leq_p:$ $p \in A\} \cup \{\sigma_p : p \in A\} \cup \{v_p : p \in A\}$ for some IS-frame $(U, \{\leq_P: P \subseteq A\}, \{\sigma_P : P \subseteq A\}, \{v_P : P \subseteq A\})$. In the following Section we present a deduction system for reasoning about properties of relations in IS-relation algebras.

## 4  Deduction in IS-Relation Algebras

The majority of deductive systems for reasoning about information relations derived from an information system are the appropriate systems of modal logics (for a survey see [5]). Modal approach enables us to study information operators, e.g., approximation operators or knowledge operators determined by information relations (see e.g. [35], [36], [37]). Here our aim is to develop a reasoning mechanism for verification of properties of plain information relations. The strategy is to design deduction rules for IS-frames and to adjoin them to the system of rules for the standard algebras of binary relations presented in Section 2, thus obtaining a deduction system for IS-relation algebras.

The formulas processed by the deduction system for IS-relation algebras are of the form $xRy$, where $x, y$ are object variables and $R$ is a term of an IS-relation algebra. For each $p \in A$ and for every $\leq_p, \sigma_p, v_p$ we assume the following rules. As usual we omit the index p in the names of the relations.

(ref $\leq$)
$$\frac{K, x \leq y, H}{K, x1'y, H, x \leq y}$$

(tran $\leq$)
$$\frac{K, x \leq y, H}{K, x \leq z, H, x \leq y \mid K, z \leq y, H, x \leq y}$$
$z$ is any object variable

(antisym $\leq$)
$$\frac{K, x(-\leq)y, y(-\leq x), H}{K, x(-1')y, H}$$

(sym $\sigma$)
$$\frac{K, x\sigma y, H}{K, y\sigma x, H}$$

(wref $\sigma$)
$$\frac{K, x\sigma y, H}{K, x1'y, H, x\sigma y \mid K, x\sigma z, H, x\sigma y}$$
$z$ is any object variable

(rIS3) $$\frac{K,x\sigma y,H}{K,x\sigma z,H,x\sigma y \mid K,z \leq y,H,x\sigma y}$$
$z$ is any object variable

(rIS4) $$\frac{K,x\sigma y,H}{K,x1'y,H,x\sigma y \mid K,x(-\leq)z,H,x\sigma y}$$
$z$ is any object variable

(sym $v$) and (wref $v$) are analogous to (sym $\sigma$) and (wref $\sigma$), respectively.

(rIS6) $$\frac{K,xvy,H}{K,z \leq x,H,xvy \mid K,zvy,H,xvy}$$
$z$ is any object variable

(rIS7) $$\frac{K,xvy,H}{K,x1'y,H,xvy \mid K,x(-\leq)z,H,xvy}$$
$z$ is any object variable

(rIS8) $$\frac{K,x \leq y,H}{K,x(-\sigma)z,H,x \leq y \mid K,z(-v)y,H,x \leq y}$$
$z$ is any object variable

(rIS9) $$\frac{K,x\sigma y,xvy,H}{K,x1'y,H,x\sigma y,xvy}$$

(rIS10) $$\frac{K,x(-v)y,H}{K,x(-v)z,H,x(-v)y \mid K,z \leq y,H,x(-v)y}$$
$z$ is any object variable

For $R \in \{\leq_P : P \subseteq A\} \cup \{\sigma_P : P \subseteq A\} \cup \{v_P : P \subseteq A\}$, the characterization of strong relations is provided by the rules (rS1), (r-S1), and the axiomatic sequence (aS2):

(rS1) $$\frac{K,xR_{P \cup Q}y,H}{K,xR_Py,H \mid K,xR_Qy,H}$$

(r-S1) $$\frac{K,x(-R_{P \cup Q})y,H}{K,x(-R_P)y,x(-R_Q)y,H}$$

(aS2)    $xR_\emptyset y$

The characterization of weak relations is given by the rules (rW1), (r-W1), and the axiomatic sequence (aW2):

(rW1) $$\frac{K,xR_{P \cup Q}y,H}{K,xR_Py,xR_Qy,H}$$

(r-W1) $$\frac{K,x(-R_{P \cup Q})y,H}{K,x(-R_P)y,H \mid K,x(-R_Q)y,H}$$

(aW2)    $x(-R_\emptyset)y$

It is easy to verify that the rules presented above are correct in view of the properties of relational constants assumed in the models. The definition of a branch model is the same as described in Section 2. A completeness theorem analogous to Theorem 1 can be proved following the principles presented in Section 2, see also the general method described in [18].

## Example 1
We show that $-\sigma$ ; $v \subseteq \leq$. Since for any binary relations $R, S$ we have $R \subseteq S$ iff $-R \cup S = 1$, we need to prove the formula (1) below:
(1) $x(-(-\sigma ; -v) \cup \leq)y$.
Applying rule ($\cup$) to (1) we get:
(2) $x(-(-\sigma ; -v))y, x \leq y$.
Applying rule ($-$ ; ) with a restricted variable $z$, and rule ($-$) to (2) we obtain:
(3) $x\sigma z, zvy, x \leq y$.
Now we apply rule (rIS8) to $x \leq y$ choosing $z$ as the new variable and we obtain two sequences (3.1) and (3.2):
(3.1) $x(-\sigma)z, x\sigma z, zvy, x \leq y$,
(3.2) $z(-v)y, x\sigma z, zvy, x \leq y$.
Both of them are axiomatic of the form (a1).

## Example 2
We show that $-v$ ; $-\sigma$ ; $-v \subseteq -v$.
(1) $x(-(-v ; -\sigma ; -v) \cup -v)y$.
We apply rule ($\cup$) and we get:
(2) $x(-(-v ; -\sigma ; -v))y, x(-v)y$.
Now we apply twice the rule ($-$ ; ) with restricted variables z and t, and then rule ($-$):
(3) $xvz, z\sigma t, tvy, x(-v)y$.
Rule (rIS10) applied to $x - vy$ with a new variable $z$ yields two sequences (3.1) and (3.2):
(3.1) $x(-v)z, xvz, z\sigma t, tvy, x(-v)y$,
(3.2) $z \leq y, xvz, z\sigma t, tvy, x(-v)y$.
Sequence (3.1) is axiomatic of the type (a1). To the sequence (3.2) we apply rule (rIS8) with a new variable $t$ and we get the following two sequences:
(3.2.1) $z(-\sigma)t, z \leq y, xvz, z\sigma t, tvy, x(-v)y$,
(3.2.2) $t(-v)y, z \leq y, xvz, z\sigma t, tvy, x(-v)y$.
Both of these sequences are axiomatic.

## Example 3
Let $\{R_P\}_{P \subseteq A}$ be a family of strong relations and let $R_p$ and $R_q$ be transitive, that is the rules (tran $R_p$) and (tran $R_q$) analogous to the rule (tran $\leq$) presented above are admitted in a proof system. For the sake of simplicity we write $R_p$ and $R_q$ instead of $R_{\{p\}}$ and $R_{\{q\}}$, respectively. We show that $R_{\{p,q\}}$ is also transitive, i.e., $R_{\{p,q\}}$ ; $R_{\{p,q\}} \subseteq R_{\{p,q\}}$. Hence, we have to prove the formula:
(1) $x(-(R_{\{p,q\}} ; R_{\{p,q\}}) \cup R_{\{p,q\}})y$
Applying rule ($\cup$) we have:
(2) $x(-(R_{\{p,q\}} ; R_{\{p,q\}}))y, xR_{\{p,q\}}y$

Now we apply rule $(-\ ;\ )$ with a restricted variable $z$:

(3)$x(-(R_{\{p,q\}})z,\ z(-(R_{\{p,q\}})y,\ xR_{\{p,q\}}y.$

Applying rule (r-S1) twice we obtain:

(4) $x(-R_p)z,\ x(-R_q)z,\ z(-R_p)y,\ z(-R_q)y,\ xR_{\{p,q\}}y.$

Now we apply rule (rS1) and we get two sequences:

(4.1) $xR_py,\ x(-R_p)z,\ x(-R_q)z,\ z(-R_p)y,\ z(-R_q)y,\ xR_{\{p,q\}}y,$

(4.2) $xR_qy,\ x(-R_p)z,\ x(-R_q)z,\ z(-R_p)y,\ z(-R_q)y,\ xR_{\{p,q\}}y.$

We apply the rule (tran $R_p$) with a new variable $z$ to the formula $xR_py$ of (4.1) and the rule (tran $R_q$) also with variable $z$ to the formula $xR_qy$ of (4.2) which yield four sequences each of which is axiomatic of the form (a1).

## 5   Relations Derived from Temporal Information Systems

A temporal information system is an information system $(U,\ \{\text{time}\},\ \text{VAL}_{\text{time}})$ whose set of attributes consists of a single attribute 'time', the set of values of this attribute is a set with a strict dense linear ordering $<$ without endpoints on it, and to every object $x$ there is associated a closed time interval $\text{time}(x) = [t,t']$, where $t < t'$. Temporal information systems are useful, for example, in temporal scenario specification of multimedia objects, where the execution of a multimedia object is usually considered to be a temporal interval.

The family of sets specific for the temporal information systems is the underlying family of time intervals:

$$\{[t,t'] : t,t' \in V_{\text{time}}, \text{time}(x) = [t,t'] \text{ for some } x \in U \text{ and } t < t'\}.$$

The typical relations defined on this family of sets are the following [1]:

$1'$ (equals):     $[t,t']\ 1'\ [u,u']$ iff $t = t'$ and $u = u'$,

$P$ (precedes):   $[t,t']\ P\ [u,u']$ iff $t' < u$,

$D$ (during):     $[t,t']\ D\ [u,u']$ iff $u < t$ and $t' < u'$,

$O$ (overlaps):   $[t,t']\ O\ [u,u']$ iff $t < u$ and $u < t'$,

$M$ (meets):      $[t,t']\ M\ [u,u']$ iff $t' = u$,

$S$ (starts):     $[t,t']\ S\ [u,u']$ iff $t = u$ and $t' < u'$,

$F$ (finishes):   $[t,t']\ F\ [u,u']$ iff $t' = u'$ and $u < t$.

By a TIS-frame (temporal information system frame) we mean a system $(U, <, 1', P, D, O, M, S, F)$, where $<$ is a strict dense linear ordering on $U$ without endpoints, and $1', P, D, O, M, S, F$ are the binary relations on the set $\{[t,t'] : t,t' \in U, t < t'\}$ as defined above.

By a *TIS-relation algebra* we understand a relation algebra generated by $1', P, D, O, M, S,$ and $F$ for some TIS-frame $(U, <, 1', P, D, O, M, S, F)$. A TIS relation algebra has 13 atoms, namely the relations from the corresponding TIS-frame and their converses. Observe that $1'^{-1} = 1'$. Any TIS-relation algebra is isomorphic to the TIS-relation algebra whose universe is the set of real numbers. Detailed discussions of relation algebras for reasoning about time (and space) can be found in [20] and [6].

## 6  Deduction in TIS-Relation Algebras

Deduction system for TIS-relation algebras processes formulas built either with relations $1', P, D, O, M, S, F$, acting on temporal intervals or with relation $<$ acting on time points. For the sake of uniformity, we can preprocess the interval formulas by replacing $1'$ (acting on temporal intervals), $P, D, O, M, S, F$ by their definitions in terms of $1'$ (acting on time points) and $<$. A deduction system for TIS-relation algebras consists of the deduction rules and axiomatic sequences for point relations presented in Section 2, the rules of the same form for interval relations, and the following specific rules:

(irref $<$)  $\dfrac{K, x(-1')y, H}{K, x < y, H, x(-1')y}$

(lin $<$)  $\dfrac{K}{K, x(-1')y \mid K, x(- <)y \mid K, y(- <)x}$,    $x, y$ are any variables

(tran $<$)  $\dfrac{K, x < y, H}{K, x < z, H, x < y \mid K, z < y, H, x < y}$,    $z$ is any variable

The following rules reflect the property that $<$ does not have endpoints and is discrete:

(nomin $<$)  $\dfrac{K}{K, x(- <)z}$,    $z$ is a restricted variable

(nomax $<$)  $\dfrac{K}{K, z(- <)x}$,    $z$ is a restricted variable

(dense)  $\dfrac{K}{K, x < y \mid K, x(- <)z, z(- <)y}$,    $z$ is a restricted variable

The rules that provide definitions of the relations $1'$ acting on the intervals $P, D, O, M, S$, and $F$ in terms of $<$ and $1'$ acting on time points are:

(1')  $\dfrac{K, [t, t']\, 1'\, [u, u'], H}{K, t1'u, H \mid K, t'1'u', H}$       ($-1'$)  $\dfrac{K, [t, t']\, (-1')\, [u, u'], H}{K, t(-1')u, t'(-1')u', H}$

(P)  $\dfrac{K, [t, t']\, P\, [u, u'], H}{K, t' < u, H}$       ($-P$)  $\dfrac{K, [t, t']\, (-P)\, [u, u'], H}{K, t'(- <)u, H}$

(D)  $\dfrac{K, [t, t']\, D\, [u, u'], H}{K, u < t, H \mid K, t' < u', H}$       ($-D$)  $\dfrac{K, [t.t']\, (-D)\, [u, u'], H}{K, u(- <)t, t'(- <)u', H}$.

(O)  $\dfrac{K, [t, t']\, O\, [u, u'], H}{K, t < u, H \mid K, u < t', H}$       ($-O$)  $\dfrac{K, [t, t']\, (-O)\, [u, u'], H}{K, t(- <)u, u(- <)t', H}$

(M)  $\dfrac{K, [t, t']\, M\, [u, u'], H}{K, t'1'u, H}$       ($-M$)  $\dfrac{[t, t']\, (-M)\, [u, u']}{K, t'(-1')u, H}$

$$(S)\ \frac{K,[t,t']\ S\ [u,u'],H}{K,t1'u,H\mid K,t'<u',H}\qquad (-S)\ \frac{K,[t,t']\ (-S)\ [u,u'],H}{K,t(-1')u,t'(-<)u',H}$$

$$(F)\ \frac{K,[t,t']\ F\ [u,u'],H}{K,t'1'u',H\mid K,u<t,H}\qquad (-F)\ \frac{K,[t,t']\ (-F)\ [u,u'],H}{K,t'(-1')u',u<t,H}$$

The decomposition rules for the compound interval relations generated by the relations listed in Section 4 with the standard relational operations and the specific rules for $1'$ on the set $\{[t,t']:t,t'\in U,t<t'\}$ and for $1'$ on $U$ are analogous to the corresponding rules in Section 2. Abusing the notation is harmless, because the arguments of the relation indicate on which set it is defined. A completeness Theorem analogous to Theorem 1 holds for the deduction system presented here. The details can be found in [2].

**Example 4**
We show that $F\ ;\ P\subseteq P$.
(1) $[t,t']\ (-(F\ ;\ P)\cup P)\ [u,u']$.
After application of rule $(\cup)$ we obtain:
(2) $[t,t']\ (-(F\ ;\ P))\ [u,u'],\ [t,t']\ P\ [u,u']$.
Now we apply rule $(-\ ;\ )$ with a restricted interval $[z,z']$:
(3) $[t.t']\ (-F)\ [z,z'],\ [z,z']\ (-P)\ [u,u'],\ [t,t']\ P\ [u,u']$.
To (3) we apply rules (P) and (-P) which yield:
(4) $[t,t']\ (-F)\ [z,z'],\ z'(-<)u,\ t'<u$.
After application of rule (-F) we have:
(5) $t'(-1')z',\ z<t,\ z'(-<)u,\ t'<u$.
Applying rule (tran $<$) with a new variable $z'$ to $z<t$ we get two sequences:
(5.1) $t'1'z',\ t'(-1')z',\ z<t,\ z'(-<)u,\ t'<u$,
(5.2) $z'<u,\ t'(-1')z',\ z<t,\ z'(-<)u,\ t'<u$.
Both of them are axiomatic of the form (a1).

## 7 Information Relations Derived from Relational Attribute Systems

Relational attribute systems [8] expand the notion of an information system in order to make explicit various conditions that are implicitly assumed in connection with information systems. By a relational attribute system (RAS) we mean a system $(U,\Omega,\{V_a:a\in\Omega\},\{\mathrm{Rel}_a:a\in\Omega\},\Delta)$, where

- $U$ is a nonempty set of objects,
- $\Omega$ is a finite nonempty set of attributes, and, for each $a\in\Omega$, $V_a$ is a nonempty set of values of attribute $a$,
- Each attribute is a function $a:U\mapsto\mathscr{P}(V_a)$.
- Each $R\in\mathrm{Rel}_a$ is a binary relation $R\subseteq U\times V_a$, and
- $\Delta$ is a set of constraints on relations from $\mathrm{Rel}_a$'s.

An appropriate choice of the families $\mathrm{Rel}_a$ of relations and constraints $\Delta$ enables us to explicitly specify various types of information structures.

For example, if an information system is assumed to be deterministic, then we postulate that for each attribute $a\in\Omega$ there is a relation $I_a\subseteq U\times V_a$ with the intuition that $xI_av$ iff $v \in a(x)$ and whenever $xI_av$ and $xI_av'$ hold, then $v = v'$. If we additionally postulate that for every $x \in U$ there is exactly one $v \in V_a$ such that $xI_av$, then such an information system does not have missing values. To represent these constraints relationally, we additionally introduce relations $1'_a$ of identity on $V_a$ and $1_a = U \times V_a$ for each $a \in \Omega$. Then, our constraints can be represented as the following two properties:

$(\Delta 1)$    $I_a^{-1}; I_a \subseteq 1'_a$,    $I_a$ is functional,
$(\Delta 2)$    $I_a; 1_a = 1_a$,    $I_a$ is total.

The family of sets specific to deterministic relational attribute systems is $\{I_a(x) : a \in \Omega, x \in U\}$.

If the system is nondeterministic, a set of attribute values assigned to an object may have several intuitive meanings, as mentioned in the Introduction. For example, to distinguish between disjunctive and conjunctive interpretation of nondeterministic information we consider relations $I_a, B_a \subseteq U \times V_a$ for each $a \in \Omega$ with the intuition that $xI_av$ iff object $x$ certainly possesses property $v$ and $xB_av$ iff object $x$ possibly possesses property $v$ (see also [8]). For every $a \in \Omega$, these relations are assumed to satisfy the constraint

$(\Delta 3)$    $I_a \cap B_a = \emptyset$,

which says that $I_a$ and $B_a$ are incompatible. The family of sets specific to nondeterministic relational attribute systems is $\{I_a(x) : a \in \Omega, x \in U\} \cup \{B_a(x) : a \in \Omega, x \in U\}$.

Concerning information relations derived from the RAS's defined above we note that the building stones are the relations $\{1', \subsetneq, \supsetneq, P, D\}$, each of which is defined on $2^{VAL}$ for $VAL = \bigcup \{V_a : a \in \Omega\}$. The relations $P$ (partial overlap) and $D$ (disjointness) are defined by

$$xPy \iff x \cap y \neq \emptyset, x \not\subseteq y, y \not\subseteq x, \text{ and } xDy \iff x \cap y = \emptyset.$$

An information relation derived from a RAS $= (U, \Omega, \{V_a : a \in \Omega\}, \{Rel_a : a \in \Omega\}, \Delta)$ is a binary relation on $U$ having the form

$$R'; \rho; S'^{-1},$$

where $\rho \in \{1', \subsetneq, \supsetneq, P, D\}$, and $R', S'$ are extensions of $R, S \in Rel_a$, defined on $U \times 2^{VAL_a}$ by

$$xR'_aA \iff R_a(x) = A.$$

Hence, $xR'; \rho; S'^{-1}y$ iff $\langle R(x), S(y) \rangle \in \rho$.

By an (abstract) RAS-frame we understand a relational system

$$(U, V, \{Rel_a : a \in A\}, 1', <, >, \pi, \delta),$$

where

- $U, V$ and $A$ are nonempty sets,
- $Rel_a \subseteq 2^{U \times V}$,
- $<, >, \pi, \delta$ are binary relations on $V$,
- $1'$ is the identity on $V$,

and the following constraints are satisfied:

(Δ4)    $1' \cup < \cup > \cup \pi \cup \delta = 1,$
(Δ5)    Any two of $1', <, >, \pi, \delta$ are disjoint,
(Δ6)    $\delta$ is irreflexive and symmetric,
(Δ7)    $<$ and $>$ are irreflexive, transitive, and $>=<^{-1}$.

Clearly, the relations $1', \subset, \supset, P, D$ satisfy the constraints $(\Delta 4), \ldots, (\Delta 7)$.

By a RAS–relation algebra we understand an algebra of relations generated by the relations of the form

$$R; \rho; S^{-1},$$

where $R, S \in \mathrm{Rel}_a$, and $\rho \in \{1', <, >, \pi, \delta\}$. Any particular class of RAS–relation algebras is obtained by specifying axiomatically the family $\{\mathrm{Rel}_a : a \in A\}$.

## 8    Deduction in RAS-Relation Algebras

The deduction system for RAS-relation algebras processes formulas built with relational terms involving both the relations from the families $\mathrm{Rel}_a, a \in A$, and relations $1', <, >, \pi, \delta$. Depending on the corresponding relations, the variables in the formulas may be either variables from $U$ or variables from $V$.

The system consists of the rules for standard algebras of binary heterogenous relations and the rules reflecting the constraints from $\Delta$. The rules for the operations of the algebras of heterogenous relations are analogous to the rules for the standard algebras of binary relations presented in Section 2 with an obvious restriction on domains of the left and right arguments of the relations. Below we present the exemplary rules reflecting the constraints discussed above. The rules for the constraints $(\Delta 1), (\Delta 2), (\Delta 3),$ $(\Delta 4),$ and $(\Delta 5)$ are as follows. We assume that $U$-variables range over elements of set $U$, and $V$-variables over elements of $V$.

(rΔ1)    $$\dfrac{K, v1'v', H}{K, xI_a v, H, v1'v' \mid K, xI_a v', H, v1'v'}$$
$x$ is any $U$-variable

(rΔ2)    $$\dfrac{K}{K, x - I_a v}$$
$x$ is any $U$-variable, $v$ is a new $V$-variable

(rΔ3)    $$\dfrac{K}{K, xB_a v, \mid K, xI_a v'}$$
$x$ is any $U$-variable, $v, v'$ are any $V$-variables

(rΔ4)    $$\dfrac{K}{K, x - 1'y \mid K, x(- <)y \mid K, x(- >)y \mid K, x - \pi y \mid K, x - \delta y}$$
where $x, y$ are any $V$-variables

(rΔ5)    $$\dfrac{K}{K, xRy \mid K, xSy}$$
where $R, S \in \{1', <, >, \pi, \delta\}, R \neq S$, and $x, y$ are any $V$-variables

The rules corresponding to the constraints $(\Delta 6)$ and $(\Delta 7)$ are analogous to the respective constraints for the relations of the preceding sections. Irreflexivity of $\delta, <,$ and $>$ is reflected by the rules obtained as the irreflexivity rule in Section 6. Symmetry of

$\delta$ is as in Section 4. The transitivity of $<$ and $>$ is as in Section 4. It is easy to extend the proof of completeness of the basic system described in Section 2 to the system expanded with the rules presented above. Namely, we check that the specific rules reflecting properties of the constants $\text{Rel}_a, a \in A$, and $1', <, >, \pi, \delta$ are correct. Next, the definition of branch model is extended to include the meaning of these constants and the branch model theorem and the satisfaction in branch model theorem are proved in a standard way. These lemmas enable us to prove soundness and completeness of the RAS-theory whose family $\{R_a : a \in \Omega\}$ consists of relations $I_a, B_a$ for $a \in \Omega$, satisfying the axioms $\Delta 1, ..., \Delta 7$. Any other RAS-theory can be obtained by choosing a family $\{R_a : a \in \Omega\}$ of relations and by postulating some axioms on the relations from that family.

**Example 5**
We show that in our exemplary RAS-theory if $I_a$ ; $1'$ ; $I_a^{-1} = 1$, then $I_a$ ; $-\delta$ ; $I_a^{-1} = 1$. We use the principle of proving entailment explained in Section 2. According to that principle we have to prove the following formula:
(1) $x(1 ; (-(I_a ; 1' ; I_a^{-1})) ; 1 \cup I_a ; -\delta ; I_a^{-1})y$.
We apply the rule $(\cup)$, then twice the rule $( ; )$ with new variables $x$ and $y$, and then twice the rule $(- ; )$ with restricted variables $z$ and $t$, and we get:
(2) $x(-I_a)z, z(-1')t, t(-I_a)y, x(I_a ; -\delta ; I_a^{-1})y$ and two axiomatic sequences containing formulas $x1x$ and $y1y$.
We apply rule $( ; )$ with new variable $z$ to the first composition and we obtain two sequences:
(2.1) containing $xI_a z$ and $x(-I_a)z$, so it is axiomatic, and
(2.2) containing $z(-\delta ; I_a^{-1})y$.
Decomposing the composition in (2.2) with rule $( ; )$ with new variable $t$ we obtain two sequences:
(2.2.1) containing $z(-\delta)t, z(-1')t$,
(2.2.2) containing $tI_a^{-1}y, yI_a t, y(-I_a)t$ which is axiomatic.
We apply rule (irref $\delta$) to (2.2.1) obtaining a sequence with $z\delta t$ and $z(-\delta)t$ which is axiomatic.

# 9    Information Relations Derived from Fuzzy Information Systems

Fuzzy information systems differ from the ordinary information systems in that the sets of properties assigned to the objects are fuzzy sets. To define fuzzy sets we need, first of all, to establish a range of fuzziness, that is an algebra whose elements serve as degrees of membership of the elements to fuzzy sets. In this paper we assume that the range of fuzziness is modelled by a class of commutative doubly residuated lattices ([27], [5]).

A commutative *doubly residuated lattice* is a structure of the form

$$(W, \vee, \wedge, \otimes, \rightarrow, \oplus, \leftarrow, 1, 0),$$

where

- $(W, \vee, \wedge, 1, 0)$ is a lattice with the top element 1 and the bottom element 0;
- $(W, \otimes, 1)$ and $(W, \oplus, 0)$ are monoids;
- $\otimes$ and $\oplus$ are commutative;

- $\rightarrow$ is a *residuum* of $\otimes$, that is $z \le x \rightarrow y$ iff $x \otimes z \le y$ for all $x, y, z \in W$;
- $\leftarrow$ is a *dual residuum* of $\oplus$, that is $x \leftarrow y \le z$ iff $y \le x \oplus z$ for all $x, y, z \in W$.

The operations $\otimes$ and $\oplus$ are referred to as *product* and *sum*, respectively. They are intended to be abstract counterparts to $t$-norms and $t$-conorms, respectively. Condition (4) is referred to as a *residuation condition* and condition (5) is a *dual residuation condition*. Clearly, $\rightarrow$ and $\leftarrow$ are uniquely determined by the residuation condition and the dual residuation condition, respectively. Furthermore, it follows from these conditions that $x \rightarrow y$ is the greatest element in the set $\{z : x \otimes z \le y\}$ and $x \leftarrow y$ is the least element in the set $\{z : y \le x \oplus z\}$.

Given a doubly residuated lattice $L = (W, \vee, \wedge, \otimes, \rightarrow, \oplus, \leftarrow, 1, 0)$ and a universe $U$ of objects, any mapping $X : U \mapsto L$ is an *L-fuzzy subset* of $U$. The family of $L$-fuzzy subsets of $U$ is denoted by $F_L \mathscr{P}(U)$. By a *fuzzy set* we understand an $L$-fuzzy set for some doubly residuated lattice $L$. The operations on $L$-fuzzy sets are defined in the following way. Let $X, Y \in F_L \mathscr{P}(U)$, then:

$$(X \cup_L Y)(x) = X(x) \oplus Y(x),$$

$$(X \cap_L Y)(x) = X(x) \otimes Y(x).$$

The empty fuzzy set is defined as $\emptyset_L(x) = 0$, and the full set is $1_L(x) = 1$.

*L-inclusion* and *L-equality* of $L$-fuzzy sets are defined as follows:

$X \subseteq_L Y$ iff for every $x \in U, X(x) \le Y(x)$, where $\le$ is the natural ordering of lattice $L$;

$X =_L Y$ iff $X \subseteq_L Y$ and $Y \subseteq_L X$.

An *L-fuzzy binary relation* on $U$ is a mapping $R : U \times U \mapsto L$. The family of all $L$-fuzzy binary relations on $U$ is denoted by $F_L \mathrm{Rel}(U)$. By a *fuzzy relation* we understand an $L$-fuzzy relation for some doubly residuated lattice $L$. Clearly, every fuzzy relation on a set $U$ is a fuzzy subset of $U \times U$, so the operations on fuzzy sets apply to fuzzy relations.

A *fuzzy information system* is a structure of the form

$$(U, L, \Omega, \{V_a : a \in \Omega\}),$$

where $U$ is a non-empty set of objects, $L$ is a commutative doubly residuated lattice, every attribute $a \in \Omega$ is a mapping $a : U \mapsto F_L \mathscr{P}(V_a)$ which assigns an L-fuzzy subset of $V_a$ to an object. Intuitively, $a(x)(v)$ is a degree to which an object $x$ assumes the value $v$ of the attribute $a$.

We define several $L$-fuzzy binary relations on a family $F_L \mathscr{P}(U)$, for any set $U$. These relations provide patterns for information relations derived from a fuzzy information system. Let $X, Y \in F_L \mathscr{P}(U)$, then:

| | |
|---|---|
| $In_L$ (*L-inclusion*): | $In_L(X, Y) = \inf\{X(x) \rightarrow Y(x) : x \in U\}$, |
| $Ni_L$ (*L-non-inclusion*): | $Ni_L(X, Y) = \sup\{Y(x) \leftarrow X(x) : x \in U\}$, |
| $Sim_L$ (*L-similarity*): | $Sim_L(X, Y) = sup\{X(x) \otimes Y(x) : x \in U\}$, |
| $Exh_L$ (*L-exhaustiveness*): | $Exh_L(X, Y) = \inf\{X(x) \oplus Y(x) : x \in U\}$, |

For a discussion of fuzzy information relations see also [30].

A specific family of fuzzy sets associated to a fuzzy information system $S = (U, L, \Omega, \{V_a : a \in \Omega\})$ is the family:

$$\{a(x) : a \in \Omega \text{ and } x \in U\}.$$

A most basic algebra of fuzzy relations is just an algebra $(F_L \mathrm{Rel}(U), \cup_L, \cap_L, \emptyset_L, 1_L)$. FIS-relation algebra (relation algebra of fuzzy information systems) is an algebra of fuzzy relations generated by the set relations $In_L, Ni_L, Sim_L$, and $Exh_L$ defined above.

A construction of a Rasiowa–Sikorski style or a Gentzen style deduction system for fuzzy relations, and in particular for FIS-relation algebras is an open problem. So the only available means of reasoning about these relations is the equational reasoning within doubly residuated lattices. For the arithmetic of doubly residuated lattices see [27].

In an abstract setting, fuzzy algebras and fuzzy relation algebras are presented and investigated, among others, in [12,14,38,39].

# References

1. Allen, J. F. (1983). Maintaining knowledge about temporal intervals. *Communications of the ACM*, 26(11): 832–843.
2. Bresolin, D., Golińska-Pilarek, J., and Orłowska, E. (2006). Relational dual tableaux for interval temporal logics. *Journal of Applied Non-Classical Logics*, 16, No 3–4: 251–277.
3. Cottrell, R. and Düntsch, I. (2005). An implementation of multivalued information systems. In Düntsch, I. and Winter, M., editors, *Proceedings of the 8th International Workshop on Relational Methods in Computer Science - RelMiCS'8*, pages 31–36, St Catharines.
4. Dallien, J. (2005). RelDT: Relational dual tableaux automated theorem prover. Available at http://logic.stfx.ca/reldt/index.html.
5. Demri, S.and Orł owska, E. (2002). *Incomplete Information: Structure, Inference, Complexity*. EATCS Monographs in Theoretical Computer Science. Springer–Verlag, Heidelberg.
6. Düntsch, I. (2005). Relation algebras and their application in temporal and spatial reasoning. *Artificial Intelligence Review*, 23:315–357.
7. Düntsch, I. and Gediga, G. (2000). Rough set data analysis. In *Encyclopedia of Computer Science and Technology*, volume 43, pages 281–301. Marcel Dekker.
8. Düntsch, I., Gediga, G., and Orłowska, E. (2001). Relational attribute systems. *International Journal of Human Computer Studies*, 55(3): 293–309.
9. Düntsch, I. and Orłowska, E. (2000). A proof system for contact relation algebras. *Journal of Philosophical Logic*, 29: 241–262.
10. Formisano, A. and Nicolosi, M. (2006). An efficient relational deductive system for propositional nonclassical logics. *Journal of Applied Non-Classical Logics*, 16, No 3–4: 367–408.
11. Frias, M.and Orłowska, E. (1995). A proof system for fork algebras and its applications to reasoning in logics based on intuitionism. *Logique et Analyse*, 150–152: 239–284.
12. Furusawa, H. (1998). *Algebraic formalisations of fuzzy relations and their representability*. PhD thesis, Kyushu University, Fukuoka.
13. Golińska-Pilarek, J. and Orłowska, E. (2006). Relational logics and their applications . In H. de Swart, E. Orłowska, M. R. and Schmidt, G., editors, *Theory and Applications of Relational Structures as Knowledge Instruments II*, Lecture Notes in Artificial Intelligence (LNAI), Nr.4342, pages 125–163. Springer, Heidelberg.

14. Kawahara, Y. and Furusawa, H. (1999). An algebraic formalization of fuzzy relations. *Fuzzy Sets and Systems*, 119: 125–135.

15. Lipski, W. (1976). Informational systems with incomplete information. In Michaelson, S. and Milner, R., editors, *Third International Colloquium on Automata, Languages and Programming*, pages 120–130, University of Edinburgh. Edinburgh University Press.

16. Lipski, W. (1981). On databases with incomplete information. *Journal of the ACM*, 28: 41–70.

17. MacCaull, W. (1998). Relational semantics and a relational proof system for the full Lambek calculus. *Journal of Symbolic Logic*, 63(2): 623–637.

18. MacCaull, W. and Orłowska, E. (2002). Correspondence results for relational proof systems with application to the Lambek calculus. *Studia Logica*, 71(3): 389–414.

19. MacCaull, W. and Orłowska, E. (2005). A logic of typed relations and its applications to relational databases. To appear in the Journal of Logic and Computation, available at `http://logic.stfx.ca/reldt/pubs.html`.

20. Maddux, R. (1993). Relation algebras for reasoning about time and space. In Nivat, M., Rattray, C., Rus, T., and Scollo, G., editors, *Algebraic Methodology and Software Technology (AMAST '93)*, Workshops in Computing, pages 27–44. Springer-Verlag, New York, NY.

21. Orłowska, E. (1991). Relational interpretation of modal logics. In Andréka, H., Nemeti, I., and Monk, D., editors, *Algebraic Logic*, volume 54 of *Colloquia mathematica Societatis János Bolyai*, pages 443–471. North-Holland, Amsterdam.

22. Orlowska, E. (1994). Relational semantics for non-classical logics: Formulas are relations. In Wolenski, J., editor, *Philosophical Logic in Poland.*, pages 167–186. Kluwer.

23. Orlowska, E. (1995). Relational proof systems for modal logics. In Wansing, H., editor, *Proof Theory of Modal Logic*, pages 55–77. Kluwer.

24. Orłowska, E. (1997). Relational formalisation of nonclassical logics. In *Relational Methods in Computer Science*, Advances in Computing Science, pages 90–105. Springer–Verlag, Wien.

25. Orłowska, E. (1998). Studying incompleteness of information: A class of information logics. In Kijania-Placek, K. and Wolenski, J., editors, *The Lvov-Warsaw School and Contemporary Philosophy*, pages 283–300. Kluwer, Dordrecht.

26. Orłowska, E. and Pawlak, Z. (1984). Representation of nondeterministic information. *Theoretical Computer Science*, 29:27–39.

27. Orłowska, E. and Radzikowska, A. (2001). Double residuated lattices and their applications. In de Swart, H., editor, *Proceedings of the 6th International Workshop on Relational Methods in Computer Science - RelMiCS'6*, volume 2561 of *Lecture Notes in Computer Science*, pages 177–198, Heidelberg. Springer–Verlag.

28. Pawlak, Z. (1981). Information systems, theoretical foundations. *Information Systems*, 6: 205–218.

29. Pawlak, Z. (1982). Rough sets. *Internat. J. Comput. Inform. Sci.*, 11: 341–356.

30. Radzikowska, A. and Kerre, E. E. (2001). On some classes of fuzzy information relations. In *IEEE International Symposium on Multiple-Valued Logic (ISMVL)*.

31. Rasiowa, H. and Sikorski, R. (1963). *The Mathematics of Metamathematics*, volume 41 of *Polska Akademia Nauk. Monografie matematyczne*. Polish Scientific Publishers, Warsaw.

32. Schmidt, G. and Ströhlein, T. (1989). *Relationen und Graphen*. Springer. English version: Relations and Graphs. Discrete Mathematics for Computer Scientists, EATCS Monographs on Comp. Sci., Springer (1993).

33. Tarski, A. (1941). On the calculus of relations. *Journal of Symbolic Logic*, 6:73–89.

34. Tarski, A. (1965). Introduction to Logic and the Methodology of Deductive Sciences. Oxford University Press, NY.

35. Vakarelov, D. (1989). Modal logics for knowledge representation systems. In Meyer, A. R. and Zalessky, M., editors, *Symposium on Logic Foundations of Computer Science, Pereslavl–Zalessky*, volume 363 of *Lecture Notes in Computer Science*, pages 257–277. Springer–Verlag.

36. Vakarelov, D. (1995). A modal logic for set relations. In *10th International Congress of Logic, Methodology and Philosophy of Science, Volume of abstracts*, Florence.

37. Vakarelov, D. (1998). Information systems, similarity relations, and modal logics. In Orłowska, E., editor, *Incomplete Information – Rough Set Analysis*, pages 492–550. Physica – Verlag, Heidelberg.

38. Winter, M. (2001). A new Algebraic Approach to *L*-Fuzzy Relations convenient to study Crispness. *Information Sciences*, 139(3–4): 233–252.

39. Winter, M. (2003). Representation Theory of Goguen Categories. *Fuzzy Sets and Systems*, 138(1): 85–126.

# Dominance-Based Rough Set Approach as a Proper Way of Handling Graduality in Rough Set Theory

Salvatore Greco[1], Benedetto Matarazzo[1], and Roman Słowiński[2]

[1] Faculty of Economics, University of Catania,
Corso Italia, 55, 95129 – Catania, Italy
[2] Institute of Computing Science, Poznań University of Technology,
60-965 Poznań, and Institute for Systems Research,
Polish Academy of Sciences, 01-447 Warsaw, Poland

**Abstract.** Referring to some ideas of Leibniz, Frege, Boole and Łukasiewicz, we represent fundamental concepts of rough set theory in terms of a generalization that permits to deal with the graduality of fuzzy sets. Our conjunction of rough sets and fuzzy sets is made using the Dominance-based Rough Set Approach (DRSA). DRSA have been proposed to take into account ordinal properties of data related to preferences. We show that DRSA is also relevant in case where preferences are not considered but a kind of monotonicity relating attribute values is meaningful for the analysis of data at hand. In general, monotonicity concerns relationship between different aspects of a phenomenon described by data, e.g.: "the larger the house, the higher its price" or "the more a tomato is red, the more it is ripe". The qualifiers, like "large house", "high price", "red" and "ripe", may be expressed either in terms of some measurement units, or in terms of degrees of membership to some fuzzy sets. In this perspective, the DRSA gives a very general framework in which the classical rough set approach based on indiscernibility relation can be considered as a particular case.

## 1 Introduction

In this paper we want to present a generalization of rough sets to handle the graduality of fuzzy sets. Since according to Pawlak [10] rough set theory refers to some ideas of Gottlob Frege (vague concepts), Gottfried Leibniz (indiscernibility), George Boole (reasoning methods), Jan Łukasiewicz (multi-valued logics), and Thomas Bayes (inductive reasoning), we want to give an account of our generalization of rough sets, justifying it by reference to some of these main ideas recalled by Pawlak.

The *identity of indiscernibles* is a principle of analytic ontology first explicitly formulated by Gottfried Leibniz in his Discourse on Metaphysics, Section 9 [6]. Two objects $x$ and $y$ are defined indiscernible if $x$ and $y$ have the same properties. The principle of identity of indiscernibles states that

$$\textit{if } x \textit{ and } y \textit{ are indiscernible, then } x = y. \tag{II1}$$

J.F. Peters et al. (Eds.): Transactions on Rough Sets VII, LNCS 4400, pp. 36–52, 2007.
© Springer-Verlag Berlin Heidelberg 2007

This can be expressed also as if $x \neq y$, then $x$ and $y$ are discernible, i.e. there is at least one property that $x$ has and $y$ does not, or vice versa. The converse of the principle of identity of indiscernibles is called *indiscernibility of identicals* and states that if $x = y$, then $x$ and $y$ are indiscernible, i.e. they have the same properties. This is equivalent to say that if there is at least one property that $x$ has and $y$ does not, or vice versa, then $x \neq y$. The conjunction of both principles is often referred to as "Leibniz's law".

Rough set theory is based on a weaker interpretation of Leibniz's law, having as objective the ability to classify objects falling under the same concept. This reinterpretation of the Leibniz's law is based on a reformulation of the principle of identity of indiscernibles as follows:

*if $x$ and $y$ are indiscernible, then $x$ and $y$ belong to the same class.*     (II2)

Let us observe that the principle of indiscernibility of identicals cannot be reformulated in analogical terms. In fact, such an analogous reformulation would amount to state that if $x$ and $y$ belong to the same class, then $x$ and $y$ are indiscernible. This principle is too strict, however, because we can well have two discernible objects $x$ and $y$ belonging to the same class. Thus, within rough set theory, the principle of indiscernibility of identicals should continue to hold in its original formulation (i.e. if $x = y$, then $x$ and $y$ are indiscernible).

For this reason, rough set theory needs a still weaker form of the principle of identity of indiscernibles. Such a principle can be formulated using the idea of *vagueness* due to Gottlob Frege. According to Frege "The concept must have a sharp boundary. To the concept without a sharp boundary there would correspond an area that had not a sharp boundary-line all around". Therefore, following this intuition, we can further reformulate the principle of identity of indiscernibles as follows:

*if $x$ and $y$ are indiscernible, then $x$ and $y$ should belong to the same class.*     (II3)

This reformulation of the principle of identity of indiscernibles implies that there is an inconsistency in the statement that $x$ and $y$ are indiscernible and $x$ and $y$ belong to different classes. We can say that the Leibniz's principle of identity of indiscernibles and the Frege's intuition about vagueness are the basic idea of the rough set concept proposed by Pawlak.

The above reconstruction of the basic idea of the Pawlak's rough set should be completed, however, by referring to another basic idea. This is the idea of Georg Boole that concerns a property which is satisfied or not satisfied. It is quite natural to weaken this principle admitting that a property can be satisfied to some degree. This idea of *graduality* can be attributed to Jan Łukasiewicz and his proposal of many-valued logic where, in addition to well-known truth values "true" and "false", other truth values representing partial degrees of truth were present. The Łukasiewicz's idea of graduality has been reconsidered, generalized and fully exploited by Zadeh [14] within fuzzy set theory, where graduality concerns membership to a set.

In this sense, any proposal of putting rough sets and fuzzy sets together can be seen as a reconstruction of the rough set concept, where the Boole's idea of binary logic is abandoned in favor of the Łukasiewicz's idea of many-valued logic, such that the Leibniz's principle of identity of indiscernibles and the Frege's intuition about vagueness are combined with the idea that a property is satisfied to some degree.

Putting aside, for the moment, the Frege's intuition about vagueness, but taking into account the concept of graduality, the principle of identity of indiscernibles can be reformulated as follows:

*if the grade of each property for x is greater than or equal to the grade of y,*
*then x belongs to the considered class in a grade at least as high as y.*    (II4)

Considering the concept of graduality together with the Frege's intuition about vagueness, one can reformulate the principle of identity of indiscernibles as follows:

*if the grade of each property for x is greater than or equal to the grade of y,*
*then x should belong to the considered class*
*in a grade at least as high as y.*    (II5)

In this paper, we show that formulation (II5) of the principle of identity of indiscernibles is perfectly concordant with the rough set concept defined within the Dominance-based Rough Set Approach (DRSA) [4]. DRSA has been proposed by the authors to deal with ordinal properties of data related to preferences in decision problems [5,12]. The fundamental feature of DRSA is that it handles monotonicity of comprehensive evaluation of objects with respect to preferences relative to evaluation of these objects on particular attributes. For example, the more preferred is a car with respect to such attributes as maximum speed, acceleration, fuel consumption, and price, the better is its comprehensive evaluation. The type of monotonicity considered within DRSA is also meaningful for problems where relationships between different aspects of a phenomenon described by data are to be taken into account, even if preferences are not considered. Indeed, monotonicity concerns, in general, mutual trends existing between different variables, like distance and gravity in physics, or inflation rate and interest rate in economics. Whenever we discover a relationship between different aspects of a phenomenon, this relationship can be represented by a monotonicity with respect to some specific measures of the considered aspects. Formulation (II5) of the principle of identity of indiscernibles refers to this type of monotonic relationships. So, in general, the monotonicity permits to translate into a formal language a primitive intuition of relationship between different concepts of our knowledge corresponding to the principle of identity of indiscernibles formulated as (II5).

Rough set approach has been proposed to approximate some relationships existing between concepts. For example, in medical diagnosis the concept of "disease Y" can be represented in terms of such concepts as "low blood pressure" and "high temperature", or "muscle pain" and "headache". The classical rough approximation is based on a very coarse representation, that is, for each aspect characterizing a concept ("low blood pressure", "high temperature", "muscle pain", etc.), only its presence or its absence is considered relevant. In this case, the rough approximation involves a very primitive idea of monotonicity related to a scale with only two values: "presence" and "absence".

Monotonicity gains importance when a finer representation of the concepts is considered. A representation is finer when for each aspect characterizing a concept, not only its presence or its absence is taken into account, but also the *degree* of its presence or absence is considered relevant. Graduality is typical for fuzzy set philosophy [14] and, therefore, a joint consideration of rough sets and fuzzy sets is worthwhile. In fact, rough sets and fuzzy sets capture the two basic complementary aspects of monotonicity: rough sets deal with relationships between different concepts and fuzzy sets deal with expression of different dimensions in which the concepts are considered. For this reason, many approaches have been proposed to combine fuzzy sets with rough sets (see e.g. [1,7,11]). Our combination of rough sets and fuzzy sets presents some important advantages with respect to competitive approaches, which are discussed below.

The main preoccupation in almost all the studies combining rough sets with fuzzy sets was related to a fuzzy extension of Pawlak's definition of lower and upper approximations using fuzzy connectives [2,3]. In fact, there is no rule for the choice of the "right" connective, so this choice is always arbitrary to some extent. Another drawback of fuzzy extensions of rough sets involving fuzzy connectives is that they are based on cardinal properties of membership degrees. In consequence, the result of these extensions is sensitive to order preserving transformation of membership degrees. For example, consider the t-conorm of Łukasiewicz as fuzzy connective; it may be used in the definition of both fuzzy lower approximation (to build fuzzy implication) and fuzzy upper approximation (as a fuzzy counterpart of a union). The t-conorm of Łukasiewicz is defined as

$$T^*(\alpha, \beta) = min(\alpha + \beta, 1), \alpha, \beta \in [0, 1].$$

$T^*(\alpha, \beta)$ can be interpreted as follows. Given two fuzzy propositions $p$ and $q$, putting $v(p) = \alpha$ and $v(q) = \beta$, $T^*(\alpha, \beta)$ can be interpreted as $v(p \vee q)$, the truth value of the proposition $p \vee q$. Let us consider the following values of arguments:

$$\alpha = 0.5, \quad \beta = 0.3, \quad \gamma = 0.2, \quad \delta = 0.1,$$

and their order preserving transformation:

$$\alpha' = 0.4, \quad \beta' = 0.3, \quad \gamma' = 0.2, \quad \delta' = 0.05.$$

The values of the t-conorm are in the two cases as follows:

$$T^*(\alpha, \delta) = 0.6, \quad T^*(\beta, \gamma) = 0.5, \quad T^*(\alpha', \delta') = 0.45, \quad T^*(\beta', \gamma') = 0.5.$$

One can see that the order of the results has changed after the order preserving transformation of the arguments. This means that the Łukasiewicz t-conorm takes into account not only the ordinal properties of the truth values, but also their cardinal properties. A natural question arises: is it reasonable to expect from truth values a cardinal content instead of ordinal only? Or, in other words, is it realistic to claim that a human is able to say in a meaningful way not only that

(a) "proposition $p$ is more credible than proposition $q$"

but even something like

(b) "proposition $p$ is two times more credible than proposition $q$"?

We claim that it is much safer to consider information of type (a), because information of type (b) is rather meaningless for a human.

Since our fuzzy generalization of rough set theory takes into account only ordinal properties of fuzzy membership degrees, we claim in this paper that it is the proper fuzzy generalization of rough set theory.

We will show, moreover, that the classical rough set approach [8,9] can be seen as a specific case of our general model. This is important for several reasons. In particular, this interpretation of DRSA gives an insight into fundamental properties of the classical rough set approach and permits its further generalization.

This article is organized as follows. Section 2 presents rough approximation of a fuzzy set based on the property of monotonicity. In section 3, we compare monotonic rough approximation of a fuzzy set with the classical rough set, and we prove that the latter is a particular case of the former. Section 4 contains conclusions.

## 2   Rough Approximations of Fuzzy Sets Based on the Property of Monotonicity

In this section, we show how the Dominance-based Rough Set Approach can be used for rough approximation of fuzzy sets.

A *fuzzy information base* is the 3-tuple $\boldsymbol{B} =< U, F, \varphi >$, where $U$ is a finite set of *objects* (universe), $F=\{f_1, f_2, ..., f_m\}$ is a finite set of *properties*, and $\varphi : U \times F \rightarrow [0, 1]$ is a function such that $\varphi(x, f_h) \in [0, 1]$ expresses the degree in which object $x$ has property $f_h$. Therefore, each object $x$ from $U$ is described by a vector

$$Des_F(x) = [\varphi(x, f_1), \ldots, \varphi(x, f_m)]$$

called *description* of $x$ in terms of the evaluations of the properties from $F$; it represents the available information about $x$. Obviously, $x \in U$ can be described in terms of any non-empty subset $E \subseteq F$ and in this case we have

$$Des_E(x) = [\varphi(x, f_h), f_h \in E].$$

Let us remark that the concept of fuzzy information base can be considered as a generalization of the concept of property system [13]. Indeed, in a property system an object may either possess a property or not, while in the fuzzy information base an object may possess a property in a given degree between 0 and 1.

With respect to any $E \subseteq F$, we can define the *dominance relation* $D_E$ as follows: for any $x, y \in U$, $x$ dominates $y$ with respect to $E$ (denoted as $xD_Ey$) if for any $f_h \in E$

$$\varphi(x, f_h) \geq \varphi(y, f_h).$$

For any $x \in U$ and for each non-empty $E \subseteq F$, let

$$D_E^+(x) = \{y \in U : yD_Ex\}, \quad D_E^-(x) = \{y \in U : xD_Ey\}.$$

Given $E \subseteq F$, for each $X \subseteq U$, we can define its *upward lower approximation* $\underline{E}^{(>)}(X)$ and its *upward upper approximation* $\overline{E}^{(>)}(X)$ as:

$$\underline{E}^{(>)}(X) = \{x \in U : D_E^+(x) \subseteq X\},$$

$$\overline{E}^{(>)}(X) = \{x \in U : D_E^-(x) \cap X \neq \emptyset\}.$$

Analogously, given $E \subseteq F$, for each $X \subseteq U$, we can define its *downward lower approximation* $\underline{E}^{(<)}(X)$ and its *downward upper approximation* $\overline{E}^{(<)}(X)$ as:

$$\underline{E}^{(<)}(X) = \{x \in U : D_E^-(x) \subseteq X\},$$

$$\overline{E}^{(<)}(X) = \{x \in U : D_E^+(x) \cap X \neq \emptyset\}.$$

Let us observe that in the above definition of rough approximations $\underline{E}^{(>)}(X)$, $\overline{E}^{(>)}(X), \underline{E}^{(<)}(X), \overline{E}^{(<)}(X)$, the elementary sets, which in the classical rough set theory are equivalence classes of the indiscernibility relation, here are positive and negative dominance cones $D_E^+(x)$ and $D_E^-(x)$. Below we prove that the rough approximations $\underline{E}^{(>)}(X)$, $\overline{E}^{(>)}(X)$, $\underline{E}^{(<)}(X)$, $\overline{E}^{(<)}(X)$ can be expressed as unions of elementary sets.

**Theorem 1.** For any $X \subseteq U$ and $E \subseteq F$

1. $\underline{E}^{(>)}(X) = \bigcup_{x \in U} \{D_E^+(x) : D_E^+(x) \subseteq X\},$

2. $\overline{E}^{(>)}(X) = \bigcup_{x \in U} \{D_E^+(x) : D_E^-(x) \cap X \neq \emptyset\},$

3. $\underline{E}^{(<)}(X) = \bigcup_{x \in U} \{D_E^-(x) : D_E^-(x) \subseteq X\},$

4. $\overline{E}^{(<)}(X) = \bigcup_{x \in U} \{D_E^-(x) : D_E^+(x) \cap X \neq \emptyset\}.$

**Proof**

1. If $D_E^+(x) \subseteq X$, then for each $y \in D_E^+(x)$, $y \in X$. But $y \in D_E^+(x)$ implies that $D_E^+(y) \subseteq D_E^+(x)$, such that $D_E^+(y) \subseteq X$, and thus $y \in \underline{E}^{(>)}(X)$. This means that

$$\bigcup_{x \in U} \{D_E^+(x) : D_E^+(x) \subseteq X\} \subseteq \underline{E}^{(>)}(X). \quad (i)$$

If $x \in \underline{E}^{(>)}(X)$, then $D_E^+(x) \subseteq X$. But $x \in D_E^+(x)$, and thus

$$\bigcup_{x \in U} \left\{ D_E^+(x) : D_E^+(x) \subseteq X \right\} \supseteq \underline{E}^{(>)}(X). \quad (ii)$$

From $(i)$ and $(ii)$ we get

$$\underline{E}^{(>)}(X) = \bigcup_{x \in U} \left\{ D_E^+(x) : D_E^+(x) \subseteq X \right\}.$$

2. Let us suppose that $D_E^-(x) \cap X \neq \emptyset$. Observe that for each $y \in D_E^+(x)$ we have $D_E^-(y) \supseteq D_E^-(x)$ which implies $D_E^-(y) \cap X \neq \emptyset$, and thus $y \in \overline{E}^{(>)}(X)$. This means that

$$\bigcup_{x \in U} \left\{ D_E^+(x) : D_E^-(x) \cap X \neq \emptyset \right\} \subseteq \overline{E}^{(>)}(X). \quad (iii)$$

If $x \in \overline{E}^{(>)}(X)$, then $D_E^-(x) \cap X \neq \emptyset$. But $x \in D_E^+(x)$, and thus

$$\bigcup_{x \in U} \left\{ D_E^+(x) : D_E^-(x) \cap X \neq \emptyset \right\} \supseteq \overline{E}^{(>)}(X). \quad (iv)$$

From $(iii)$ and $(iv)$ we get

$$\overline{E}^{(>)}(X) = \bigcup_{x \in U} \left\{ D_E^+(x) : D_E^-(x) \cap X \neq \emptyset \right\}.$$

3. and 4. can be proved analogously.    □

The rough approximations $\underline{E}^{(>)}(X)$, $\overline{E}^{(>)}(X)$, $\underline{E}^{(<)}(X)$, $\overline{E}^{(<)}(X)$ can be used to analyze data relative to gradual membership of objects to some concepts representing properties of objects and their assignment to decision classes. This analysis takes into account the following monotonicity principle: "the greater the degree to which an object has properties from $E \subseteq F$, the greater its degree of membership to a considered class". This principle can be formalized as follows. Let us consider a fuzzy set $X$ in $U$, characterized by the membership function $\mu_X : U \to [0, 1]$. This fuzzy set represents a class of interest, such that function $\mu$ specifies a graded membership of objects from $U$ to considered class $X$. For each cutting level $\alpha \in [0, 1]$ we can consider the following sets

– weak upward cut of fuzzy set $X$:

$$X^{\geq \alpha} = \{ x \in U : \mu(x) \geq \alpha \},$$

– strict upward cut of fuzzy set $X$:

$$X^{> \alpha} = \{ x \in U : \mu(x) > \alpha \},$$

– weak downward cut of fuzzy set $X$:

$$X^{\leq \alpha} = \{ x \in U : \mu(x) \leq \alpha \},$$

– strict downward cut of fuzzy set $X$:

$$X^{<\alpha} = \{x \in U : \mu(x) < \alpha\}.$$

Let us remark that, for any fuzzy set $X$ and for any $\alpha \in [0,1]$, we have that

$$U - X^{\geq \alpha} = X^{<\alpha}, \quad U - X^{\leq \alpha} = X^{>\alpha},$$

$$U - X^{>\alpha} = X^{\leq \alpha}, \quad U - X^{<\alpha} = X^{\geq \alpha}.$$

Given a set of fuzzy sets $\mathbf{X} = \{X_1, X_2, ...., X_p\}$ on $U$, whose respective membership functions are $\mu_1, \mu_2, ..., \mu_p$, let $P^>(\mathbf{X})$ be the set of all the sets obtained as a union and intersection of weak and strict upward cuts of component fuzzy sets. Analogously, let $P^<(\mathbf{X})$ be the set of all the sets obtained as a union and intersection of weak and strict downward cuts of component fuzzy sets.

$P^>(\mathbf{X})$ and $P^<(\mathbf{X})$ are closed under set union and set intersection operations, i.e. for all $Y_1, Y_2 \in P^>(\mathbf{X})$, $Y_1 \cup Y_2$ and $Y_1 \cap Y_2$ belong to $P^>(\mathbf{X})$, as well as for all $W_1, W_2 \in P^<(\mathbf{X})$, $W_1 \cup W_2$ and $W_1 \cap W_2$ belong to $P^<(\mathbf{X})$. Observe, moreover, that the universe $U$ and the empty set $\emptyset$ belong both to $P^>(\mathbf{X})$ and to $P^<(\mathbf{X})$ because, for any fuzzy set $X_i \in \mathbf{X}$,

$$U = X_i^{\geq 0} = X_i^{\leq 1}$$

and

$$\emptyset = X_i^{>1} = X_i^{<0}.$$

Let us remark that due to the distributive property of $\cup$ with respect to $\cap$, as well as that of $\cap$ with respect to $\cup$, we can always represent any $Y \in P^>(\mathbf{X})$ in its disjunctive form

$$Y = \bigcup_{i=1}^{p} \left( \bigcap_{j_i=1}^{m_i} X_i^{*_{j_i}\alpha_{j_i}} \right)$$

or in its conjunctive form

$$Y = \bigcap_{i=1}^{p} \left( \bigcup_{j_i=1}^{m_i} X_i^{*_{j_i}\alpha_{j_i}} \right)$$

where, for any $j_i = 1, ..., m_i$ and $i = 1, ..., p$, $*_{j_i} \in \{\geq, >\}$ and $\alpha_{j_i} \in [0,1]$. Of course, also the elements of $P^<(\mathbf{X})$ can be represented in a conjunctive or a disjunctive form, as above, with the only exception that, for any $j_i = 1, ..., m_i$ and $i = 1, ..., p$, $*_{j_i} \in \{\leq, <\}$.

Let us observe, moreover, that for any $Y \in P^>(\mathbf{X})$ we have that $U - Y \in P^<(\mathbf{X})$ and, viceversa, for any $Y \in P^<(\mathbf{X})$ we have that $U - Y \in P^>(\mathbf{X})$. This can be explained as follows. Let us consider a set $Y \in P^>(\mathbf{X})$ represented in its disjunctive form

$$Y = \bigcup_{i=1}^{p} \left( \bigcap_{j_i=1}^{m_i} X_i^{*_{j_i}\alpha_{j_i}} \right).$$

For De Morgan property we have

$$U - Y = U - \bigcup_{i=1}^{p} \left( \bigcap_{j_i=1}^{m_i} X^{*j_i \alpha_{j_i}} \right) =$$

$$\bigcap_{i=1}^{p} \left( \bigcup_{j_i=1}^{m_i} U - X_i^{*j_i \alpha_{j_i}} \right) = \bigcap_{i=1}^{p} \left( \bigcup_{j_i=1}^{m_i} X_i^{\star j_i \alpha_{j_i}} \right),$$

where for any $j_i = 1, ..., m_i$ and $i = 1, ..., n$,

$$\star_{j_i} = \text{"} < \text{"} \quad \text{if} \quad *_{j_i} = \text{"} \geq \text{"}$$

and

$$\star_{j_i} = \text{"} \leq \text{"} \quad \text{if} \quad *_{j_i} = \text{"} > \text{"}.$$

Thus, $\bigcap_{i=1}^{p} \left( \bigcup_{j_i=1}^{m_i} X_i^{\star j_i \alpha_{j_i}} \right)$ is clearly a conjunctive form of an element $Y' \in P^{<}(\mathbf{X})$. Analogous reasoning can be done for any $Y \in P^{<}(\mathbf{X})$.

Let us remark that we can rewrite the rough approximations $\underline{E}^{(>)}(Y), \overline{E}^{(>)}(Y)$, $\underline{E}^{(<)}(Y)$ and $\overline{E}^{(<)}(Y)$ as follows:

$$\underline{E}^{(>)}(Y) = \{x \in U : \forall w \in U, \ wD_E x \Rightarrow w \in Y\},$$

$$\overline{E}^{(>)}(Y) = \{x \in U : \exists w \in U \text{ such that } xD_E w \text{ and } w \in Y\},$$

$$\underline{E}^{(<)}(Y) = \{x \in U : \forall w \in U, \ xD_E w \Rightarrow w \in Y\},$$

$$\overline{E}^{(<)}(Y) = \{x \in U : \exists w \in U \text{ such that } wD_E x \text{ and } w \in Y\}.$$

The following theorem states some important properties of the dominance-based rough approximations.

**Theorem 2**

1. For any $Y \in P^{>}(\mathbf{X})$ and for any $W \in P^{<}(\mathbf{X})$ and for any $E \subseteq F$,

$$\underline{E}^{(>)}(Y) \subseteq Y \subseteq \overline{E}^{(>)}(Y), \quad \underline{E}^{(<)}(W) \subseteq W \subseteq \overline{E}^{(<)}(W).$$

2. For any $E \subseteq F$,

$$\underline{E}^{(>)}(\emptyset) = \overline{E}^{(>)}(\emptyset) = \emptyset, \quad \underline{E}^{(<)}(\emptyset) = \overline{E}^{(<)}(\emptyset) = \emptyset,$$

$$\underline{E}^{(>)}(U) = \overline{E}^{(>)}(U) = U, \quad \underline{E}^{(<)}(U) = \overline{E}^{(<)}(U) = U.$$

3. For any $E \subseteq F$, for any $Y_1, Y_2 \in P^>(\mathbf{X})$ and for any $W_1, W_2 \in P^<(\mathbf{X})$,

$$\overline{E}^{(>)}(Y_1 \cup Y_2) = \overline{E}^{(>)}(Y_1) \cup \overline{E}^{(>)}(Y_2),$$

$$\overline{E}^{(<)}(W_1 \cup W_2) = \overline{E}^{(<)}(W_1) \cup \overline{E}^{(<)}(W_2).$$

4. For any $E \subseteq F$, for any $Y_1, Y_2 \in P^>(\mathbf{X})$ and for any $W_1, W_2 \in P^<(\mathbf{X})$,

$$\underline{E}^{(>)}(Y_1 \cap Y_2) = \underline{E}^{(>)}(Y_1) \cap \underline{E}^{(>)}(Y_2),$$

$$\underline{E}^{(<)}(W_1 \cap W_2) = \underline{E}^{(<)}(W_1) \cap \underline{E}^{(<)}(W_2).$$

5. For any $E \subseteq F$, for any $Y_1, Y_2 \in P^>(\mathbf{X})$ and for any $W_1, W_2 \in P^<(\mathbf{X})$,

$$Y_1 \subseteq Y_2 \Rightarrow \underline{E}^{(>)}(Y_1) \subseteq \underline{E}^{(>)}(Y_2),$$

$$W_1 \subseteq W_2 \Rightarrow \underline{E}^{(<)}(W_1) \subseteq \underline{E}^{(<)}(W_2).$$

6. For any $E \subseteq F$, for any $Y_1, Y_2 \in P^>(\mathbf{X})$ and for any $W_1, W_2 \in P^<(\mathbf{X})$,

$$Y_1 \subseteq Y_2 \Rightarrow \overline{E}^{(>)}(Y_1) \subseteq \overline{E}^{(>)}(Y_2),$$

$$W_1 \subseteq W_2 \Rightarrow \overline{E}^{(<)}(W_1) \subseteq \overline{E}^{(<)}(W_2).$$

7. For any $E \subseteq F$, for any $Y_1, Y_2 \in P^>(\mathbf{X})$ and for any $W_1, W_2 \in P^<(\mathbf{X})$,

$$\underline{E}^{(>)}(Y_1 \cup Y_2) \supseteq \underline{E}^{(>)}(Y_1) \cup \underline{E}^{(>)}(Y_2),$$

$$\underline{E}^{(<)}(W_1 \cup W_2) \supseteq \underline{E}^{(<)}(W_1) \cup \underline{E}^{(<)}(W_2).$$

8. For any $E \subseteq F$, for any $Y_1, Y_2 \in P^>(\mathbf{X})$ and for any $W_1, W_2 \in P^<(\mathbf{X})$,

$$\overline{E}^{(>)}(Y_1 \cap Y_2) \subseteq \overline{E}^{(>)}(Y_1) \cap \overline{E}^{(>)}(Y_2),$$

$$\overline{E}^{(<)}(W_1 \cap W_2) \subseteq \overline{E}^{(<)}(W_1) \cap \overline{E}^{(<)}(W_2).$$

9. For any $E \subseteq F$, for any $Y \in P^>(\mathbf{X})$ and for any $W \in P^<(\mathbf{X})$,

$$\underline{E}^{(<)}(U - Y) = U - \overline{E}^{(>)}(Y),$$

$$\underline{E}^{(>)}(U - W) = U - \overline{E}^{(<)}(W).$$

10. For any $E \subseteq F$, for any $Y \in P^>(\mathbf{X})$ and for any $W \in P^<(\mathbf{X})$,

$$\overline{E}^{(<)}(U - Y) = U - \underline{E}^{(>)}(Y),$$

$$\overline{E}^{(>)}(U - W) = U - \underline{E}^{(<)}(W).$$

11. For any $E \subseteq F$, for any $Y \in P^>(\mathbf{X})$ and for any $W \in P^<(\mathbf{X})$,

$$\underline{E}^{(>)}[\underline{E}^{(>)}(Y)] = \overline{E}^{(>)}[\underline{E}^{(>)}(Y)] = \underline{E}^{(>)}(Y),$$

$$\underline{E}^{(<)}[\underline{E}^{(<)}(W)] = \overline{E}^{(<)}[\underline{E}^{(<)}(W)] = \underline{E}^{(<)}(W).$$

12. For any $E \subseteq F$, for any $Y \in P^>(\mathbf{X})$ and for any $W \in P^<(\mathbf{X})$,

$$\overline{E}^{(>)}[\overline{E}^{(>)}(Y)] = \underline{E}^{(>)}[\overline{E}^{(>)}(Y)] = \overline{E}^{(>)}(Y),$$

$$\overline{E}^{(<)}[\overline{E}^{(<)}(W)] = \underline{E}^{(<)}[\overline{E}^{(<)}(W)] = \overline{E}^{(<)}(W).$$

## Proof

In the following we shall consider only upward rough approximations $\underline{E}^{(>)}(Y)$ and $\overline{E}^{(>)}(Y)$. Analogous proof holds for the downward rough approximations $\underline{E}^{(<)}(W)$ and $\overline{E}^{(<)}(W)$.

1a. If $x \in \underline{E}^{(>)}(Y)$, then $D^+(x) \subseteq Y$, but $x \in D^+(x)$, hence $x \in Y$, and $\underline{E}^{(>)}(Y) \subseteq Y$.

1b. If $x \in Y$, then $D^-(x) \cap Y \neq \emptyset$ (because $x \in D^-(x) \cap Y$), hence $x \in \overline{E}^{(>)}(Y)$, and $Y \subseteq \overline{E}^{(>)}(Y)$.

2a. From 1, $\underline{E}^{(>)}(\emptyset) \subseteq \emptyset$ and $\emptyset \subseteq \underline{E}^{(>)}(\emptyset)$ (because the empty set is included in every set), thus $\underline{E}^{(>)}(\emptyset) = \emptyset$.

2b. Assume that $\overline{E}^{(>)}(\emptyset) \neq \emptyset$. Then, there exists $x$ such that $x \in \overline{E}^{(>)}(\emptyset)$. Hence, $D^-(x) \cap \emptyset \neq \emptyset$, but $D^-(x) \cap \emptyset = \emptyset$, what contradicts the assumption, thus $\overline{E}^{(>)}(\emptyset) = \emptyset$.

2c. From 1, $\underline{E}^{(>)}(U) \subseteq U$. In order to show that $U \subseteq \underline{E}^{(>)}(U)$, let us observe that if $x \in U$, then $D^+(x) \subseteq U$, hence $x \in \underline{E}^{(>)}(U)$, thus $\underline{E}^{(>)}(U) = U$.

2d. From 1, $\overline{E}^{(>)}(U) \supseteq U$ and, obviously, $\overline{E}^{(>)}(U) \subseteq U$, thus $\overline{E}^{(>)}(U) = U$.

3. $x \in \overline{E}^{(>)}(Y_1 \cup Y_2)$ iff $D^-(x) \cap (Y_1 \cup Y_2) \neq \emptyset$ iff $D^-(x) \cap Y_1 \cup D^-(x) \cap Y_2 \neq \emptyset$ iff $D^-(x) \cap Y_1 \neq \emptyset \vee D^-(x) \cap Y_2 \neq \emptyset$ iff $x \in \overline{E}^{(>)}(Y_1) \vee x \in \overline{E}^{(>)}(Y_2)$ iff $x \in \overline{E}^{(>)}(Y_1) \cup \overline{E}^{(>)}(Y_2)$. Thus, $\overline{E}^{(>)}(Y_1 \cup Y_2) = \overline{E}^{(>)}(Y_1) \cup \overline{E}^{(>)}(Y_2)$.

4. $x \in \underline{E}^{(>)}(Y_1 \cap Y_2)$ iff $D^+(x) \subseteq Y_1 \cap Y_2$ iff $D^+(x) \subseteq Y_1 \wedge D^+(x) \subseteq Y_2$ iff $x \in \underline{E}^{(>)}(Y_1) \cap \underline{E}^{(>)}(Y_2)$. Thus $\underline{E}^{(>)}(Y_1 \cap Y_2) = \underline{E}^{(>)}(Y_1) \cap \underline{E}^{(>)}(Y_2)$.

5. Because $Y_1 \subseteq Y_2$ iff $Y_1 \cap Y_2 = Y_1$, by virtue of 4 we have $\underline{E}^{(>)}(Y_1 \cap Y_2) = \underline{E}^{(>)}(Y_1)$ iff $\underline{E}^{(>)}(Y_1) \cap \underline{E}^{(>)}(Y_2) = \underline{E}^{(>)}(Y_1)$, which gives $\underline{E}^{(>)}(Y_1) \subseteq \underline{E}^{(>)}(Y_2)$.

6. Because $Y_1 \subseteq Y_2$ iff $Y_1 \cup Y_2 = Y_2$, by virtue of 3 we have $\overline{E}^{(>)}(Y_1 \cup Y_2) = \overline{E}^{(>)}(Y_2)$ iff $\overline{E}^{(>)}(Y_1) \cup \overline{E}^{(>)}(Y_2) = \overline{E}^{(>)}(Y_2)$, which gives $\overline{E}^{(>)}(Y_1) \subseteq \overline{E}^{(>)}(Y_2)$.

7. Since $Y_1 \subseteq Y_1 \cup Y_2$ and $Y_2 \subseteq Y_1 \cup Y_2$, by virtue of 6 we have $\underline{E}^{(>)}(Y_1) \subseteq \underline{E}^{(>)}(Y_1 \cup Y_2)$ and $\underline{E}^{(>)}(Y_2) \subseteq \underline{E}^{(>)}(Y_1 \cup Y_2)$ and thus $\underline{E}^{(>)}(Y_1) \cup \underline{E}^{(>)}(Y_2) \subseteq \underline{E}^{(>)}(Y_1 \cup Y_2)$.

8. Since $Y_1 \cap Y_2 \subseteq Y_1$ and $Y_1 \cap Y_2 \subseteq Y_2$, by virtue of 5 we have $\overline{E}^{(>)}(Y_1 \cap Y_2) \subseteq \overline{E}^{(>)}(Y_1)$ and $\overline{E}^{(>)}(Y_1 \cap Y_2) \subseteq \overline{E}^{(>)}(Y_2)$ and thus $\overline{E}^{(>)}(Y_1 \cap Y_2) \subseteq \overline{E}^{(>)}(Y_1) \cap \overline{E}^{(>)}(Y_2)$.

9a. Let us suppose that $Y \in P^{>}(\mathbf{X})$. $x \in \underline{E}^{(<)}(U - Y)$ iff $D^{-}(x) \subseteq U - Y$ iff $D^{-}(x) \cap Y = \emptyset$ iff $x \notin \overline{E}^{(>)}(Y)$, hence $\underline{E}^{(<)}(U - Y) = U - \overline{E}^{(>)}(Y)$.

9b. Analogously to 9a, if $W \in P^{<}(\mathbf{X})$, we get $\underline{E}^{(>)}(U - W) = U - \overline{E}^{(<)}(W)$.

10a. Putting $Y = U - W$ in 9 we get $\underline{E}^{(>)}(Y) = U - \overline{E}^{(<)}(U - Y)$ and, in consequence, $U - \underline{E}^{(>)}(Y) = \overline{E}^{(<)}(U - Y)$, which is the thesis.

10b. Analogously to 10a, putting $W = U - Y$ in 9 we get $U - \underline{E}^{(<)}(W) = \overline{E}^{(>)}(U - W)$ which is the thesis.

11a. From 1, $\underline{E}^{(>)}[\underline{E}^{(>)}(Y)] \subseteq \underline{E}^{(>)}(Y)$, thus we have to show that $\underline{E}^{(>)}(Y) \subseteq \underline{E}^{(>)}[\underline{E}^{(>)}(Y)]$. If $x \in \underline{E}^{(>)}(Y)$, then $D_E^{+}(x) \subseteq Y$, hence $\underline{E}^{(>)}[D_E^{+}(x)] \subseteq \underline{E}^{(>)}(Y)$ but $\underline{E}^{(>)}[D_E^{+}(x)] = D_E^{+}(x)$, thus $D_E^{+}(x) \subseteq \underline{E}^{(>)}(Y)$ and $x \in \underline{E}^{(>)}[\underline{E}^{(>)}(Y)]$, that is $\underline{E}^{(>)}(Y) \subseteq \underline{E}^{(>)}[\underline{E}^{(>)}(Y)]$.

11b. From 1, $\underline{E}^{(>)}(Y) \subseteq \overline{E}^{(>)}[\underline{E}^{(>)}(Y)]$, thus we have to show that $\underline{E}^{(>)}(Y) \supseteq \overline{E}^{(>)}[\underline{E}^{(>)}(Y)]$. If $x \in \overline{E}^{(>)}[\underline{E}^{(>)}(Y)]$, then $D_E^{-}(x) \cap \underline{E}^{(>)}(Y) \neq \emptyset$, i.e. there exists $y \in D_E^{-}(x)$ such that $y \in \underline{E}^{(>)}(Y)$, hence $D_E^{+}(y) \subseteq Y$ but $D_E^{+}(x) \subseteq D_E^{+}(y)$, thus $D_E^{+}(x) \subseteq Y$ and $x \in \underline{E}^{(>)}(Y)$, that is $\underline{E}^{(>)}(Y) \supseteq \overline{E}^{(>)}[\underline{E}^{(>)}(Y)]$.

12a. From 1, $\overline{E}^{(>)}(Y) \subseteq \overline{E}^{(>)}[\overline{E}^{(>)}(Y)]$. We have to show that $\overline{E}^{(>)}(Y) \supseteq \overline{E}^{(>)}[\overline{E}^{(>)}(Y)]$. If $x \in \overline{E}^{(>)}[\overline{E}^{(>)}(Y)]$, then $D_E^{-}(x) \cap \overline{E}^{(>)}(Y) \neq \emptyset$ and for some $y \in D_E^{-}(x)$, $y \in \overline{E}^{(>)}(Y)$, hence $D_E^{-}(y) \cap Y \neq \emptyset$, but $D_E^{-}(y) \subseteq D_E^{-}(x)$, thus $D_E^{-}(x) \cap Y \neq \emptyset$, i.e. $x \in \overline{E}^{(>)}(Y)$, which gives $\overline{E}^{(>)}(Y) \supseteq \overline{E}^{(>)}[\overline{E}^{(>)}(Y)]$.

12b. From 1, $\underline{E}^{(>)}[\overline{E}^{(>)}(Y)] \subseteq \overline{E}^{(>)}(Y)$. We have to show that $\underline{E}^{(>)}[\overline{E}^{(>)}(Y)] \supseteq \overline{E}^{(>)}(Y)$. If $x \in \overline{E}^{(>)}(Y)$, then $D_E^{-}(x) \cap Y \neq \emptyset$. Hence, $D_E^{+}(x) \supseteq \overline{E}^{(>)}(Y)$ (because if $y \in D_E^{+}(x)$, then $D_E^{-}(x) \subseteq D_E^{-}(y)$ and $D_E^{-}(x) \cap Y \neq \emptyset$ implies $D_E^{-}(y) \cap Y \neq \emptyset$, i.e. $y \in \overline{E}^{(>)}(Y)$) and $x \in \underline{E}^{(>)}[\overline{E}^{(>)}(Y)]$, which gives $\underline{E}^{(>)}[\overline{E}^{(>)}(Y)] \supseteq \overline{E}^{(>)}(Y)$. $\square$

Let us remark that the results given in Theorem 2 correspond to well known properties of classical rough sets (see the original properties numbered in the same way in [9]), however, with the noticeable exception of properties 9 and 10 characterizing the specific nature of complementarity relations within the Dominance-based Rough Set Approach.

## 3   Classical Rough Set as a Particular Case of the Monotonic Rough Approximation of a Fuzzy Set

In this section, we show that the classical rough approximation is a particular case of the rough approximation of a fuzzy set presented in the previous section.

Let us remember that in classical rough set approach [8,9], the original information is expressed by means of an *information system*, that is the 4-tuple $S = \langle U, Q, V, \phi \rangle$, where $U$ is a finite set of *objects* (universe), $Q=\{q_1, q_2, ..., q_m\}$ is a finite set of *attributes*, $V_q$ is the set of values of the attribute $q$, $V = \bigcup_{q \in Q} V_q$ and $\phi : U \times Q \to V$ is a total function such that $\phi(x, q) \in V_q$ for each $q \in Q$, $x \in U$, called *information function*.

Therefore, each object $x$ from $U$ is described by a vector $Des_Q(x) = [\phi(x, q_1), \phi(x, q_2), ..., \phi(x, q_m)]$, called *description* of $x$ in terms of the evaluations of the attributes from $Q$; it represents the available information about $x$. Obviously, $x \in U$ can be described in terms of any non-empty subset $P \subseteq Q$.

With every (non-empty) subset of attributes $P$ there is associated an *indiscernibility relation* on $U$, denoted by $I_P$:

$$I_P = \{(x, y) \in U \times U : \phi(x, q) = \phi(y, q), \forall q \in P\}.$$

If $(x, y) \in I_P$, it is said that the objects $x$ and $y$ are $P$-indiscernible. Clearly, the indiscernibility relation thus defined is an equivalence relation (reflexive, symmetric and transitive). The family of all the equivalence classes of the relation $I_P$ is denoted by $U|I_P$, and the equivalence class containing an element $x \in U$ by $I_P(x)$, i.e.

$$I_P(x) = \{y \in U : \phi(y, q) = \phi(x, q), \forall q \in P\}.$$

The equivalence classes of the relation $I_P$ are called *P-elementary sets*.

Let $S$ be an information system, $X$ a non-empty subset of $U$ and $\emptyset \neq P \subseteq Q$. The *P-lower approximation* and the *P-upper approximation* of $X$ in $S$ are defined, respectively, as:

$$\underline{P}(X) = \{x \in U : I_P(x) \subseteq X\},$$

$$\overline{P}(X) = \{x \in U : I_P(x) \cap X \neq \emptyset\}.$$

The elements of $\underline{P}(X)$ are all and only those objects $x \in U$ which belong to the equivalence classes generated by the indiscernibility relation $I_P$, *contained* in $X$; the elements of $\overline{P}(X)$ are all and only those objects $x \in U$ which belong to the equivalence classes generated by the indiscernibility relation $I_P$, *containing at least one* object $x$ belonging to $X$. In other words, $\underline{P}(X)$ is the largest union of the $P$-elementary sets included in $X$, while $\overline{P}(X)$ is the smallest union of the $P$-elementary sets containing $X$.

Now, we prove that any information system can be expressed in terms of a specific type of an information base (see section 2). An *information base* is called *Boolean* if $\varphi : U \times F \to \{0,1\}$. A partition $\boldsymbol{F}=\{F_1,\ldots,F_r\}$ of the set of properties $F$, with $card(F_k) \geq 2$ for all $k = 1,\ldots,r$, is called *canonical* if, for each $x \in U$ and for each $F_k \subseteq \boldsymbol{F}$, $k = 1,\ldots,r$, there exists only one $f_j \in F_k$ for which $\varphi\,(x, f_j) = 1$ (and, therefore, for all $f_i \in F_k - \{f_j\}$, $\varphi(x, f_i)=0$). The condition $card(F_k) \geq 2$ for all $k = 1,\ldots,r$, is necessary because, otherwise, we would have at least one element of the partition $F_k=\{f'\}$ such that $\varphi(x,f')=1$ for all $x \in U$, and this would mean that property $f'$ gives no information and can be removed. We can observe now that any *information system* $\boldsymbol{S}=< U, Q, V, \phi >$ can be transformed to a Boolean information base $\boldsymbol{B}=< U, F, \varphi >$ assigning to each $v \in V_q$, $q \in Q$, one property $f_{qv} \in F$ such that $\varphi(x, f_{qv}) = 1$ if $\phi(x, q) = v$, and $\varphi(x, f_{qv}) = 0$, otherwise. Let us remark that $\boldsymbol{F}=\{F_1,\ldots,F_r\}$, with $F_q = \{f_{qv}, v \in V_q\}$, $q \in Q$, is a canonical partition of $F$. The opposite transformation, from a Boolean information base to an information system, is not always possible, i.e. there may exist Boolean information bases which cannot be transformed into information systems because their sets of properties do not admit any canonical partition, as shown by the following example.

**Example.** Let us consider a Boolean information base, such that $U = \{x_1, x_2, x_3\}$, $F = \{f_1, f_2\}$ and function $\varphi$ is defined by Table 1. One can see that $\boldsymbol{F}=\{\{f_1, f_2\}\}$ is not a canonical partition because $\varphi(x_3, f_1) = \varphi(x_3, f_2) = 1$, while definition of canonical partition $\boldsymbol{F}$ does not allow that for an object $x \in U$, $\varphi(x, f_1) = \varphi(x, f_2) = 1$. Therefore, this Boolean information base has no equivalent information system. Let us remark that also the Boolean information base presented in Table 2, where $U = \{x_1, x_2, x_4\}$ and $F = \{f_1, f_2\}$, cannot be transformed to an information system because partition $\boldsymbol{F}=\{\{f_1, f_2\}\}$ is not canonical. Indeed, $\varphi(x_4, f_1) = \varphi(x_4, f_2) = 0$, while definition of canonical partition $\boldsymbol{F}$ does not allow that for an object $x \in U$, $\varphi(x, f_1) = \varphi(x, f_2) = 0$.

The above says that consideration of rough approximation in the context of a Boolean information base is more general than the same consideration in the context of an information system. This means, of course, that the rough approximation considered in the context of a fuzzy information base is yet more general.

| **Table 1.** Information base $B$ | | |
|---|---|---|
| | $f_1$ | $f_2$ |
| $x_1$ | 0 | 1 |
| $x_2$ | 1 | 0 |
| $x_3$ | 1 | 1 |

| **Table 2.** Information base $B'$ | | |
|---|---|---|
| | $f_1$ | $f_2$ |
| $x_1$ | 0 | 1 |
| $x_2$ | 1 | 0 |
| $x_4$ | 0 | 0 |

It is worth stressing that the Boolean information bases $B$ and $B'$ are not Boolean information systems. In fact, on one hand, a Boolean information base provides information about absence ($\varphi(x,f) = 0$) or presence ($\varphi(x,f) = 1$) of properties $f \in F$ in objects $x \in U$. On the other hand, a Boolean information system provides information about values assigned by attributes $q \in Q$, whose sets of values are $V_q = \{0,1\}$, to objects $x \in U$, such that $\phi(x,q) = 1$ or $\phi(x,q) = 0$ for all $x \in U$ and $q \in Q$. Observe, therefore, that to transform a Boolean information system $S$ into a Boolean information base $B$, each attribute $q$ of $S$ corresponds to two properties $f_{q0}$ and $f_{q1}$ of $B$, such that for all $x \in U$

- $\varphi(x, f_{q0}) = 1$ and $\varphi(x, f_{q1}) = 0$ if $\phi(x,q) = 0$,
- $\varphi(x, f_{q0}) = 0$ and $\varphi(x, f_{q1}) = 1$ if $\phi(x,q) = 1$.

Thus, the Boolean information base $B$ in Table 1 and the Boolean information system $S$ in Table 3 are different, despite they could seem identical. In fact, the Boolean information system $S$ in Table 3 can be transformed into the Boolean information base $B''$ in Table 4, which is clearly different from $B$.      ◇

**Table 3.** Information system $S$

|       | $q_1$ | $q_2$ |
|-------|-------|-------|
| $x_1$ | 0     | 1     |
| $x_2$ | 1     | 0     |
| $x_3$ | 1     | 1     |

**Table 4.** Information base $B''$

|       | $f_{q_1 0}$ | $f_{q_1 1}$ | $f_{q_2 0}$ | $f_{q_2 1}$ |
|-------|-------------|-------------|-------------|-------------|
| $x_1$ | 1           | 0           | 0           | 1           |
| $x_2$ | 0           | 1           | 1           | 0           |
| $x_3$ | 0           | 1           | 0           | 1           |

What are the relationships between the rough approximation considered in the context of a fuzzy information base and the classical definition of rough approximation in the context of an information system? The following theorem is useful for answering this question.

**Theorem 3.** Let us consider an information system and the corresponding Boolean information base. For each $P \subseteq Q$, let $E^P$ be the set of all the properties $f_{qv}$ corresponding to values $v \in V_q$ of each particular attribute $q \in P$. For each $X \subseteq U$ we have

$$\underline{E^{P(>)}}(X) = \underline{E^{P(<)}}(X) = \underline{P}(X),$$
$$\overline{E^{P(>)}}(X) = \overline{E^{P(<)}}(X) = \overline{P}(X).$$

**Proof**
Since the information base is Boolean, then $D^+_{E^P}(x) = I_P(x)$ and $D^-_{E^P}(x) = I_P(x)$. In fact,

$$D_{EP}^+(x) =$$
$$\{y \in U : \varphi(y, f_{qv}) \geq \varphi(x, f_{qv}) \text{ for all } f_{qv}, \text{ such that } v \in V_q \text{ and } q \in P\} =$$
$$\{y \in U : \varphi(y, f_{qv})=1 \text{ if } \varphi(x, f_{qv})=1 \text{ for all } f_{qv}, \text{ such that } v \in V_q \text{ and } q \in P \}=$$
$$\{y \in U : f(y, q) = v \text{ if } f(x, q) = v \text{ for all } v \in V_q \text{ and } q \in P \}=$$
$$\{y \in U : f(y, q) = f(x, q) \text{ for all } q \in P\} = I_P(x),$$

$$D_{EP}^-(x) =$$
$$\{y \in U : \varphi(y, f_{qv}) \leq \varphi(x, f_{qv}) \text{ for all } f_{qv}, \text{ such that } v \in V_q \text{ and } q \in P\} =$$
$$\{y \in U : \varphi(y, f_{qv})=0 \text{ if } \varphi(x, f_{qv})=0 \text{ for all } f_{qv}, \text{ such that } v \in V_q \text{ and } q \in P \}=$$
$$\{y \in U : f(y, q) \neq v \text{ if } f(x, q) \neq v \text{ for all } v \in V_q \text{ and } q \in P \}=$$
$$\{y \in U : f(y, q) = f(x, q) \text{ for all } q \in P\} = I_P(x).$$

Thus, for all $X \subseteq U$, we have that

$$\underline{E^{P^{(>)}}}(X) = \{x \in U : D_{EP}^+(x) \subseteq X\} = \{x \in U : I_P(x) \subseteq X\} = \underline{P}(X),$$

$$\overline{E^{P^{(>)}}}(X) = \{x \in U : D_{EP}^-(x) \cap X \neq \emptyset\} = \{x \in U : I_P(x) \cap X \neq \emptyset\} = \overline{P}(X),$$

$$\underline{E^{P^{(<)}}}(X) = \{x \in U : D_{EP}^-(x) \subseteq X\} = \{x \in U : I_P(x) \subseteq X\} = \underline{P}(X),$$

$$\overline{E^{P^{(<)}}}(X) = \{x \in U : D_{EP}^+(x) \cap X \neq \emptyset\} = \{x \in U : I_P(x) \cap X \neq \emptyset\} = \overline{P}(X). \qquad \square$$

The above theorem proves that the rough approximation of a set considered within a Boolean information base admitting a canonical partition is equivalent to the classical rough approximation of the same set considered within the corresponding information system. Therefore, the classical rough approximation of a set is a particular case of the rough approximation of this set considered within a fuzzy information system.

## 4 Conclusions

We presented a general model of rough approximations based on ordinal properties of membership functions of fuzzy sets. In this very general framework, the classical rough set theory can be considered as a particular case. This result gives us a certainty that the fuzzy generalization of rough approximations based on Dominance-based Rough Set Approach is a real generalization of the original rough set concept. We have shown, moreover, that this generalization is a natural continuation of some ideas given by Leibniz, Frege, Boole, Łukasiewicz and Pawlak. As it exploits only ordinal properties of membership degrees and monotonic relationships between them, without using any (arbitrary) fuzzy connective, we claim that this is the most appropriate fuzzy generalization of the rough set concept. Finally, this result opens a promising research direction in which we envisage further developments, in particular, concerning algebraic properties of the proposed rough approximations.

**Acknowledgements.** The research of the first two authors has been supported by the Italian Ministry of University and Scientific Research (MUR). The third author wishes to acknowledge financial support from the Polish Ministry of Science and Higher Education (grant no. 3T11F 02127).

# References

1. Dubois, D., Prade, H., Putting rough sets and fuzzy sets together, in: R. Słowiński (ed.), *Intelligent Decision Support. Handbook of Applications and Advances of the Sets Theory*, Kluwer, Dordrecht, 1992, pp. 203-232
2. Fodor, J., Roubens, M., *Fuzzy Preference Modelling and Multicriteria Decision Support*, Kluwer, Dordrecht, 1994
3. Klement, E. P., Mesiar, R., Pap, E., *Triangular Norms*, Kluwer, Dordrecht, 2000
4. Greco, S., Matarazzo, B., Słowiński, R., Rough sets theory for multicriteria decision analysis, *European Journal of Operational Research*, 129 (2001) 1-47
5. Greco, S., Matarazzo, B., Słowiński, R., Decision rule approach, in: J. Figueira, S. Greco, M. Erghott (eds.) *Multiple Criteria Decision Analysis: State of the Art Surveys*, Springer, Berlin, 2005, pp. 507-563
6. Loemker, L., (ed. and trans.), Leibniz, G. W., *Philosophical Papers and Letters*, 2nd ed., Dordrecht, D. Reidel, 1969
7. Pal, S.K., Skowron, A. (eds.), *Rough-Fuzzy Hybridization: A new trends in decision making*, Springer-Verlag, Singapore, 1999
8. Pawlak, Z., Rough Sets, *International Journal of Computer and Information Sciences*, 11 (1982) 341-356
9. Pawlak, Z., *Rough Sets. Theoretical Aspects of Reasoning about Data*, Kluwer, Dordrecht, 1991
10. Pawlak, Z., Rough Set Theory, *Kunstliche Intelligenz*, 3 (2001) 38-39
11. Peters, J. F., Skowron, A., Dubois, D., Grzymala-Busse, J., Inuiguchi, M., Polkowski, L. (eds.), *Rough Sets and Fuzzy Sets*, Transactions on Rough Sets II, LNCS 3135, Springer-Verlag, 2005
12. Słowiński, R., Greco, S., Matarazzo, B., Rough set based decision support, chapter 16 in: E.K. Burke and G. Kendall (eds.), *Search Methodologies: Introductory Tutorials in Optimization and Decision Support Techniques*, Springer-Verlag, New York, 2005, pp. 475-527
13. Vakarelov, D., Consequence Relations and Information Systems, in: R. Słowiński (ed.), *Intelligent Decision Support. Handbook of Applications and Advances of the Rough Sets Theory*, Kluwer, Dordrecht, 1992, pp. 391-399
14. Zadeh, L., Fuzzy Set, *Information Control*, 8 (1965) 338-353

# Deriving Belief Networks and Belief Rules from Data: A Progress Report

Jerzy W. Grzymała-Busse[1,2], Zdzisław S. Hippe[1], and Teresa Mroczek[1]

[1] University of Information Technology and Management,
ul. Sucharskiego 2, 35-225 Rzeszów, Poland
{zhippe,tmroczek}@wsiz.rzeszow.pl
[2] Department of Electrical Engineering and Computer Science,
University of Kansas, Lawrence KS 66045-7523, USA
jerzy@ku.edu

**Abstract.** An in-house developed computer program *Belief*-**SEEKER**, capable to generate belief networks and also to generate sets of belief rules, has been presented in this paper. This system has a modular architecture, and consists of the following modules: *Knowledge Discovery Module* (**KDM**, an intelligent agent or pre-processor), *Belief Network Development Module* (**BDM**, generates belief networks), *Belief Network Training Module* (**BTM**, shows the distribution of conditional probabilities using a two-dimensional graph, together with some hints extracted from the investigated data), *Belief Network Conversion Module* (**BCM**, converts generated belief networks into relevant sets of belief rules of the type **IF...THEN**), and *Probability Reasoning Module* (**PRM**, checks the correctness of developed learning models as the "prediction of future" in classification of unseen examples).

**Keywords:** belief networks, belief rules, *Belief*SEEKER.

## 1 Introduction

Belief network and belief rule induction system *Belief*SEEKER has been developed at the University of Information Technology and Management (Rzeszow, Poland), in cooperation with the University of Kansas. The first application of *Belief*SEEKER was applied to the classification and prediction of melanocytic skin lesions [11]. Other applications were reported in [1,2,4,5,6,14,18]. The current version of *Belief*SEEKER, implemented in RAD-Delphi (Borland Software Corporation), searches for knowledge hidden in data and then uses it for classification of various objects, physical or abstract, such as ideas, concepts and/or processes. Moreover, a version of the system described here contains a new pre-processor that enables

- detecting and correcting typographical errors in the source decision table,
- creating truncated representation of the inputted data (in a sense of reduced representation of all concepts, using the most representative cases), and
- applying specific discretization schemes to deal with numerical attributes.

J.F. Peters et al. (Eds.): Transactions on Rough Sets VII, LNCS 4400, pp. 53–69, 2007.

Additionally, special mechanisms were developed for handling missing attribute values and for loading external numerical data, discretized using advanced algorithms elaborated at the University of Kansas [9].

The main aim of the paper is to describe internal features of *Belief*SEEKER that have never been published so far. Therefore, a general description of the system has been included.

## 2  Decision Table

Formally, a decision table is an ordered set DT = (U, C, D, V, f), where

- U, C, D are nonempty sets of elements, U is a set of objects (or cases), C is a set of descriptive attributes, D is a set of decision attributes,
- C, D $\subseteq$ A, C $\cup$ D = A, C $\cap$ D = $\emptyset$, where A is finite set of attributes,
- $V = \bigcup_{a \in A} V_a$ ($V_a$ is a domain of attribute a $\in$ A), and
- f: U $\times$ A $\rightarrow$ V is an information function, such that $\forall x \in U, a \in A[f(x, a) \in V_a]$.

In addition, specific values of the decision attribute d, d $\in$ D, correspond to classes or concepts.

The *Belief*SEEKER system accepts input data in the form of text files, or decision tables, specified by Pawlak [15], i.e., consisting of some descriptive attributes, and only one decision attribute located in the rightmost column. Thus, decision tables processed by the system are of the type 2a [17], see Table 1.

**Table 1.** An example decision table

| U | A | | | | | |
|---|---|---|---|---|---|---|
| | C | | | | | D |
| | Color | Size | Shape | Taste | Weight | Fruit |
| x1 | green | small | round | sweet | 0.1 | Banana |
| x2 | green | medium | round | sweet | 0.5 | Apple |
| x3 | green | big | round | sweet | 0.4 | Apple |
| x4 | green | medium | round | sour | 0.3 | Apple |
| x5 | green | small | round | sweet | 0.1 | Grape |
| x6 | yellow | medium | elongated | sweet | 0.3 | Banana |
| x7 | green | medium | elongated | sweet | 0.2 | Banana |
| x8 | green | big | elongated | sweet | 0.3 | Banana |
| x9 | maroon | small | round | sweet | 0.2 | Grape |
| x10 | green | small | round | sour | 0.1 | Grape |
| x11 | red | medium | round | sweet | 0.3 | Apple |
| x12 | red | medium | round | sour | 0.4 | Apple |
| x13 | maroon | small | round | sweet | 0.2 | Grape |

Table 1 provides the description of selected fruits (decision), using five relevant features (descriptive attributes). It contains 13 cases, and may be presented as DT = (U, C, D, V, f), where

$U = \{x_1, x_2, x_3, x_4, x_5, x_6, x_7, x_8, x_9, x_{10}, x_{11}, x_{12}, x_{13}\}$,

$C = \{$Color, Size, Shape, Taste, Weight$\}$,

$D = \{$Fruit$\}$,

$V_A = V_{Color} \cup V_{Size} \cup V_{Shape} \cup V_{Taste} \cup V_{Weight}$ and $V_D = V_{Fruit}$,

$V_{Color} = \{$green, maroon, red, yellow$\}$,

$V_{Size} = \{$small, medium, big$\}$,

$V_{Shape} = \{$elongated, round$\}$,

$V_{Taste} = \{$sour, sweet$\}$,

$V_{Weight} = \{0.1, 0.2, 0.3, 0.4, 0.5\}$,

$V_{Fruit} = \{$Apple, Banana, Grape$\}$,

f: $U \times A \rightarrow V$, e.g.: f(1, Color) = green, f(3, Shape) = round, etc.

## 3 Inconsistent (Contradictory) Data

When data in decision table are *inconsistent*, *Belief*SEEKER activates internal mechanisms based on the theory of rough sets. The main advantage of this theory, introduced by Z. Pawlak in 1982 [16], is that it does not need any preliminary or additional information about data (like grade of membership in fuzzy set theory). In rough set theory approach inconsistencies are not removed from consideration. Instead, *lower* and *upper* approximations of the concept are computed. On the basis of these approximations, *Belief*SEEKER computes two corresponding sets of belief networks (*certain* and *possible*), as well as two sets of belief rules (*certain* and *possible*). Some basic notions and ideas of the rough set theory are quoted from [3]. In a decision table any subset of the set of all examples, defined by the same value of the decision is called a *concept*. Let $U$ denote the set of all examples of the decision table and let $P$ be a subset of the set $A$ of all variables, i.e., attributes and decisions. Let $P$ be a subset of $A$. An

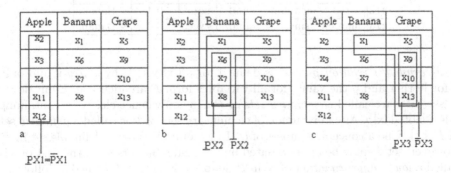

**Fig. 1.** Lower and upper approximations for concepts: a) Fruit = *Apple*, b) Fruit = *Banana*, c) Fruit = *Grape*

**Table 2.** $\underline{P}X1 = \overline{P}X1$

| U | A | | | | | D |
|---|---|---|---|---|---|---|
| | C | | | | | |
| | Color | Size | Shape | Taste | Weight | Fruit |
| x1 | green | small | round | sweet | 0.1 | ~ Apple |
| x2 | green | medium | round | sweet | 0.5 | Apple |
| x3 | green | big | round | sweet | 0.4 | Apple |
| x4 | green | medium | round | sour | 0.3 | Apple |
| x5 | green | small | round | sweet | 0.1 | ~ Apple |
| x6 | yellow | medium | elongated | sweet | 0.3 | ~ Apple |
| x7 | green | medium | elongated | sweet | 0.2 | ~ Apple |
| x8 | green | big | elongated | sweet | 0.3 | ~ Apple |
| x9 | maroon | small | round | sweet | 0.2 | ~ Apple |
| x10 | green | small | round | sour | 0.1 | ~ Apple |
| x11 | red | medium | round | sweet | 0.3 | Apple |
| x12 | red | medium | round | sour | 0.4 | Apple |
| x13 | maroon | small | round | sweet | 0.2 | ~ Apple |

**Table 3.** $\underline{P}X2$

| U | A | | | | | D |
|---|---|---|---|---|---|---|
| | C | | | | | |
| | Color | Size | Shape | Taste | Weight | Fruit |
| x1 | green | small | round | sweet | 0.1 | ~ Banana |
| x2 | green | medium | round | sweet | 0.5 | ~ Banana |
| x3 | green | big | round | sweet | 0.4 | ~ Banana |
| x4 | green | medium | round | sour | 0.3 | ~ Banana |
| x5 | green | small | round | sweet | 0.1 | ~ Banana |
| x6 | yellow | medium | elongated | sweet | 0.3 | Banana |
| x7 | green | medium | elongated | sweet | 0.2 | Banana |
| x8 | green | big | elongated | sweet | 0.3 | Banana |
| x9 | maroon | small | round | sweet | 0.2 | ~ Banana |
| x10 | green | small | round | sour | 0.1 | ~ Banana |
| x11 | red | medium | round | sweet | 0.3 | ~ Banana |
| x12 | red | medium | round | sour | 0.4 | ~ Banana |
| x13 | maroon | small | round | sweet | 0.2 | ~ Banana |

*indiscernibility relation* $\rho$ on $U$ is defined for all x, y $\in$ U by $x \rho y$ if and only if for both $x$ and $y$ the values for all variables form $P$ are identical. Equivalence classes of $\rho$ are called *elementary sets* of $P$. An equivalence class of $\rho$ containing $x$ is denoted $[x]\rho$. Any finite union of elementary sets of $P$ is called a *definable* set in $P$. Let X be a concept, a subset of $U$. In general, $X$ is not a definable set in $P$. However, set $X$ may be approximated by two definable sets in $P$, the first one is called a *lower approximation* of $X$ in $P$ denoted by $\underline{P}X$ and defined as follows:

$$\{x \in U \mid [x]_\rho \subseteq X\}.$$

**Table 4.** $\overline{P}X2$

| U | A | | | | | |
|---|---|---|---|---|---|---|
| | C | | | | | D |
| | Color | Size | Shape | Taste | Weight | Fruit |
| x1 | green | small | round | sweet | 0.1 | Banana |
| x2 | green | medium | round | sweet | 0.5 | ~ Banana |
| x3 | green | big | round | sweet | 0.4 | ~ Banana |
| x4 | green | medium | round | sour | 0.3 | ~ Banana |
| x5 | green | small | round | sweet | 0.1 | Banana |
| x6 | yellow | medium | elongated | sweet | 0.3 | Banana |
| x7 | green | medium | elongated | sweet | 0.2 | Banana |
| x8 | green | big | elongated | sweet | 0.3 | Banana |
| x9 | maroon | small | round | sweet | 0.2 | ~ Banana |
| x10 | green | small | round | sour | 0.1 | ~ Banana |
| x11 | red | medium | round | sweet | 0.3 | ~ Banana |
| x12 | red | medium | round | sour | 0.4 | ~ Banana |
| x13 | maroon | small | round | sweet | 0.2 | ~ Banana |

**Table 5.** $\underline{P}X3$

| U | A | | | | | |
|---|---|---|---|---|---|---|
| | C | | | | | D |
| | Color | Size | Shape | Taste | Weight | Fruit |
| x1 | green | small | round | sweet | 0.1 | ~ Grape |
| x2 | green | medium | round | sweet | 0.5 | ~ Grape |
| x3 | green | big | round | sweet | 0.4 | ~ Grape |
| x4 | green | medium | round | sour | 0.3 | ~ Grape |
| x5 | green | small | round | sweet | 0.1 | ~ Grape |
| x6 | yellow | medium | elongated | sweet | 0.3 | ~ Grape |
| x7 | green | medium | elongated | sweet | 0.2 | ~ Grape |
| x8 | green | big | elongated | sweet | 0.3 | ~ Grape |
| x9 | maroon | small | round | sweet | 0.2 | Grape |
| x10 | green | small | round | sour | 0.1 | Grape |
| x11 | red | medium | round | sweet | 0.3 | ~ Grape |
| x12 | red | medium | round | sour | 0.4 | ~ Grape |
| x13 | maroon | small | round | sweet | 0.2 | Grape |

The second set is called an *upper approximation* of $X$ in $P$, denoted by $\overline{P}X$ an defined as follows:

$$\{x \in U \mid [x]_\rho \cap X \neq \emptyset\}$$

The *lower approximation* of $X$ in $P$ is the greatest definable set in $P$, contained in $X$. The *upper approximation* of $X$ in $P$ least definable set in $P$ containing $X$. A rough set of $X$ is the family of all subsets of $U$ having the same *lower* and the same *upper approximations* of $X$. In our example, classification based on the set C of attributes is given by the following partition:

$$\{\{x_1, x_5\}, \{x_2\}, \{x_3\}, \{x_4\}, \{x_6\}, \{x_7\}, \{x_8\}, \{x_9, x_{13}\}, \{x_{10}\}, \{x11\}, \{x_{12}\}\}.$$

The set D of decisions contains only one element. As it follows from Table 1 cases $x_1$ and $x_5$ are contradictory. In addition, Table 1 contains three concepts (X1 - Fruit $=$ *Apple*, X2 - Fruit $=$ *Banana*, X3 - Fruit $=$ *Grape*), where

$$X1 = \{x_2, x_3, x_4, x_{11}, x_{12}\},$$

$$X2 = \{x_1, x_6, x_7, x_8\},$$

$$X3 = \{x_5, x_9, x_{10}, x_{13}\}.$$

Therefore:

1. $\underline{P}X1 = \{x_2, x_3, x_4, x_{11}, x_{12}\}$ , $\overline{P}X1 = \{x_2, x_3, x_4, x_{11}, x_{12}\}$,
2. $\underline{P}X2 = \{x_6, x_7, x_8\}$, $\overline{P}X2 = \{x_1, x_5, x_6, x_7, x_8\}$,
3. $\underline{P}X3 = \{x_9, x_{10}, x_{13}\}$, $\overline{P}X3 = \{x_1, x_5, x_9, x_{10}, x_{13}\}$.

Graphical interpretation of these sets is shown on Fig. 1.

Lower and upper approximations of concepts are presented in Tables 2–6.

In the next step, *Belief*SEEKER derives a collection of nets, labeled as certain belief networks (for $\underline{P}X$) or possible belief networks (for $\overline{P}X$), among them an optimum network, featuring the smallest classification error.

Table 6. $\overline{P}X3$

| U | A | | | | | |
|---|---|---|---|---|---|---|
| | C | | | | | D |
| | Color | Size | Shape | Taste | Weight | Fruit |
| x1 | green | small | round | sweet | 0.1 | Grape |
| x2 | green | medium | round | sweet | 0.5 | ~ Grape |
| x3 | green | big | round | sweet | 0.4 | ~ Grape |
| x4 | green | medium | round | sour | 0.3 | ~ Grape |
| x5 | green | small | round | sweet | 0.1 | Grape |
| x6 | yellow | medium | elongated | sweet | 0.3 | ~ Grape |
| x7 | green | medium | elongated | sweet | 0.2 | ~ Grape |
| x8 | green | big | elongated | sweet | 0.3 | ~ Grape |
| x9 | maroon | small | round | sweet | 0.2 | Grape |
| x10 | green | small | round | sour | 0.1 | Grape |
| x11 | red | medium | round | sweet | 0.3 | ~ Grape |
| x12 | red | medium | round | sour | 0.4 | ~ Grape |
| x13 | maroon | small | round | sweet | 0.2 | Grape |

# 4   Architecture of *Belief*SEEKER

*Belief*SEEKER, with a modular architecture based on five units: KDM, BDM, BTM, BCM, and PRM, is depicted on Fig. 2.

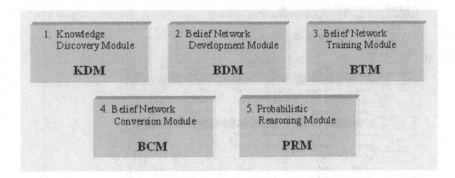

**Fig. 2.** Simplified architecture of *Belief*SEEKER

*Knowledge Discovery Module (KDM)* plays a role of an intelligent agent or pre-processor, aimed at preliminary processing of data collected in a form of decision table (in text format), fulfilling requirements described in [13]. The **KDM** module checks thoroughly the consistency of semantics used for the description of data, allows to handle missing data (both numeric or symbolic) and additionally has some internal mechanisms to create reliable image of extremely large data sets, i.e., develops the truncated representation of source data. The **KDM** module is also responsible for quantization of continuous data and visualization of the developed belief networks. Thus, the main task of **KDM** module is related with intense analysis of the source decision table. Some ideas of this process were already discussed in the introductory part of the paper. Within a decision table, **KDM** module seeks for consistent rows, rows that are inconsistent, and for redundant rows (Fig. 3).

An interesting feature of the discussed module is the ability to develop the meaning of symbols "*" and "?". In the first case, valid for symbolic attributes only, new cases are automatically generated. The number of cases added equals to the number of values of the attribute for which symbol "*" did appear. Say, for a row

| green | * | round | sweet | 0.1 | Banana |
|-------|---|-------|-------|-----|--------|

three new cases are developed because the attribute *Size* (see Table 1) has three different values *big*, *medium*, and *small*.

| green | small | round | sweet | 0.1 | Banana |
|-------|--------|-------|-------|-----|--------|
| green | medium | round | sweet | 0.1 | Banana |
| green | big | round | sweet | 0.1 | Banana |

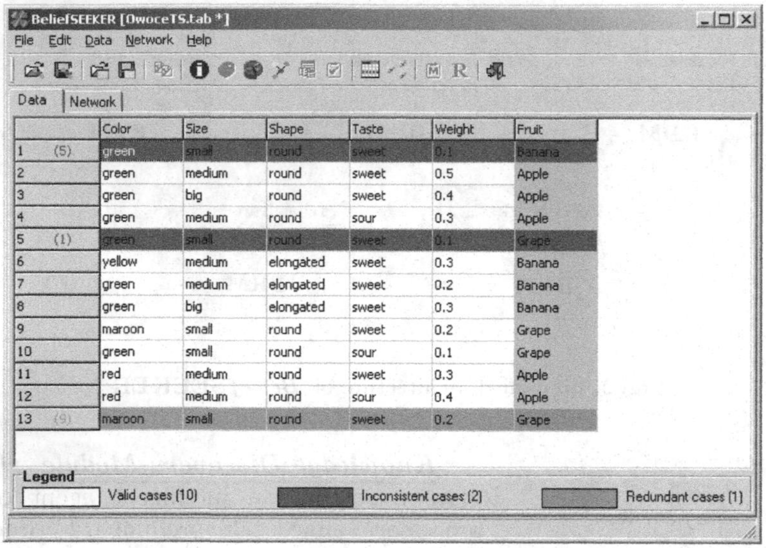

**Fig. 3.** Decision table - a dialogue window of *Belief*SEEKER. Object (row) 1 is inconsistent whit object 5; object 13 is exactly the same as the object 9, i.e., is redundant. Additionally, any entry in the table can be easily modified (corrected) at hand.

In the second case (?), two different outcomes are possible, depending on the attribute type. For symbolic attributes, question mark is replaced by the most frequent value for a given column. For continuous attribute, "?" is replaced by the arithmetic mean of all values, taken from vectors (rows) that are most similar to the considered row. However, the final decision which of the mechanisms will be selected is up to the user. The last step of the pre-processing with **KDM** module is devoted to a visualization of the developed belief networks (Fig. 4). Here, individual nodes of the network (green for numerical and yellow for symbolic attributes) are located in a circle with the decision node (gray) located in the center.

On the left side of Fig. 4 graphical information is displayed, indicating the frequency of occurrence of a given value for each attribute.

2. Belief Network
Development Module

**BDM**

***Belief Network Development Module
(BDM)*** Within this module the process of development of belief networks is steadily controlled by a specific parameter, known as marginal likelihood (*ML*) [13], showing the maximum dependence between variables:

$$ML = \prod_{i=1}^{v} \prod_{j=1}^{q_i} \frac{\Gamma\left(\alpha_{ij}\right)}{\Gamma\left(\alpha_{ij} + n_{ij}\right)} \prod_{k=1}^{c_i} \frac{\Gamma\left(\alpha_{ijk} + n_{ijk}\right)}{\Gamma\left(\alpha_{ijk}\right)}, \tag{1}$$

where $i = 1, ..., v$, where v is the number of nodes in the network, $j = 1, ..., q_i$, where $q_i$ is the number of possible combinations of parents of the node $X_i$ (if a

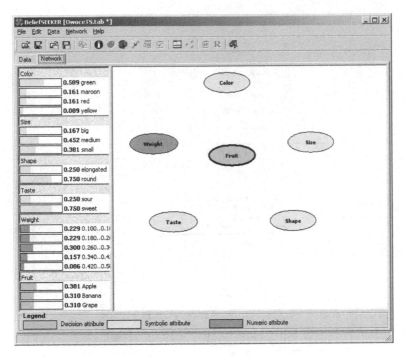

**Fig. 4.** Preliminary visualization of the developed belief network (see explanation in the text)

given attribute does not contain nodes of the type "parent", $q_i = 1, k = 1, ..., c_i$, where $c_i$ is the number of classes within the attribute $X_i$, $n_{ijk}$ - is the number of rows in the database, for which parents of the attribute $X_i$ have value j, and this attribute has the value of k, and $\alpha_{ijk}, \alpha_{ij}$ - are parameters of the initial Dirichlet's distribution [10].

Note that the calculation of Dirichlet's parameter ($DP$) has been favorably optimized through reducing the number of iteration steps, owing to the application of a special algorithm for elimination of variables. Currently, *Belief*SEEKER allows to develop a single (optimal) belief network (for any distinct value of $DP$), or can generate a set of belief networks for incrementally increased value of $DP$. In a separate process of global optimization, only dissimilar networks are kept for further processing, i.e., generation of belief rules and/or classification of unseen cases. Belief networks can be developed using various basic algorithms like K2, K2 augmented for exhaustive searching of all possible connections of nodes, Naive Bayesian Classifier, and Reversed Naive Bayesian Classifier [13]. An outline of the improved version of K2 algorithm [12] is shown on Fig. 5. Belief networks developed within the **BDM** module are displayed in a very special way, facilitating their understanding.

Although application of Bayes theory itself allows for correct processing of contradictory data, the current version of *Belief*SEEKER contains additional mechanisms to deal with such data. The approach used is based on Pawlak's

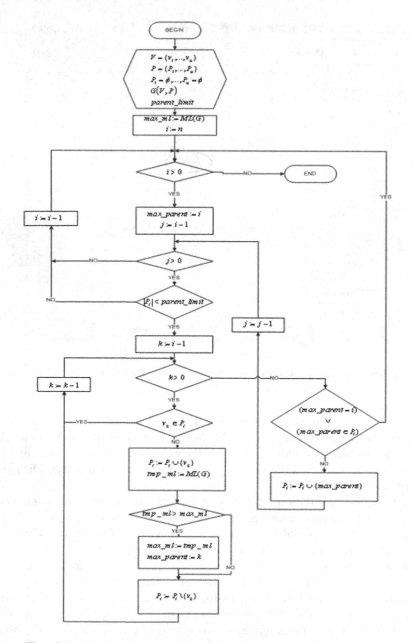

**Fig. 5.** Flowchart of the improved version of K2 algorithm. Explanatory notes: n - number of nodes in a network, $V = \{v_1, ..., v_n\}$ - set of network's nodes, $P = P_1, ..., P_n$ - sets of parents of individual nodes; at the beginning this set is empty (i.e., a network has no connections), G(V, P) - a graph of the network, described by the set of nodes and the set of parents of individual nodes, parent_limit - maximum allowed number of parents of a node, $ML$ - marginal likelihood, $|Pi|$ - number of elements in the set of parents of the i-th node.

theory of rough sets, which has already been discussed previously. Thus for inconsistent data, the system produces two types of belief networks, namely *certain networks* and *possible networks*. Similarly, also two types of rules (*certain rules* and *possible rules*) can be created (see description of **BCM** module).

*Belief Network Training Module*

3. Belief Network
Training Module

**BTM**

*(BTM)* supports the development process of belief networks. Within this module, Bayesian belief network is split into two main components, tentatively called *qualitative component* and *quantitative component*. The first component (qualitative) is formally represented by a two-dimensional graph, showing mutual dependences between all variables. In other words, this component reveals internal links within independent variables and their global influence onto the dependent variable (i.e., decision). The second component (quantitative) stores information about distribution of conditional probabilities (in the network) for all entities of the type "parent-child". The distribution of conditional probabilities can be estimated using the 2D-graph (earlier referred to) and some hints extracted from the investigated data. Finally, for each node of the generated network, a special array is automatically created, to store information about influence of a given variable on other descriptive variables in the network.

One of the most important task of the discussed module is searching for the best learning model, depicted in a form of belief network. The learning process begins by the estimation of marginal likelihood (*ML*) for a network without arcs, and terminates when *ML* reaches the maximum value. This means that in consecutive steps of building the learning model, arcs connecting nodes can be created only when their addition causes maximization of matching function. The learning process can be conveniently displayed in *statu nascendi*, i.e., while the network is developed. The final network for *Fruit* data is shown on Fig. 6.

Arrows on arcs connecting given nodes depict relations of the type "parent" $\rightarrow$ "child" represented by the conditional probability table (**CPT**) $P(A|B_1,...,B_n)$, where $B_1,...,B_n$ are variables of the type "parent" for the node A. In order to estimate that distribution we need to know (i) distribution of a priori probability (equation (2), $c_i$ - is here a number of unique values of a given attribute), and (ii) distribution of probability for unconnected network (equation (3)), where $\pi$ is an a priori probability of attribute $X_i$,
$\alpha$ - is a constant (usually equal to 1),
$x_i$ - is a number of instantions of attribute $X_i$, and
$n$ - is a number of cases in a data set, where

$$\pi = \frac{1}{c_i}, \tag{2}$$

and

$$P(X_i) = \frac{\pi \cdot \alpha + x_i}{\alpha + n}. \tag{3}$$

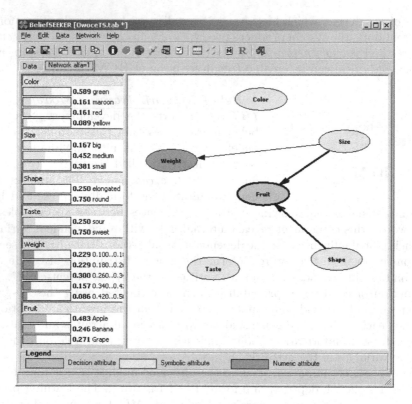

**Fig. 6.** Belief network generated for *Fruit* data. Bold arcs outgoing from nodes represent attributes with the highest influence on the decision. Attributes *Color* and *Taste* are less significant for classification of *Fruit* data.

Distribution of a prior probability for *Fruit* data is presented on Fig. 7. Therefore, distribution of probability for nodes $X_i$ having parents $Y_i$ can be expressed by function (4)

$$P(X_i|Y_i) = \frac{\pi \cdot \alpha_i + n_i}{\alpha_i + n}, \tag{4}$$

where $\pi$ is an a priori probability of attribute $X_i$,
$\alpha_i = \frac{\alpha}{q_i}$ - ($q_i$ is the number of possible combinations of values, assigned to parents of the regarded node),
$n_i$ - is a number of cases in the decision table for given combination of attribute values, and
$n$ - is a number of all expected values of the parent node.

The conditional probability $P(X_i|Y_i)$ describes a "cause"→"result" relation and allows to estimate the influence of "parent node" onto a "child node", taking into account all possible values of variables. Results of the computation are kept in conditional probability tables. An example of such a table, showing

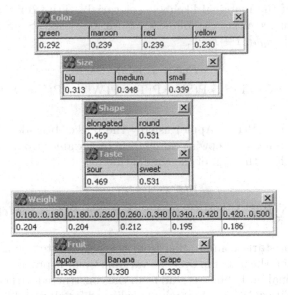

**Fig. 7.** Distribution of a priori probability for Fruit data

| Shape | Size | Apple | Banana | Grape |
|-------|------|-------|--------|-------|
| elongated | big | 0.048 | 0.905 | 0.048 |
| round | big | 0.905 | 0.048 | 0.048 |
| elongated | medium | 0.026 | 0.949 | 0.026 |
| round | medium | 0.973 | 0.013 | 0.013 |
| elongated | small | 0.333 | 0.333 | 0.333 |
| round | small | 0.011 | 0.204 | 0.785 |

**Fig. 8.** Conditional probability table for dependent variable *Fruit*

the influence of variable *Size* and *Shape* onto the dependent variable *Fruit* is presented on Fig. 8.

The topology of Bayesian network defines distribution of probabilities in an inclusive way, with probability specification for each event. It represents a complete specification of the developed model. An inclusive probability distribution for the network given on Fig. 6 is

$$P = P(Color) * P(Size) * P(Shape) * P(Taste) * (Weight|Size) * P(Fruit|Size, Shape).$$

Apart from probabilities defined so far, the topology of Bayesian network allows to define marginal distribution of all variables. The distribution of randomly labeled variable X, computed from inclusive probability distribution is called a marginal likelihood of X.

Calculation of the marginal likelihood of variable Fruit for the variable Apple can be conducted using the following codes: C→*Color*, S→*Size*, Sh→*Shape*, T→*Taste*, W→*Weight*, F→*Fruit*,

$$P(F = Apple) =$$
$$\sum_{C,S,Sh,T,W} P(C) * P(S) * P(Sh) * P(T) * P(W|S) * P(F = Apple|S, Sh).$$

For calculation of $P(F = Apple)$ usually the bucket brigade algorithm is applied. Its action relies on stepwise elimination of variables from the equation by shifting them before the sign of the summation

$$P(F = Apple) =$$
$$\sum_C P(C) \sum_T P(T) \sum_S P(S) \sum_W P(W|S) \sum_{Sh} P(Sh)P(F = Apple|S, Sh).$$

The algorithm starts computation from the most nested sum, i.e., from the right to left. When the last sum over the variable Color is processed, the required value of conditional probability is read from the constructed earlier tables of conditional probabilities. Values of marginal likelihood for all variables are displayed in graphical form on the left side of the main screen (see Fig. 6).

> 4. Belief Network
> Conversion Module
>
> **BCM**

**Belief Network Conversion Module (BCM).** Current version of **Belief SEEKER** provides special means for specific interpretation of the developed learning model. Namely, the classic belief network can be converted into relevant sets of belief rules (both certain and possible) of the type **IF...THEN**. All algorithms responsible for the conversion are included in the **BCM** module. The conversion process is controlled by a specific parameter called *belief factor* **BF**. This factor reveals indirectly the influence of the most significant descriptive attributes on the dependent variable (i.e., decision). Also, to facilitate the preliminary evaluation of generated rules, an additional mechanism supports the calculation of their specificity, strength, generality and accuracy. In the extended research based on the application of **Belief SEEKER** in various domains of data (business, chemistry, medicine, etc.) it has been found that sets of rules, developed using qualitative component of the belief network, were more accurate in classification of unseen cases than rules, generated with the use of qualitative component of the network. Note, however, that development of rules begins—in both cases—using the most significant variables, and is then stepwise augmented using variables (and rules) of the next generation.

In our example (*Fruit* data), the dependent variable is preceded by one generation only, therefore, the final set of belief rules—for BF = 0.900—consists of four rules (Fig. 9). However, selecting smaller BF, say 0.4, results in more rules, while the error rate may decrease.

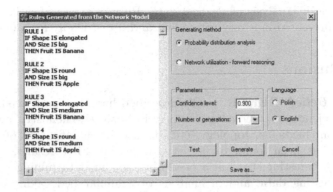

**Fig. 9.** Belief network conversion module - a dialogue window

**Rules Testing Results**

Testing database: "DwoceTS.tab"     Number of cases: 13

| | Color | Size | Shape | Taste | Weight | Fruit | Classification | Rules fitted |
|---|---|---|---|---|---|---|---|---|
| 1 | green | small | round | sweet | 0.1 | Banana | ? | |
| 2 | green | medium | round | sweet | 0.5 | Apple | Apple | 4 |
| 3 | green | big | round | sweet | 0.4 | Apple | Apple | 2 |
| 4 | green | medium | round | sour | 0.3 | Apple | Apple | 4 |
| 5 | green | small | round | sweet | 0.1 | Grape | ? | |
| 6 | yellow | medium | elongated | sweet | 0.3 | Banana | Banana | 3 |
| 7 | green | medium | elongated | sweet | 0.2 | Banana | Banana | 3 |
| 8 | green | big | elongated | sweet | 0.3 | Banana | Banana | 1 |
| 9 | maroon | small | round | sweet | 0.2 | Grape | ? | |
| 10 | green | small | round | sour | 0.1 | Grape | ? | |
| 11 | red | medium | round | sweet | 0.3 | Apple | Apple | 4 |
| 12 | red | medium | round | sour | 0.4 | Apple | Apple | 4 |
| 13 | maroon | small | round | sweet | 0.2 | Grape | ? | |

Number of correct classified cases: 8 (61.54%)
Number of incorrect classified cases: 0 (0.00%)
Number of unclassified cases: 5 (38.46%)

| Class.\Origi. | Apple | Banana | Grape |
|---|---|---|---|
| Apple | 5 | 0 | 0 |
| Banana | 0 | 3 | 0 |
| Grape | 0 | 0 | 0 |

Legend     Incorrect classified cases     Unclassified cases     Save as...     Statistics     OK

**Fig. 10.** Confusion matrix and auxiliary information - a dialogue window of *Belief*SEEKER during classification. Clicking on the Statistics at bottom displays information about most important parameters of corresponding rules, and case(s) classificated by such rules.

5. Probabilistic
Reasoning Module

**PRM**

***Probability Reasoning Module   (PRM)***
This module checks the correctness of developed learning models in the "prediction of future", i.e., in classification of unseen examples. Note that **PRM** module checks correctness of belief networks and belief rules as well. Results gathered inside this module are applied to construct the confusion matrix, containing auxiliary information about all false-negative and false-positive cases. This feature of the module plays an extremely

important role in analysis of medical data. Results of classification by **PRM** module are presented in a very convenient and user-friendly form (see Fig. 10)

## 5  Conclusions

The system *Belief*SEEKER, briefly described here, has been used in various domains of science and technology, namely in business, chemistry, environmental protection, medicine, topology, etc. Its use in classification of melanoma skin lesions has been found to be very successful. First results, quite satisfactory and interesting, have also been obtained in data mining experiments on hop processing data, and in analysis of granular bed caking during hop extraction [6,7,8].

## References

1. Błajdo, P., Grzymała-Busse, J.W., Hippe, Z.S., Knap, M., Marek, T., Mroczek, T., Wrzesień, M.: A suite of machine learning systems for mining of information and knowledge from data. In: Tadeusiewicz R., Ligęza A., Szymkat M. (Eds.), *Computer Methods and Systems. Scientific & Technical Programs.* Cracow, Poland (2003) 479–484 (in Polish).
2. Błajdo, P., Grzymała-Busse, J.W., Hippe, Z.S., Knap, M., Marek, T., Mroczek, T., Wrzesień , M.: A suite of machine learning programs for data mining: chemical applications. In: Dębska B., Fic G. (Eds.), *Information Systems in Chemistry* 2. University of Technology Editorial Office, Rzeszow, Poland (2004) 7–14.
3. Grzymała-Busse, J.W.: A new version of the rule induction system. *Fundamenta Informaticae* 31 (1997) 27–39.
4. Grzymała-Busse, J.W., Hippe, Z.S., Knap, M., Mroczek, T.: New computer program systems for knowledge engineering and machine learning. Comparison of some models of hidden uncertain knowledge. In: Bubnicki Z., Grzech A. (Eds.), *Knowledge Engineering and Expert Systems.* University of Technology Publishing. Office, Wrocław, Poland, Vol. 1 (2003) 239–247 (in Polish).
5. Grzymała-Busse, J.W., Hippe, Z.S., Mroczek, T.: Belief rules vs. decision rules: A preliminary appraisal to the problem. In: Kłopotek M.A., Wierzchoń S.T., Trojanowski K. (Eds.), *Advances in Soft Computing (Intelligent Information Processing and Web Mining).* Springer-Verlag, Heidelberg (2003) 431–435.
6. Grzymała-Busse, J.W., Hippe, Z.S., Mroczek, T.: System BeliefSEEKER — A new approach to induction of belief networks and belief rules. In: Burczyński T., Cholewa W., Moczulski W. (Eds.), *Artificial Intelligence Methods (AI-METH).* Silesian University of Technology Edit. Office, Gliwice, Poland (2005) 59–60.
7. Grzymała-Busse, J.W., Hippe, Z.S., Mroczek, T., Rój, E. Skowroński, B.: Data mining analysis of granular bed caking during hop extraction. Proceedings of the ISDA'2005, Fifth International Conference on Intelligent System Design and Applications, IEEE Computer Society, Wroclaw, Poland (2005) 426–431 .
8. Grzymała-Busse, J.W., Hippe, Z.S., Mroczek, T., Rój, E. Skowroński, B.: Data mining experiments on hop processing data. Proceedings of the HIS'2005, Fifth International Conference on Hybrid Intelligent Systems, IEEE Computer Society, Rio de Janeiro, Brazil (2005) 175–180 .

9. Grzymała-Busse, J.W., Santoso, S.: Experiments on data with three interpretations of missing attribute values—A rough set approach. Proc. Intern. Conference on New Trends in Intelligent Information Processing and Web Mining, Ustroń, Poland (2006) 143–152.
10. Heckerman, D.: A tutorial on learning Bayesian networks. Technical report MSR-TR-95-06: http://research.microsoft.com/research/pubs/view.aspx?msr_tr_id=MSR-TR-95-06 (1995)
11. Hippe, Z.S., T. Mroczek, T.: Melanoma classification and prediction using belief networks. In: Kurzyński M., Puchała E., Woźniak M. (Eds.), *Computer Recognition Systems*. University of Technology Publishing Office, Wrocław, Poland (2003) 337–342.
12. Hippe, Z.S. , Mroczek, T.: Belief networks and belief rules—Promising tools for mining hidden knowlegde in chemistry?. Proc. 3rd International Conference Information Systems in Chemistry (SIC), Bezmiechowa, Poland (in printing).
13. Jensen F.V.: *Bayesian Networks and Decision Graphs*. Springer-Verlag, Berlin-Heidelberg (2001).
14. Mroczek, T., Grzymała-Busse, J.W., Hippe, Z.S.: Rules from belief networks: a rough set approach. In: Tsumoto S., Słowiński R., Komorowski J., Grzymała-Busse J.W. (Eds.), *Rough Sets and Current Trends in Computing*. Springer-Verlag, Berlin-Heidelberg (2004) 483–487.
15. Pawlak, Z.: Knowledge and Rough Sets. In: Traczyk W. (Ed.) *Problems of Artificial Intelligence*. Wiedza i Życie, Warsaw, Poland (1995) 9–21 (in Polish).
16. Pawlak, Z.: Rough Sets. *Int. Journ. Computer and Information Sci.* 11 (1982) 341–356.
17. Varmuza, K.: Chemometrics - multivariate view on chemical problems. In: Schleyer P.v.R., Allinger N.L., Clark T., Gasteiger J., Kollman H.F., Schafer III F., Schreiner P.R. (Eds.), *The Encyclopedia of Computational Chemistry*. J. Willey & Sons Ltd., Chichester, Vol. 1 (1998) 346–366.
18. Varmuza, K., Grzymała-Busse, J.W., Hippe, Z.S., Mroczek T.: Comparison of consistent and inconsistent models in biomedical domain: a rough set approach to melanoma data. In: Burczyński T., Cholewa W., Moczulski W. (Eds.), *Artificial Intelligence Methods (AI-METH)*. Silesian University of Technology Publishing Office, Gliwice, Poland (2003) 323–328.

# Selection of Important Attributes for Medical Diagnosis Systems

Grzegorz Ilczuk[1] and Alicja Wakulicz-Deja[2]

[1] Siemens AG Medical Solutions,
Allee am Roethelheimpark 2, 91052 Erlangen, Germany
Grzegorz.Ilczuk@ilczuk.com
[2] Institut of Informatics University of Silesia,
Bedzinska 39, 41-200 Sosnowiec, Poland
wakulicz@us.edu.pl

**Abstract.** Success of machine learning algorithms is usually dependent on a quality of a dataset they operate on. For datasets containing noisy, inadequate or irrelevant information these algorithms may produce less accurate results. Therefore a common pre-processing step in data mining domain is a selection of highly predictive attributes. In this case study we select subsets of attributes from medical data using filter feature selection algorithms. To validate the algorithms we induce decision rules from the selected subsets of attributes and compare classification accuracy on both training and test datasets. Additionally medical relevance of the selected attributes is checked with help of domain experts.

## 1 Introduction

Feature selection is often an essential data processing step prior to applying a learning algorithm. If processed information contains irrelevant, unreliable or redundant data then a process of knowledge discovery is more difficult and achieved results are complicate to analyze. One way to remove the unneeded information is a selection of a subset of attributes from an original dataset for further processing. Depending on purposes of data mining this selection can be focused on:

- finding the minimally sized feature subset that is necessary and sufficient to the target concept
- improving classification accuracy or decreasing a number of selected features without significantly decreasing the prediction accuracy of a selected classifier
- approximating an original class distribution with a smaller subset of features

In our research we are trying to join both last goals, so that a classification accuracy after removal of some attributes does not significantly decrease and the resulting class distribution is close to the original class distribution calculated for all attributes. Main area of interest in our research is a complete vertical solution which is able to extract knowledge in form of decision rules from raw, medical data. This solution is built from a number of processing modules/subsystems:

J.F. Peters et al. (Eds.): Transactions on Rough Sets VII, LNCS 4400, pp. 70–84, 2007.

1. Import subsystem-responsible for importing data from medical information systems into our storage subsystem
2. Data recognition subsystem-this subsystem transforms raw data to a form suited for further data processing. Additionally noise and redundant data are removed based on a statistical analysis. Partial results were described in [16]
3. Feature subset selection - responsible for selecting an optimal set of attributes for a generation of decision rules.
4. Rule induction subsystem - uses based on Rough Sets [13,14] MLEM2 algorithm [23] for generating decision rules. Early research on this area was described in [15,19].
5. Visualization of the collected knowledge in a form easily understandable by humans. Partial results which based on decision trees were published in [20]

An block diagram of the described modules is shown at figure 1.

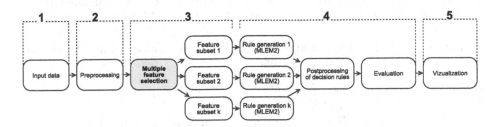

**Fig. 1.** Modules of the data exploration

In this paper we extend our initial research on a feature selection process and present results achieved with three different *filter* algorithms. Resulted feature subsets were used for rule generation using rough set MLEM2 algorithm. Validation of the subsets was done based on both prediction accuracy of decision rules generated from these sets and medical relevance proved by domain experts.

## 2 Rough Sets - Basic Notions

Developed by Pawlak and presented in 1982 Rough Sets theory is a mathematical approach to handle imprecision and uncertainty [11,24]. The main goal of rough set analysis is to synthesize approximation of concepts from the acquired data. Some basic definitions are presented below.

*Information system* is a pair $\mathbf{A} = (U, A)$ where $U$ is a non-empty, finite set called the universe and $A$ is a non-empty, finite set of *attributes*, i.e. $a : U \to V_a$ for $a \in A$, where $V_a$ is called the *value set* of attribute $a$. Elements of $U$ are called *objects*.

The special case of information systems called *decision system* is defined as $\mathbf{A} = (U, A \cup \{d\})$, where $d \notin A$ is a distinguished attribute called *decision* and elements of $A$ are called *conditions*.

A *decision rule* is defined as $r = (a_{i1} = v_1) \wedge \ldots \wedge (a_{im} = v_m) \Rightarrow (d = k)$ where $1 \leq i_1 < \ldots < i_m \leq |A|, v_i \in V_{ai}$. We say an object matches a rule if its attributes satisfy all *atomic formulas* $(a_{ij} = v_j)$ of the rule. A rule is called minimal consistent with **A** when any decision rule $r'$ created from $r$ by removing one of atomic formula of $r$ is not consistent with **A**.

We assume that the set $V_d$ of values of the decision $d$ is equal to $\{1, \ldots, r(d)\}$. The decision $d$ determines the partition $\{C_1, \ldots, C_{r(d)}\}$ of the universe $U$, where $C_k = \{x \in U : d(x) = k\}$ for $1 \leq k \leq r(d)$. The set $C_k$ is called k-th decision class of **A**.

Let $\mathbf{A} = (U, A)$ be an information system. With any subset of attribute $B \subseteq A$, an equivalence relation, denoted by $IND(B)$ called the *B-indiscernibility relation*, is associated and defined by:

$$IND(B) = \{(x, y) \in U \times U : \bigvee_{a \in B}(a(x) = a(y))\}$$

Objects $x, y$ satisfying relation $IND(B)$ are indiscernible by attributes from $B$. By $[x]_{IND(B)}$ we denote the equivalence class of $IND(B)$ defined by $x$. A minimal subset $B$ of $A$ such that $IND(A) = IND(B)$ is called a *reduct* of **A**.

If $\mathbf{A} = (U, A)$ is an information system, $B \subseteq A$ is a set of attributes and $X \subseteq U$ is a set of objects then the sets

$\underline{B}X = \{x \in U : [x]_{IND(B)} \subseteq X\};$ and
$\overline{B}X = \{x \in U : [x]_{IND(B)} \cap X \neq \emptyset\}$
are called *B-lower* and *B-upper approximation* of $X$ in **A**, respectively.

In our Data Exploration system we use a modified version of LEM2 algorithm - MLEM2 to generate decision rules. LEM2 (Learning from Examples Module, version 2) algorithm was firstly presented in [21,22] and then implemented in [23]. Successful examples of medical appliance of LERS in medical diagnostic systems were described among other in [25,17].

LEM2 induces a rule set by exploring the space of blocks of attribute-value pairs to generate a local covering. Afterwards the found local covering is converted into the rule set. To define local covering following definitions must be quoted [18].

For a variable (attribute or decision) $x$ and its value $v$, a block $[(x, v)]$ of a variable-value pair $(x, v)$ is the set of all cases for which variable $x$ has value $v$.

Let $B$ be a nonempty lower or upper approximation of a concept represented by a decision-value pair $(d, w)$. Set $B$ *depends* on a set $T$ of attribute-value pairs $(a, v)$ if and only if

$$\emptyset \neq [T] = \bigcap_{(a,v) \in T} [(a, v)] \subseteq B. \tag{1}$$

Set $T$ is a *minimal complex* of $B$ if and only if $B$ depends on $T$ and no proper subset $T'$ of $T$ exists such that $B$ depends on $T'$. Let **T** be a nonempty collection of nonempty sets of attribute-value pairs. Then **T** is a *local covering* of $B$ if and only if the following conditions are satisfied:

- each member $T$ of $\mathbf{T}$ is a minimal complex of $B$,
- $\bigcup_{T \in \mathbf{T}}[T] = B$, and
- $\mathbf{T}$ is minimal, i.e., $\mathbf{T}$ has the smallest possible number of members.

Modified LEM2 (MLEM2) proposed by Grzymala-Busse in [18] in compare to LEM2 induces decision rules from data containing numerical attributes without a need of a separate discretization step. Our implementation of MLEM2 algorithm generates decision rules from both *lower approximation* (certain rules) and *upper approximation* (possible rules). This technique allows us reasoning from "real" data, which contains uncertain, noisy and redundant information.

## 3   Feature Selection

*Feature Selection* is a process that attempts to select a subset of features, satisfying a combination of application and methodology-dependent criteria: minimizing the cardinality of the feature subset; ensuring classification accuracy does not significantly decrease; and approximating the original class distribution with the class distribution given the selected features.

Attribute selection techniques can be categorized using different criteria. One commonly used categorization describes a nature of a metric used to evaluate the worth of attributes and divides it into two different approaches. One approach called *wrapper* uses a statistical re-sampling technique (e.g. cross validation) together with a target learning algorithm to estimate an accuracy of feature subsets [8]. As it is discussed in literature *wrapper* approach should result in a better prediction accuracy on new, unseen data. The main limitation of the approach is an additional computational cost resulting from frequently repeated cross-validation. This process which is used for a validation of each feature subset makes this method inapplicable for processing of large datasets.

The second approach called *filter* uses a general characteristics of data to filter out undesirable features independently of a learning algorithm and without the knowledge of the classifying properties of the data[10]. Since *filters* are not coped with classifier systems or other learner they use quality metric to evaluate features. Four main types of such metric are commonly used:

1. Distance metrics. During feature selection a distance between samples is kept maximal to improve separability.
2. Information measures. They measure information gain for each feature (or subsets of features) and attempt to maximize it.
3. Dependence/Correlation metrics. These metrics identify redundant features by calculating the correlation between them and other features. Afterwards redundant features are removed.
4. Consistency metrics. This new class of metrics employs the training data to assess their consistency, given the subset of features currently evaluated.

*Filters* have proven to be significantly faster then *wrappers* and this advantage makes then suitable for analyzing huge datasets. Due to the fact,

A block diagram of a typical feature selection algorithm

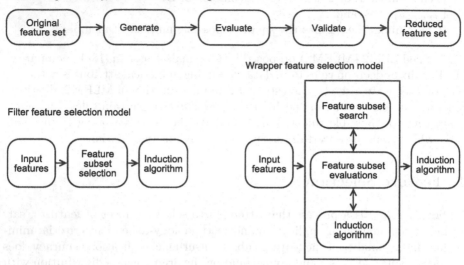

**Fig. 2.** Block diagram of a FS algorithm and two FS models: filter and wrapper

that in medical domain datasets containing information about several thousand of patients described with hundreds of attributes are common we concentrate in our research on appliance of *filter* methods for feature selection. In this study we present results achieved with three different FS algorithms: QUICKREDUCT, CFS (Correlation-based Feature Selection) and $\chi^2$ test.

**QUICKREDUCT.** QUICKREDUCT defines a rough set-based attribute reduction family of feature selection algorithms based on concepts developed by Ziarko [26] and Modrzejewski [27]. For selection of relevant attributes a dependency metric defined as follows is used.

Assuming $P$ and $Q$ are equivalence relations in $U$, the important concept *positive region* $POS_P(Q)$ is defined as:

$$POS_P(Q) = \bigcup_{X \in Q} \underline{P}X. \tag{2}$$

A positive region contains all patterns in $U$ that can be classified in attribute set $Q$ using the information in attribute set $P$.

*The degree of dependency* $\gamma_P(Q)$ of a set $P$ of variables with respect to a set $Q$ of class labelings is defined as:

$$\gamma_P(Q) = \frac{|POS_P(Q)|}{|U|}. \tag{3}$$

Where $|S|$ denotes the cardinality of set S.

The degree of dependency provides a measure of how important $P$ is in mapping the dataset examples into $Q$. If $\gamma_P(Q) = 0$, then classification $Q$ is independent of the attributes in $P$, hence the decision attributes are of no use to this classification. If $\gamma_P(Q) = 1$, then $Q$ is completely dependent on $P$, hence the attributes are indispensable. Values $0 < \gamma_P(Q) < 1$ denote partial dependency, which shows that only some of the attributes in P may be useful.

Given a classification task mapping a set of variables $C$ to a set of labelings $D$, a *reduct* is defined as any subset $R \subseteq C$, such that $\gamma(C, D) = \gamma(R, D)$.

Given a classification task mapping a set of variables $C$ to a set of labelings $D$, a *reduct set* is defined with respect to the power set $\mathbf{P}(C)$ as the set $\mathcal{R} \subseteq \mathbf{P}(C)$ such that $\mathcal{R} = \{A \in \mathbf{P}(C) : \gamma(A, D) = \gamma(C, D)\}$. That is, the reduct set is the set of all possible reducts of the equivalence relation denoted by $C$ and $D$.

Given a classification task mapping a set of variables $C$ to a set of labelings $D$, and $\mathcal{R}$, the reduct set for this problem space, a *minimal reduct* is defined as any reduct $R$ such that $|R| \leq |A|, \forall A \in \mathcal{R}$. That is, the minimal reduct is the reduct of least cardinality for the equivalence relation denoted by $C$ and $D$.

Using the presented definitions a QUICKREDUCT algorithm using forward searching hill climbing is provided as algorithm 1. The algorithm starts with an empty set of variables. Following heuristic is used to add variables to the initial set: the next variable chosen to be added to the candidate reduct is the variable that adds the most to the candidate reducts dependency. The hill climb ends when the dependency reaches one, or when no more variables are left. It must be mentioned, that QUICKREDUCT algorithm does not always generate a reduct. For some cases, the resulting attribute set will be a superreduct, i.e. it will be possible to reduce it further.

---

**Algorithm 1.** QUICKREDUCT(C,D)

---

**Input** :
C-set of all feature attributes
D-set of class attributes
**Output**:
R- attribute reduct, R $\subseteq$ C

R $\leftarrow$ {};
**repeat**
    $T \leftarrow$ R;
    **foreach** $x \in (C - R)$ **do**
        **if** $\gamma_{R \cup \{x\}}(D) > \gamma_T(D)$ **then**
            $T \leftarrow R \cup \{x\}$;
    $R \leftarrow T$;
**until** $\gamma_R(D) = \gamma_C(D)$ ;
**return** R

---

QUICKREDUCT in its search does not compromise with reducts offering a near-perfect consistency. It looks for a strict reduct which is not always desirable. During processing of real data with noise it could be advantageous to relax the concept of a reduct.

Given a classification task mapping a set of variables $C$ to a set of labelings $D$, an *approximate reduct* is defined as any subset $R \subseteq C$, such that $0 \leq \gamma(C, D) - \gamma(R, D) \leq \epsilon \leq 1$, where $\epsilon$ is a suitably chosen neighborhood.

In this paper we present results for different values of $\epsilon$. Results achieved for QUICKREDUCT are compared with other two filter algorithms: CFS and Chi-square test.

Feature selection can be based on selection of reducts or approximations of reducts. It is worthwhile to mention that reducts selected randomly can be of low quality, i.e., classifiers induced from such reducts can have the low classification quality [34,35,36]. There have been developed methods for selection of relevant reducts making it possible to induce classifiers with the high classification quality. Among them are the method based on dynamic reducts (see, e.g., [32,37,38]); the method for selection reducts using as a criterion for selection the number of rules generated by reducts (see, e.g., [28]), and the selection method with application of ensembles of reducts (see, e.g., [31]).

**CFS: Correlation-based Feature Selection.** Described by Hall Correlation-based Feature Selection (CFS) uses a heuristic which measures the usefulness of individual features for predicting the class label along with the level of intercorrelation among them [7]. This defined by equation 4 heuristic should filter out: irrelevant features because they are poor predictors of the class and redundant features which should be ignored because of their high correlation with each other[2].

$$G_s = \frac{k\overline{r_{ci}}}{\sqrt{k + k(k-1)\overline{r_{ii'}}}} \tag{4}$$

$k$ is the number of features in the subset, $\overline{r_{ci}}$ is the mean feature correlation with the class, and $\overline{r_{ii'}}$ is the average feature intercorrelation. For computing the correlations necessary for equation 4 a number of information based measures of association were proposed such as: the uncertainty coefficient, the gain ratio or the minimum description length principle [6,7,4]. The best results however were achieved with the gain ratio used for feature-class correlations and symmetrical uncertainty coefficient used for feature intercorrelations. Following equations define these terms:

$$H(Y) = \sum_y p(y) log_2(p(y)) \tag{5}$$

$$H(Y \mid X) = \sum_x p(x) \sum_y p(y \mid x) log_2(p(y \mid x)) \tag{6}$$

$$gain = H(Y) - H(Y \mid X) = H(X) - H(X \mid Y) = H(Y) + H(X) - H(X,Y) \tag{7}$$

$$gain\ ratio = \frac{gain}{H(X)} \tag{8}$$

$$symmetrical\ uncertainty = 2.0 \times \left[\frac{gain}{H(Y) + H(X)}\right] \tag{9}$$

In our research we use CFS heuristic with Best First search strategy [1]. This strategy starts with an empty set of attributes and generates all single feature expansions which are possible. Then the subset of attributes with the highest value of the evaluation function is chosen and the procedure repeats. A stop criteria is defined as a number of subsets that results in no improvement.

**Chi-square test.** Chi-square ($\chi^2$) is a non-parametric test of statistical significance for bivariate tabular analysis (crossbreaks) [12]. The $\chi^2$ test is defined by equation 10.

$$\chi^2 = \sum \frac{(f_0 - f)^2}{f} \tag{10}$$

Where $f_0$ is an observed frequency and $f$ is an expected frequency. Essentially the $\chi^2$ test is commonly used for testing independence and/or correlation between two vectors. The test compares observed frequencies with the corresponding expected frequencies. Value 0 of $\chi^2$ means that the corresponding two vectors are statistically independent with each other. At a certain threshold value (e.g., 3.84 at the 95% significance level [12]) an independence assumption between two vectors can be rejected. It can be said, that the higher value $\chi^2$ takes the higher the correlation between the corresponding vectors. In our work we test correlations between each attribute and a class label and select for further processing only attributes with the highest $\chi^2$ value.

## 4 Dataset Preparation and Experimental Environment

Data used in our research was obtained from the Cardiology Department of Silesian Medical Academy in Katowice - the leading Electrocardiology Department in Poland specializing in hospitalization of severe heart diseases. For our experiments we took a data of 4318 patients hospitalized in this Department between 2003 and 2005. The data were imported from a clinical information system and transformed to binary attributes using our, based on regular expressions, hierarchical dictionary algorithm. Afterwards with a help of domain experts these binary attributes were joined into 13 grouped attributes. Following joined attributes were created:

**ID:0 AVBL**:Atrioventricular block, value range:[0,1,2,3,4,5,6,7]
**ID:1 DIAB**:Diabetes, value range:[0,1]
**ID:2 PTACH**:Paroxysmal tachycardia, value range:[0,1,2,3,4,5,6,7]
**ID:3 HYPERCHOL**:Hypercholesterolaemia, value range:[0,1]
**ID:4 CARDIOMYO**:Cardiomyopathy, value range:[0,1]
**ID:5 ATHEROSC**:Atherosclerosis, value range:[0,1]
**ID:6 AFF**:Atrial fibrillation and flutter, value range:[0,1,2,3]
**ID:7 HYPERTEN**:Hypertension (High Blood Pressure), value range:[0,1]
**ID:8 CIHD**:Chronic ischaemic heart disease, value range:[0,1,2,3]
**ID:9 OBESITY**:Obesity, value range:[0,1]

**ID:10 SSS**:Sick Sinus Syndrome, value range:[0,1,2,3,4,5,6,7]
**ID:11 TYROIDG**:Disorders of thyroid gland, value range:[0,1,2,3]
**ID:12 MIOLD**:Myocardial infarction in past, value range:[0,1]

For our experiments we selected also three decision attributes representing a decision about implementation of the specific type of pace-maker. If a specified pace-maker was not implanted then the decision attribute took a value 0. Following decision attributes were selected:

**DDD**, value range:[0,1]
**SSI**, value range:[0,1]
**VDD**, value range:[0,1]

For each decision attribute we created a dataset containing all 13 joined attributes and a decision attribute. Then this dataset was divided into two parts: a train dataset containing 66% of objects from the entry dataset and a test dataset containing the rest 33%. In the next step for each decision attribute we created totally 1830 copies of each training dataset and applied on each copy a different combination of noise reduction algorithms. Afterwards we used QUICKREDUCT, CFS and $\chi^2$ on these datasets to select subsets of attributes.

We integrated all described in this study algorithms in our Data Exploration system written in Java v.1.5. To compare the achieved results we chose two classifiers from a powerful open-source machine learning workbench Weka (Waikato Environment for Knowledge Analysis) v.3.4.3. These were: Decision Table - an implementation of Decision Table Majority (DTM) algorithm used successfully in discrete spaces by Kohavi [9] and J48 Decision tree (C4.5 release 8) - a TDIDT (top-down induction of decision trees) approach derived from Quinlan's ID3 induction system [3].

## 5   Results

In this section we present results of attribute selection done with different FS algorithms. To measure the quality of the selected subsets of attributes we compare classification accuracy of decision rules generated with these sets and check their medical relevance. Decision rules were tested on two datasets:a test dataset containing 33% of the input data and at training dataset with rest 66%. Judging the medical importance of selected attributes was done with a help of domain experts from the Cardiology Department of Silesian Medical Academy in Katowice/Poland. They prepared a list of important attributes taken into consideration if a specified pace-maker type should be implanted. Afterwards they valuated an importance of attributes for each decision attribute using three symbolic categories:(+) is an important attribute for classification,(+/-) a less important attribute, which under circumstances can be used additionally for classification (-) an unimportant attribute. Table 1 shows the described classification (numbers identify attributes as described in section 4). Summary results presented in tables 2,3,4 use the following column description:

**Selected attrib**-a selected set of attributes. Each number equals to an attribute ID as described in section 4. A keyword 'unreduced' means a dataset with 13 attributes

**Method**-a method used for selecting attributes or in case of DT and J48 a method used for comparison

**Sel. Attr**-a number of selected attributes.

**% corretly. classified**-percentage of correctly classified objects to a number of all objects in a dataset; a number in brackets shows results of classification for a training set

**Correctly classified**-a sum of correct classifications for both classes; (in brackets classification for a training dataset)

**Rules**-a number of generated decision rules from a training dataset, in case of J48 it is a tree size

**Attr (+), (+/-), (-)**-a number of important (+), less important (+/-) and unimportant (-) attributes

Table 2 presents results achieved for the DDD decision attribute. From this table it can be seen that both feature selection methods (CFS and $\chi^2$) were able to significantly reduce a set of attributes from initial 13 attributes to 3-5 attributes in case of CFS and to 4-5 in case of $\chi^2$. These results were achieved without a loose of prediction accuracy on the contrary the number of correctly classified objects is increased. In is also noticeable, that these results were achieved with strongly reduced (in most cases more then ten times) number of decision rules. This simplify a validation process by domain experts and therefore determines a medical appliance of the proposed methods. In case of QUICKREDUCK achieved results strongly depends on the number of selected attributes. Specifying low values of $\epsilon$ in a definition of approximate reduct allows to generate subsets with lower number of attributes. In this case classification accuracy and number of generated decision rules is similar to results achieved for CFS and $\chi^2$. With an increasing number of selected attributes an overfitting effect is visible. The generated rules show a high prediction accuracy on the same test dataset as the training set but are less able to recognize new cases. Two, selected for comparison, methods from Weka package show results, which are not significantly different from our methods. The highest medical relevance of the selected attribute shown $\chi^2$ algorithm. Four of five important attributes were recognized and the number of (-) attributes was the lowest. CFS was able to recognize at least 3 medically important attributes but the number of attributes marked by domain experts as unimportant was higher. Number of relevant attributes

**Table 1.** Medical importance of attributes for selection of pace-maker type

| Pace-maker type | (+) | (+/-) | (-) |
|---|---|---|---|
| DDD | 0, 2, 4, 6, 10 | 5, 8, 11, 12 | 1, 3, 7, 9 |
| SSI | 0, 2, 6, 10 | 5, 8, 11, 12 | 1, 3, 4, 7, 9 |
| VDD | 0, 2, 6, 10 | 4, 5, 8, 11, 12 | 1, 3, 7, 9 |

**Table 2.** Summary results for DDD recognition

| Selected attrib. | Method | Sel. Attr. | % corretly. classified | Correctly classified | Rules | Attr. (+) | Attr. (+/-) | Attr. (-) |
|---|---|---|---|---|---|---|---|---|
| **Train set 66%, Test set 33% - C:1 570 objects, C:0 899 objects** | | | | | | | | |
| unreduced | - | - | 71.14 (82.70) | 1045 (2356) | 252 | - | - | - |
| unreduced | J48 | - | 75.28 (76.93) | 1106 (2192) | 28 | - | - | - |
| unreduced | DT | - | 75.08 (76.76) | 1103 (2187) | 55 | - | - | - |
| 0,2,7,11 | QRD | 4 | 75.96 (76.29) | 1109 (2162) | 21 | 2 | 1 | 1 |
| 0,2,7,9,11 | QRD | 5 | 75.48 (76.53) | 1102 (2169) | 41 | 2 | 1 | 2 |
| 0,2,7,8,9,11 | QRD | 6 | 73.15 (77.31) | 1068 (2191) | 58 | 2 | 2 | 2 |
| 0,1,2,3,7,8,9,11 | QRD | 8 | 73.08 (78.83) | 1067 (2234) | 115 | 2 | 2 | 4 |
| 0,1,2,3,7,8,9,10,11,12 | QRD | 10 | 71.85 (80.52) | 1049 (2282) | 171 | 3 | 3 | 4 |
| 0,1,2,3,4,7,8,9,10,11,12 | QRD | 11 | 71.51 (81.26) | 1044 (2303) | 205 | 4 | 3 | 4 |
| 2,6,10 | CFS | 3 | 75.56 (76.59) | 1110 (2182) | 15 | 3 | 0 | 0 |
| 2,6,7,10 | CFS | 4 | 75.02 (76.76) | 1102 (2187) | 16 | 3 | 0 | 1 |
| 2,6,9,10 | CFS | 4 | 75.63 (76.66) | 1111 (2184) | 18 | 3 | 0 | 1 |
| 2,3,6,7,10 | CFS | 5 | 74.88 (76.83) | 1100 (2189) | 17 | 3 | 0 | 2 |
| 0,2,6,10 | $\chi^2$ | 4 | 75.22 (76.69) | 1105 (2185) | 34 | 4 | 0 | 0 |
| 0,2,6,8,10 | $\chi^2$ | 5 | 74.88 (77.22) | 1100 (2200) | 54 | 4 | 1 | 0 |
| 0,2,3,6,10 | $\chi^2$ | 5 | 75.29 (76.87) | 1106 (2190) | 42 | 4 | 0 | 1 |

selected by QUICKREDUCT was about 2 for small sets (4-6 attributes) and reached 4 for a set of 11 attributes. The number of (+/-) and (-) attributes was also the highest what placed this algorithm at the last place.

Data presented in table 3 shows results achieved for the SSI decision attribute. These results show at a first view a good recognition accuracy (about 80%) for rules generated from datasets with reduced attributes. Especially the results achieved with a small number of decision rules for new cases are promising. The higher recognition accuracy for this type of pace-maker as for the DDD type can be explainable with an unequal class distribution in the training set. More then 66% of objects belong to class 0 and only 33 % were classified as class 1. This inequality led to a low coverage rate for class 1, which oscillated between 60-67% (50-53% for J48 and DT). To handle such unbalanced class distributions either an additional pre-processing step is needed before the data can be analyzed or a classification algorithm must be extended. Nevertheless the small number of strong decision rules can be used for analysis the most typical factors, which play a role in the decision about implantation of this kind of pace-maker. The highest medical importance shown attribute subsets selected with $\chi^2$ (3-4 important attributes) followed by subsets generated with CFS. For both methods the number of medically unimportant attributes was oscillating between 0 and 2. QUICKREDUCT selected only 2 (+) attributes and completed the generated subsets with 1-3 (+/-) attributes. In can be observed, that although the dependency metrics used in implementation of QUICKREDUCT shows very similar recognition accuracy as the other two methods it selects often attributes with less medical importance.

**Table 3.** Summary results for SSI recognition

| Selected attrib. | Method | Sel. Attr. | % correctly. classified | Correctly classified | Rules | Attr. (+) | Attr. (+/-) | Attr. (-) |
|---|---|---|---|---|---|---|---|---|
| **Train set 66%, Test set 33% - C:1 444 objects, C:0 1025 objects** | | | | | | | | |
| unreduced | - | - | 76.45 (85.43) | 1123 (2434) | 253 | - | - | - |
| unreduced | J48 | - | 80.46 (79.92) | 1182 (2277) | 30 | - | - | - |
| unreduced | DT | - | 80.12 (80.09) | 1177 (2282) | 83 | - | - | - |
| 0,2,5,7 | QRD | 4 | 81.23 (79.22) | 1186 (2245) | 18 | 2 | 1 | 1 |
| 0,2,5,7,11 | QRD | 5 | 81.03 (79.36) | 1183 (2249) | 27 | 2 | 2 | 1 |
| 0,1,2,3,5,7,8,9,11 | QRD | 9 | 78.84 (82.36) | 1151 (2334) | 156 | 2 | 3 | 4 |
| 0,1,2,3,5,7,8,9,11,13 | QRD | 10 | 78.49 (83.06) | 1146 (2354) | 185 | 2 | 3 | 5 |
| 0,1,2,3,5,7,8,9,10,11,13 | QRD | 11 | 77.47 (83.80) | 1131 (2375) | 203 | 3 | 3 | 5 |
| 1,2,3,6 | CFS | 4 | 81.01 (79.36) | 1190 (2261) | 8 | 2 | 0 | 2 |
| 1,2,6,10 | CFS | 4 | 80.94 (79.40) | 1189 (2262) | 12 | 3 | 0 | 1 |
| 1,2,3,6,10 | CFS | 5 | 80.60 (79.50) | 1184 (2265) | 22 | 3 | 0 | 2 |
| 0,1,2,6,10 | CFS | 5 | 79.99 (79.85) | 1175 (2275) | 48 | 4 | 0 | 1 |
| 0,2,6,10 | $\chi^2$ | 4 | 80.74 (79.57) | 1186 (2267) | 23 | 4 | 0 | 0 |
| 0,1,2,6,10 | $\chi^2$ | 5 | 79.99 (79.85) | 1175 (2275) | 48 | 4 | 0 | 1 |
| 0,1,3,6,10 | $\chi^2$ | 5 | 80.33 (79.75) | 1180 (2272) | 55 | 3 | 0 | 2 |

**Table 4.** Summary results for VDD recognition

| Selected attrib. | Method | Sel. Attr. | % correctly. classified | Correctly classified | Rules | Attr. (+) | Attr. (+/-) | Attr. (-) |
|---|---|---|---|---|---|---|---|---|
| **Train set 66%, Test set 33% - C:1 96 objects, C:0 1373 objects** | | | | | | | | |
| unreduced | - | - | 92.24 (94.42) | 1355 (2690) | 253 | - | - | - |
| unreduced | J48 | - | 93.46 (93.57) | 1373 (2666) | 30 | - | - | - |
| unreduced | DT | - | 93.46 (93.57) | 1373 (2666) | 83 | - | - | - |
| 0,7,8,11 | QRD | 4 | 95.21 (92.38) | 1390 (2618) | 11 | 1 | 2 | 1 |
| 0,7,8,9,11 | QRD | 5 | 95.21 (92.41) | 1390 (2619) | 12 | 1 | 2 | 2 |
| 0,3,7,8,9,11 | QRD | 6 | 94.93 (92.55) | 1386 (2623) | 22 | 1 | 2 | 3 |
| 0,1,3,7,8,9,11 | QRD | 7 | 94.38 (92.77) | 1378 (2629) | 43 | 1 | 2 | 4 |
| 0,3,5,6,10 | CFS | 5 | 92.65 (93.58) | 1361 (2666) | 15 | 3 | 1 | 1 |
| 0,5,6,9,10 | CFS | 5 | 93.12 (93.72) | 1368 (2670) | 12 | 3 | 1 | 1 |
| 0,2,6,8,10,12 | CFS | 6 | 93.19 (93.58) | 1369 (2666) | 10 | 4 | 2 | 0 |
| 0,2,6,8,10 | $\chi^2$ | 5 | 93.19 (93.58) | 1369 (2666) | 10 | 4 | 1 | 0 |
| 0,5,6,8,10 | $\chi^2$ | 5 | 93.19 (93.58) | 1369 (2666) | 11 | 3 | 2 | 0 |
| 0,6,8,9,10 | $\chi^2$ | 5 | 92.99 (93.72) | 1366 (2670) | 15 | 3 | 1 | 1 |
| 0,1,6,9,10 | $\chi^2$ | 5 | 92.85 (93.65) | 1364 (2668) | 16 | 3 | 0 | 2 |
| 0,6,8,9,10 | $\chi^2$ | 5 | 92.99 (93.72) | 1366 (2670) | 15 | 3 | 1 | 1 |

Results for a decision attribute VDD, shown in table 4, presents a similar picture as results achieved for SSI. The overall correct classification rate is very high (between 92-93%). Due to a very unbalanced class distribution (96% class 0, 4% class 1) a coverage for class 1 is very low (2-14 %), so that almost all correctly classified objects belong to class 0. This effect was also observed with

J48 and Decision Tables from Weka, which were unable to classify a single object belonging to class 1. From the medical point of view again subsets selected by $\chi^2$ and CFS contained the most important attributes. QUICKREDUCT managed to select only one attribute from (+) category and 2 attributes from (+/-) group. Unfortunately it selected also the highest number of attributes marked by domain experts as unimportant.

# 6   Conclusions

In the case study presented in this paper we showed results of feature selection for medical datasets. These sets contain a lot of noise and redundant information, which should be filtered out before next machine learning algorithms are used. Additional advantage of feature selection is a reduction of search space, which, as presented in this paper and our entry research [16], reduce a number of decision rules (sometimes by factor 10) without compromising prediction accuracy. This fact is very important in medical domain where achieved results must be explainable and verifiable by experts. In this paper we showed results for three feature selection algorithms: QUICKREDUCT, CFS and $\chi^2$ all belonging to the *filter* category. There are two main advantages of *Filter* algorithms over *Wrappers* based ones: they require significantly less computational effort and the achieved results do not depend on a specific learning algorithm.

In our experiments we selected subsets of attributes both from original training sets and from training sets after applying different noise reduction algorithms (over 1800 combinations pro decision attribute). The selected subsets were than used to generate decision rules using MLEM2 algorithm. Afterwards both the test set and the training set were used for classification. The accuracy of classification was compared with J48 and Decision Table from Weka package. In all cases decision rules generated with MLEM2 were comparable with results achieved by DT and J48. Current implementation of the classification system showed some problems with classifying datasets with unbalanced class distribution (SSI and VDD). An additional work on this topic is needed to improve a handling of such situations in the future.

In this paper we presented not only numerical results of classification but also compared generated subsets for their medical relevance. This comparison shown differences between selected algorithms in ability to find medically important attributes. In this competition $\chi^2$ and CFS were the winner. The degree of dependency metric used by QUICKREDUCT showed a tendency to select attributes with less medical relevance. Results presented in this paper and research on our Data Exploration system led us to a conclusion, that it will be advantageous to join strength of all presented algorithms. A possible solution could be a measurement of an importance of attributes based on decision rules generated from the reduced datasets, for example in form of a rule importance as proposed by Li in [5].

# Acknowledgment

We would like to thank Rafal Mlynarski from Cardiology Department of Silesian Medical Academy in Katowice, Poland for providing us the data and giving us feedbacks.

# References

1. Rich, E. and Knight, K.: Artificial Intelligence. McGraw-Hill Science, New York (1990)
2. Ghiselli, E.: Theory of Psychological Measurement. McGraw-Hill Book, New York (1964)
3. Quinlan, J. R.: C4.5: Programs for Machine Learning. Morgan Kaufmann Publishers, Inc,Los Altos, California (1993)
4. Kononenko, I.: On Biases in Estimating Multi-Valued Attributes. In: International Joint Conference on Artificial Intelligence,Montreal (1995) 1034–1040
5. Li, J and Cercone, N: Introducing A Rule Importance Measure. In: LNCS Transactions on Rough Sets, Springer Verlag, 5 (2006) 171–194
6. Quinlan, J. R: Induction of Decision Trees. In: Mach. Learn., (2003) 81–106
7. Hall, M.: Correlation-based Feature Selection for Machine Learning. Ph.D diss. Hamilton, NZ: Waikato University, Department of Computer Science, 1998
8. Kohavi, R. and John, G.: Wrappers for Feature Subset Selection. In: Artif. Intell., (97) 273–324
9. Kohavi, R.: The Power of Decision Tables. In: Lecture Notes in Artificial Intelligence, (914) 174–189
10. John, G. and Kohavi, R. and Pfleger, K.: Irrelevant Features and the Subset Selection Problem. In: International Conference on Machine Learning, New Jersey (1994), 121–129
11. Pawlak, Z.: Rough sets. International Journal of Computer and Information Science 11 (1982) 341–356
12. Everitt, B. S.: The analysis of contingency tables. Chapman and Hall, London (1977)
13. Pawlak, Z.: Knowledge and Uncertainty: A Rough Set Approach. SOFTEKS Workshop on Incompleteness and Uncertainty in Information Systems (1993) 34–42
14. Pawlak, Z. and Grzymala-Busse, J. W. and Slowinski, R. and Ziarko, W.: Rough Sets. Commun. ACM 38 (1995) 88–95
15. Ilczuk, G. and Wakulicz-Deja, A.: Rough Sets Approach to Medical Diagnosis System. In: AWIC 2005, Lodz (2005) 204–210
16. Ilczuk, G. and Wakulicz-Deja, A.: Attribute Selection and Rule Generation Techniques for Medical Diagnosis Systems. In: RSFDGrC 2005, Regina (2005) 352–361
17. Wakulicz-Deja, A. and Paszek, P.: Applying Rough Set Theory to Multi Stage Medical Diagnosing. Fundam. Inform. 54 (2003) 387–408
18. Grzymala-Busse, J. W.: MLEM2 - Discretization During Rule Induction. In: IIS 2003, Zakopane (2003) 499–508
19. Ilczuk, G. and Mlynarski, R. and Wakulicz-Deja, A. and Drzewiecka, A. and Kargul, W.: Rough Sets Techniques for Medical Diagnosis Systems. In: Computers in Cardiology 2005, Lyon (2005) 837–840
20. Mlynarski, R. and Ilczuk, G. and Wakulicz-Deja, A. and Kargul, W.: Automated Decision Support and Guideline Verification in Clinical Practice. In: Computers in Cardiology 2005, Lyon (2005) 375–378

21. Chan, C. C. and Grzymala-Busse, J. W.: On the two local inductive algo-rithms: PRISM and LEM2. Foundations of Computing and Decision Sciences **19** (1994) 185–203
22. Chan, C. C. and Grzymala-Busse, J. W.: On the attribute redundancy and the learning programs ID3, PRISM, and LEM2.Department of Computer Science, University of Kansas,TR-91-14, (1991)
23. Grzymala-Busse, J. W.: A new version of the rule induction system LERS. Fundam. Inform. **31** (1997) 27–39
24. Komorowski, H. J. and Pawlak, Z. and Polkowski, L. T. and Skowron, A.: Rough Sets: A Tutorial. Springer-Verlag, Singapore (1999)
25. Paszek, P. and Wakulicz-Deja, A.: The Application of Support Diagnose in Mito-chondrial Encephalomyopathies. Rough Sets and Current Trends in Computing. **2475** (2002) 586–593
26. Ziarko, W.: The discovery, analysis and representation of data dependencies in databases. In G. Piatesky-Shapiro and W. J. Frawley, editors, Knowledge Discovery in Databases. MIT Press, (1991)
27. Modrzejewski, M.: Feature selection using rough sets theory. In Pavel B. Brazdil, editor, Proceedings of the European Conference on Machine Learning (ECML-93) **667** (1993) 213–226
28. J. Bazan, H. S. Nguyen, S. H. Nguyen, P. Synak, J. Wróblewski. Rough set algo-rithms in classification problems. In: Polkowski et al. [29], pp. 49–88.
29. L. Polkowski, T. Y. Lin, S. Tsumoto (Eds.). Rough Set Methods and Applica-tions: New Developments in Knowledge Discovery in Information Systems, Studies in Fuzziness and Soft Computing, vol. 56. Springer-Verlag/Physica-Verlag, Heidel-berg, 2000.
30. S. K. Pal, L. Polkowski, A. Skowron (Eds.). Rough-Neural Computing: Techniques for Computing with Words. Cognitive Technologies, Springer-Verlag, Heidelberg, 2004.
31. J. Wróblewski. Adaptive aspects of combining approximation spaces. In: Pal et al. [30], pp. 139–156.
32. J. G. Bazan. A comparison of dynamic and nondynamic rough set methods for extracting laws from decision tables. In: Polkowski and Skowron [33], pp. 321–365.
33. L. Polkowski, A. Skowron (Eds.). Rough Sets in Knowledge Discovery 1: Method-ology and Applications, Studies in Fuzziness and Soft Computing, vol. 18. Physica-Verlag, Heidelberg, 1998.
34. Z. Pawlak, A. Skowron, Rudiments of rough sets. Information Sciences. An Inter-national Journal. Elsevier (2006) 177(1) (2007) 3-27.
35. Z. Pawlak, A. Skowron, Rough sets: Some extensions. Information Sciences. An International Journal. Elsevier (2006) 177(1) (2007) 28-40.
36. Z. Pawlak, A. Skowron, Rough sets and Boolean reasoning. Information Sciences. An International Journal. Elsevier (2006) 177(1) (2007) 41-73.
37. R. Swiniarski, A. Skowron. Rough set methods in feature selection and extraction. Pattern Recognition Letters 24(6) (2003) 833–849.
38. R. W. Swiniarski, A. Skowron. Independent component analysis, principal compo-nent analysis and rough sets in face recognition. In: Peters and Skowron [39], pp. 392–404.
39. J. F. Peters, A. Skowron (Eds.). Transactions on Rough Sets I: Journal Subline, Lecture Notes in Computer Science, vol. 3100. Springer, Heidelberg, 2004.

# Using Approximate Reduct and LVQ in Case Generation for CBR Classifiers[*]

Yan Li[1,2], Simon Chi-Keung Shiu[2],
Sankar Kumar Pal[3], and James Nga-Kwok Liu[2]

[1] College of Mathematics and Computer, Hebei University,
Baoding City, 071002, Hebei Province, China
polyuliyan@yahoo.com.cn
[2] Department of Computing, Hong Kong Polytechnic University,
Kowloon, Hong Kong
{csckshiu, csnkliu}@comp.polyu.edu.hk
[3] Machine Intelligence Unit, Indian Statistical Institute,
Kolkata, 700 035, India
sankar@isical.ac.in

**Abstract.** Case generation is a process of extracting representative cases to form a compact case base. In order to build competent and efficient CBR classifiers, we develop a case generation approach which integrates fuzzy sets, rough sets and learning vector quantization (LVQ). If the feature values of the cases are numerical, fuzzy sets are firstly used to discretize the feature spaces. Secondly, a fast rough set-based feature selection method is applied to identify the significant features. Different from the traditional discernibility function-based methods, the feature reduction method is based on a new concept of approximate reduct. The representative cases (prototypes) are then generated through LVQ learning process on the case bases after feature selection. LVQ is the supervised version of self-organizing map (SOM), which is more suitable to classification problems. Finally, a few of prototypes are generated as the representative cases of the original case base. These prototypes can be also considered as the extracted knowledge which improves the understanding of the case base. Three real life data are used in the experiments to demonstrate the effectiveness of this case generation approach. Several evaluation indices, such as classification accuracy, the storage space, case retrieval time and clustering performance in terms of intro-similarity and inter-similarity, are used in these testing.

## 1  Introduction

Case-based Reasoning (CBR) is a reasoning methodology that is based on prior experience and examples. When a CBR reasoner is presented with a problem, it searches it memory of past cases (called the case base) and attempts to retrieve a case or multiple cases that most closely match the case under analysis [1,2].

---

[*] This work is supported by the Hong Kong government CERG research grant BQ-496.

J.F. Peters et al. (Eds.): Transactions on Rough Sets VII, LNCS 4400, pp. 85–102, 2007.
© Springer-Verlag Berlin Heidelberg 2007

Compared with rule-based systems, CBR systems usually require significantly less knowledge acquisition, since it involves collecting a set of past experiences without the added necessity of extracting a formal domain model from these cases. The CBR systems have been widely used in classification problems [3,4,5] (to determine if an object is a member of a class or not, or which of several classes), which are called CBR classifiers.

The performance of CBR classifiers, in terms of classification accuracy and case retrieval speed, closely depends on the competence and size of the case base. The competence of a case base is the range of unseen cases which the case base can correctly classify. In general, the more competent the case base, the higher the classification accuracy of the CBR classifier. On the other hand, it is obvious that the larger the size of a case base, the lower the case retrieval speed. It is difficult to achieve the maximal classification accuracy and the least case retrieval time simultaneously. If the size of a case base is reduced, the competence of the case base may be hurt because of the removal of some important cases. In this research, we attempt to make a trade-off by selecting the most representative cases to reduce the case retrieval speed and preserve the competence of the case bases. That is, to build both compact and competent case bases for CBR classifiers. In this paper, we achieve this goal by developing a rough learning vector quantization (LVQ)-based case generation approach. A few of prototypes are generated to represent the entire case base without loss of the competence of the original case base.

As a necessary preprocessing of LVQ-based case generation, a fast rough set-based method is developed to select the relevant features and eliminate the irrelevant ones. The feature selection process can modify the similarity among cases and achieve better clustering performance. Rough sets [6] allow the most informative features to be detected and then selected through the reduct computation. There is much research work in rough set-based feature selection [7,8], the effectiveness of which have been demonstrated in many different domains. There are two main groups of rough set-based feature selection methods: discernibility function-based [9] and attribute dependency-based [10]. Such methods are computational intensive, i.e., in the former, during the generation of the discernibility matrix and in the latter, during the discovery of the positive regions.

To reduce the computational load inherent in such methods, a new concept called approximate reduct is introduced. It is a generalization of the classical concept of reduct, and can be computed by counting the distinct rows in some decision sub-tables. The computational load is linear with the number of cases and features. Other primary concepts in rough sets, such as dispensable (indispensable) attribute and core, are also modified. Using these extended concepts, we develop a fast rough set based approach to finding the approximate reduct. Our feature selection approach can be considered as a generalization of the original attribute dependency-based or discernibility function-based techniques. In this feature selection process, fuzzy sets are used to discretize the numerical attribute values to generate indiscernibility relation and equivalence classes of the given case base. Triangular membership functions are applied in the discretization of feature spaces.

Learning vector quantization is then applied to extract the representative cases (also called prototypes) to represent the entire case base. LVQ is a kind of competitive algorithms. Some of them can be considered to be a supervised version of the Self-Organizing Map (SOM) algorithm. The SOM algorithm [11,12] constructs a stable topology preserving mapping from the high-dimensional space onto map units in such a way that relative distances between data points are preserved. Mangiameli et al. [13] demonstrated that SOM is a better clustering algorithm than hierarchical clustering with regard to clustering data with overlapped dispersion, irrelevant variables, outliers or different sized populations. Their study also proved that SOM is insensitive to learning rates which vary in the self-organizing process, and the clusters resulted from SOM are robust. Pal et al. used SOM to extract prototypical cases in [14] and reported a compact representation of data. However, Kohonen pointed out in [11], "the SOM has not been meant for statistical pattern recognition; it is a clustering, visualization, and abstraction method. Anybody wishing to implement decision and classification processes should use Learning Vector Quantization (LVQ) instead of SOM".

Since we focus on the classification problems in this paper, LVQ is used to generate prototypical cases. LVQ is introduced by Teuvo Kohonen in 1986, which inherits almost all the features of SOM, except that it is a supervised learning algorithm. LVQ is able to summarise or reduce large datasets to a smaller number of representative vectors suitable for classification or visualization. LVQ has similar advantages of SOM, such as the robustness with noise and missing information.

After applying the case selection approach, the original case base can be reduced to a few prototypes which can be directly used to predict the class label of the unseen cases. These prototypes can be regarded as the specific domain knowledge which is extracted from the case base. This will speed up the case retrieval and make the case base be more easily understood. On the other hand, since the most representative cases are generated, case base competence can be also preserved. Therefore, using our developed rough LVQ-based case generation approach, the retrieval speed, clustering performance, and the understanding of the case base are all improved without decreasing the classification accuracy.

The remainder of this paper is organized as follows. In Section 2, fuzzy sets are applied to discretize the continuous-valued attributes of the cases. Three triangular membership functions - "low", "medium", and "high" - are used to describe each attribute. This is followed by Section 3 in which we describe the feature selection method based on the concept of approximate reduct computation. Some primary concepts in rough sets are also generalized in this section. In Section 4, the supervised learning process of LVQ is presented to generate prototypical cases for the given case base. To validate the developed rough LVQ case generation approach, section 5 presents the experimental results on three real life data. The classification accuracy, case retrieval speed, intra- similarity and inter- similarity are used as the indices to evaluate the performance of our approach. Comparisons are made among the developed rough LVQ approach, LVQ, SOM, and Random case generation methods.

## 2   Fuzzy Discretization of Feature Space

The rough set-based feature selection methods are all built on the basis of indiscernibility relation. If the attribute values are continuous, the feature space needs to be discretized for defining the indiscernibility relations and equivalence classes on different subset of attribute sets. In this paper, fuzzy sets are used for the discretization by partition each attribute into three levels: Low ($L$), Medium ($M$), and High ($H$). Finer partitions may lead to better accuracy at the cost of higher computational load. The use of fuzzy sets has several advantages over the traditional "hard" discertizations, such as handling the overlapped clusters and linguistic representation of data [14].

Triangular membership functions are used to define the fuzzy sets: $L$, $M$ and $H$. There are three parameters $C_L$, $C_M$, and $C_H$ for each attribute which should be determined beforehand. They are considered as the centers of the three fuzzy sets. Noted that, the determination of the parameter values is problem-dependent, and there are different methods that can be used. In this paper, we do not intend to compare these methods. Here, the center of fuzzy set $M$ for a given attribute $a$ is the average value of all the values occurring in the domain of $a$. Assume $V_a$ is the domain of attribute $a$, then $C_M = \sum_{y \in V_a} y / |V_a|$, where $|*|$ is the cardinality of set $*$.

$C_L$ and $C_H$ are computed as
$C_L = (C_M - Min_a)/2$ and $C_H = (Max_a - C_M)/2$,
where $Min_a = min\{y|y \in V_a\}$ and $Max_a = max\{y|y \in V_a\}$.

More formally, these membership functions are denoted by $\mu_L$, $\mu_M$, and $\mu_H$, which are defined by the following equations.

$$\mu_L(x) = \begin{cases} 1, & Min_a \leq x \leq C_L \\ \frac{C_M - x}{C_M - C_L}, & C_L \leq x \leq C_M \\ 0, & x > C_M \end{cases}$$

$$\mu_M(x) = \begin{cases} 0, & x \leq C_L \\ \frac{x - C_L}{C_M - C_L}, & C_L < x \leq C_M \\ \frac{C_H - x}{C_H - C_M}, & C_M < x \leq C_H \\ 0, & x > C_H \end{cases}$$

$$\mu_H(x) = \begin{cases} 0, & x \leq C_M \\ \frac{x-C_M}{C_H-C_M}, & C_M < x \leq C_H \\ 1, & x > C_H \end{cases}$$

where $\mu_*(x)$ is the membership value of case $x$ to fuzzy set $*$.

For an attribute $a$, the membership functions of $L$, $M$, and $H$ are illustrated in Fig.1.

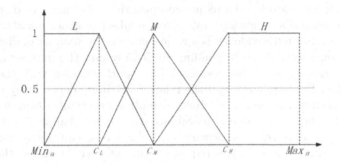

**Fig. 1.** Membership functions of $L$, $M$ and $H$ for attribute $a$

## 3    Feature Selection Based on Approximate Reduct

The purpose of rough set-based feature selection is to identify the most significant attributes and eliminate the irrelevant ones to form a good feature subset for classification tasks. It reduces the running time of classification processes and increases the accuracy of classification models.

In this section, we develop a fast feature reduction method based on the concept of approximate reduct. Before we present the proposed algorithm, the rough set-based feature selection methods are briefly reviewed.

### 3.1    Traditional Rough Set-Based Feature Selection Methods

The discernibility function-based reduct computation algorithms belong to the traditional rough set-based feature selection methods. They are built on the concepts of indiscernibility relation, set approximations, dispensable and indispensable attributes, reduct and core. Readers may refer to [6] for details.

The reduct computation is directly based on discernibility matrix [6]. Assume $IS$ is an information system which can be represented by a triplet $IS = (U, A, f)$, where $U$ is a finite nonempty set of $n$ objects $\{x_1, x_2, ..., x_n\}$; $A$ is a finite nonempty set of $m$ attributes (features) $\{a_1, a_2, ..., a_m\}$; $f_a : U \to V_a$ for any $a \in A$, where $V_a$ is called the domain of attribute $a$. The discernibility matrix is defined as a $n \times n$ matrix represented by $(dm_{ij})$, where $dm_{ij} = \{a \in A : f_a(x_i) \neq f_a(x_j)\}$ for $i, j = 1, 2, ..., n$.

The discernibility matrix of an IS completely depicts the identification capability of the system and all reducts are therefore hidden in some discernibility function induced by the discernibility matrix. Due to the computation of discernibility matrix, if there are $n$ objects in the IS, $m$ attributes in $A \cup \{d\}$, the computation complexity of these methods is $O(n^2 \times m)$.

## 3.2 Relative Dependency-Based Reduct Computation

To reduce the computational load of the discernibility function-based methods, Han et al. [15] have developed a reduct computation approach based on the concept of relative attribute dependency. Given a subset of condition attributes, $B$, the relative attribute dependency is a ratio between the number of distinct rows in the decision sub-table corresponding to $B$ only and the number of distinct rows in the decision sub-table corresponding to $B$ together with the decision attributes, i.e., $B \cup \{d\}$. The larger the relative attribute dependency value (i.e., close to 1), the more useful is the subset of condition attributes $B$ in discriminating the decision attribute values. To evaluate the generated approximate reducts, Bazan et al. [16] used a quality measure for reducts based on the number of rules generated by the reducts. In this paper, we do not need to obtain the optimal approximate reduct, which requires much more computational effort. In the following sections, a feature selection method will be built based on the work of Han et al. Some pertinent concepts are defined as:

**Definition 1.** (Projection) [15]
Let $P \subseteq A \cup D$, where $D = \{d\}$. The projection of $U$ on $P$, denoted by $\Pi_P(U)$, is a sub table of $U$ and is constructed as follows:

    1) remove attributes $A \cup D - P$; and
    2) merge all indiscernible rows.

**Definition 2.** (Consistent Decision Table)
A decision table $DT$ or $U$ is consistent when $\forall x, y \in U$, if $f_D(x) \neq f_D(y)$, then $\exists a \in A$ such that $f_a(x) \neq f_a(y)$.

Table 1 provides an example of consistent decision table. Here $U = \{c_1, c_2, ..., c_8\}$, $A = \{a, b, c, d\}$ and $D = \{e\}$. In Table 1, for every two objects in $U$, if they have the same attribute values for all the attributes, their decision attribute must be equal. In contrast, Table 2 shows an example of inconsistent table which derived from Table 1. It is obvious that $c_8$ and $c_9$ have the same condition attribute values, 1, 1, 1, 2, but different decision attribute value, 1 for $c_8$ and 2 for $c_9$.

**Definition 3.** (Relative Dependency Degree)
Let $B \subseteq A$, $A$ be the set of conditional attributes. $D$ is the set of decision attributes. The *relative dependency degree* of $B$ w.r.t. $D$ is defined as

**Table 1.** A consistent decision table

| ID | a | b | c | d | e |
|----|---|---|---|---|---|
| $c_1$ | 1 | 1 | 2 | 1 | 2 |
| $c_2$ | 1 | 2 | 1 | 2 | 1 |
| $c_3$ | 2 | 2 | 2 | 1 | 2 |
| $c_4$ | 3 | 1 | 2 | 2 | 1 |
| $c_5$ | 3 | 2 | 2 | 1 | 1 |
| $c_6$ | 1 | 2 | 2 | 1 | 2 |
| $c_7$ | 3 | 2 | 1 | 2 | 1 |
| $c_8$ | 1 | 1 | 1 | 2 | 1 |

**Table 2.** An inconsistent decision table

| ID | a | b | c | d | e |
|----|---|---|---|---|---|
| $c_1$ | 1 | 1 | 2 | 1 | 2 |
| $c_2$ | 1 | 2 | 1 | 2 | 1 |
| $c_3$ | 2 | 2 | 2 | 1 | 2 |
| $c_4$ | 3 | 1 | 2 | 2 | 1 |
| $c_5$ | 3 | 2 | 2 | 1 | 1 |
| $c_6$ | 1 | 2 | 2 | 1 | 2 |
| $c_7$ | 3 | 2 | 1 | 2 | 1 |
| $c_8$ | 1 | 1 | 1 | 2 | 1 |
| $c_9$ | 1 | 1 | 1 | 2 | 2 |

$\delta_B^D$, $\delta_B^D = |\Pi_B(U)|/|\Pi_{B \cup D}(U)|$, where $|\Pi_X(U)|$ is the number of equivalence classes in $U/IND(X)$.

The relative dependency degree $\delta_B^D$ implies how well subset $B$ discerns the objects in $U$ relative to the original attribute set $A$. It can be computed by counting the number of equivalence classes induced by $B$ and $B \cup D$, i.e., the distinct rows in the projections of $U$ on $B$ and $B \cup D$.

Take Table 3 for example to show the process of the computation of relative dependency degree $\delta_B^D$. Here $B = \{a, c, d\}$, $D = \{e\}$.

Since
$$U/IND(B) = \{\{c_1, c_6\}, \{c_2, c_8\}, \{c_3\}, \{c_4\}, \{c_5\}, \{c_7\}\}$$
$$= U/IND(B \cup D) = \{\{c_1, c_6\}, \{c_2, c_8\}, \{c_3\}, \{c_4\}, \{c_5\}, \{c_7\}\},$$

$$\delta_B^D = \frac{|\Pi_B(U)|}{|\Pi_{B \cup D}(U)|} = \frac{6}{6} = 1.$$

We can say that the attribute subset $B$ has preserved the discernibility ability of the original feature set $A$.

Let $P = \{a, c, d\}$, according to Definition 1, the projection of $U$ on $P$, $\Pi_P(U)$ can be described by Table 3 which is a sub-table of Table 1.

**Table 3.** An example of projection

| ID | a | c | d | e |
|----|---|---|---|---|
| $c_1$ | 1 | 2 | 1 | 2 |
| $c_2$ | 1 | 1 | 2 | 1 |
| $c_3$ | 2 | 2 | 1 | 2 |
| $c_4$ | 3 | 2 | 2 | 1 |
| $c_5$ | 3 | 2 | 1 | 1 |
| $c_6$ | 1 | 2 | 1 | 2 |
| $c_7$ | 3 | 1 | 2 | 1 |
| $c_8$ | 1 | 1 | 2 | 1 |

It can be easily induced that $\delta_A^D = 1$ when $U$ is consistent. A attribute subset $B \subseteq A$ is found to be a reduct, if the decision sub-table is still consistent after removing the attributes in $A - B$. This is given as Theorem 1.

**Theorem 1.** If $U$ is consistent, $B \subseteq A$ is a reduct of $A$ w.r.t. $D$, if and only if $\delta_B^D = \delta_A^D = 1$ and for $\forall Q \subset B$, $\delta_Q^D \neq \delta_A^D$. (See [15] for the proof)

Theorem 1 gives the necessary and sufficient conditions for reduct computation and implies that the reduct can be generated by only counting the distinct rows in some projections. The computational load is linear to the number of cases, $n$, and the number of attributes, $m$.

Here we use the example in Table 1 to illustrate Han's method. As mentioned previously, the consistent decision table consists of 8 cases, 5 attributes including 4 conditional attributes, $A = \{a, b, c, d\}$, and 1 decision attribute, $D = \{e\}$. We have the following computations:

Since $\delta_{A-\{a\}} = \frac{|\Pi_{\{b,c,d\}}(U)|}{|\Pi_{\{b,c,d,e\}}(U)|} = 5/6$, and $\delta_{A-\{b\}} = \frac{6}{6} = 1$,

$b$ is considered to be dispensable and therefore removed from $A$, i.e., $A = \{a, c, d\}$.

Next, because $\delta_{A-\{c\}} = \frac{|\Pi_{\{a,d\}}(U)|}{|\Pi_{\{a,d,e\}}(U)|} = \frac{5}{5} = 1$, $c$ is then removed from $A$, $A = \{a, d\}$.

However, this method can not directly used on inconsistent decision table such as that shown in Table 2. For every subset of condition attributes $B \subseteq A$, we always have $\delta_B^D = \frac{|\Pi_B(U)|}{|\Pi_{B \cup D}(U)|} < 1$. Therefore, the reduct cannot be found. In next section, we will introduce the concept of approximate reduct to overcome this problem.

### 3.3   Feature Selection Based on Approximate Reduct

From Section 3.2, we notice that, although the relative dependency-based reduct computation is fast, $U$ is always assumed to be consistent in theorem 1. This assumption is not necessarily true in real life applications. In this section, we relax this condition by finding approximate reduct instead of exact reduct. The use of a relative dependency degree in reduct computation is extended to inconsistent information systems. Some new concepts, such as the $\beta$-*dispensable attribute*, $\beta$-*indispensable attribute*, $\beta$-*reduct* (i.e., approximate reduct), and $\beta$-*core* are introduced to modify the traditional concepts in rough set theory. The parameter is used as the consistency measurement to evaluate the goodness of the subset of attributes currently under consideration. These are explained as follows.

**Definition 4.** ($\beta$-dispensable attribute and $\beta$-indispensable attribute)
If $a \in A$ is an attribute that satisfies $\delta^D_{A-\{a\}} \geq \beta \cdot \delta^D_A$, $a$ is called a $\beta$-*dispensable* attribute in $A$. Otherwise, $a$ is called a $\beta$-*indispensable* attribute.

The parameter $\beta$, $\beta \in [0,1]$, is called the consistency measurement.

For example, in Table 1, we have $\delta_{A-\{a\}} = \frac{5}{6} < 1$, $\delta_{A-\{d\}} = \frac{3}{4} < 1$. If $\beta$ is set as 0.75, then the attributes $a$ and $d$ are both considered as $\beta$-indispensable attribute. If $\beta$ is set as 0.8, then only the attribute $d$ is $\beta$-indispensable attribute.

**Definition 5.** ( $\beta$-reduct/approximate reduct and $\beta$-core)
$B$ is called a $\beta$-*reduct* or *approximate reduct* of conditional attribute set $A$ if $B$ is the minimal subset of $A$ such that $\delta^D_B \geq \beta \cdot \delta^D_A$. The $\beta$-*core* of $A$ is the set of $\beta$-indispensable attributes.

The relationship between $\beta$-reduct and $\beta$-core is similar to the relationship between the traditional reduct and core, which is described in theorem 2.

**Theorem 2 (Relationship of $\beta$-reduct and $\beta$-core)**
$\beta$-core can be computed as the interaction of all approximate reducts, i.e., $\beta$-core $= \cap_i reduct_i$, where $reduct_i$ is the i-th approximate reduct.

Proof: The proof is divided into two parts.

(1) For every attribute $a \in \beta$-core, $a$ is a $\beta$-indispensable attribute, i.e., $\delta_{A-\{a\}} < \beta \cdot \delta^D_A$. According to definition 5, an approximate reduct implies that $a \in \cap_i reduct_i$. This can be proved using the method of contradiction as follows:

If $\exists i$, $a \notin reduct_i$, then $reduct_i \subseteq A - \{a\}$ and $\delta^D_{reduct_i} < \delta^D_{A-\{a\}} < \beta \cdot \delta^D_A$. This result contradicts the assumption that $reduct_i$ is an approximate reduct. Therefore, $a \in \cap_i reduct_i$, and then $A \subseteq B$ holds.

(2) Let an attribute $a \in \cap_i reduct_i$. If we assume $a \notin \beta$-core, that is, $a$ is a dispensable attribute, then $\exists i$, such that $a \notin reduct_i$. This is not possible since $a \in \cap_i reduct_i$. Therefore, $a \in \beta$-core.

This completes the proof.                                        $\square$

The consistency measurement $\beta$ reflects the relationship of the approximate reduct and the exact reduct. The larger the value of $\beta$, the more similar is the approximate reduct to the exact reduct computed using the traditional discernibility function-based methods. If $\beta = 1$ (i.e., attains its maximum), the two reducts are equal (according to theorem 1). The reduct computation is implemented by counting the distinct rows in the decision sub-tables of some attribute subsets. $\beta$ controls the end condition of the algorithm and therefore controls the size of reduced feature set. It can be determined beforehand by experts or can be learned during the feature selection process. Based on Definitions 4-5, the rough set-based feature selection algorithm in our developed approach is given as follows.

### Feature Selection Algorithm

Input: $U$ - the entire case base; $A$ - the entire condition attribute set; $D$ - the decision attribute set.
Output: $R$ - the approximate reduct of $A$.

Step 1. Initialize $R$=empty set;
Step 2. Compute the approximate reduct.

> While $A$ is not empty
> 1. For each attribute $a \in A$
>    Compute the significance of $a$;
> 2. Add the most significant one, $q$, to $R$: $R = R \cup \{q\}$;
>    $A = A - q$;
> 3. Compute the relative dependency degree $\delta_R^D$ for current $R$;
> 4. If $\delta_R^D > \beta$ , return $R$ and stop.

Notice that the significance of an attribute $a$ can be evaluated in many ways using different evaluation criteria such as information gain (IG), frequency of occurrence (often used in text documents),and dependency factors (in rough set-based methods). In this paper, we use the dependency factors to compute the significance of the attributes.

Since the computation of approximate reduce does not increase the computational load to the method of Han et al., the computation complexities of the feature selection algorithms is also $O(n \times m)$, where $m$ is the number of features in $A \cup D$, $n$ is the number of objects in $U$. This can be explained as follows. In each iteration of the feature selection algorithm, one attribute is added to R (Reduct) and the corresponding $\delta$ value (i.e., $\delta_R^D > \beta$) is computed. The computation of $\delta$ is to count the rows in the projection, which needs $|U| = n$ computations. Since that it is required maximally m iterations, the overall time complexity of the FR algorithm is $O(m \times n)$. It is linear to the number of attributes and cases.

## 4   LVQ-Based Case Generation

After the approximate reduct-based feature selection, the supervised learning process of LVQ is used for generating prototypes which represent the entire case base.

## 4.1   Learning Vector Quantization

Vector quantization is one example of competitive learning. The goal is to discover structure in the data by finding how the data is clustered. The results can be used for data encoding and compression. LVQ is a supervised version of vector quantization. Classes are predefined and we have a set of labelled data. A set of prototypes are then determined to best represent each class.

Another method of competitive learning is SOM, which is unsupervised and can serve as a clustering tool of high-dimensional data. For classification problems, supervised learning LVQ should be superior to SOM since the information of classification results is incorporated to guide the learning process. LVQ is more robust to redundant features and cases, and more insensitive to the learning rate. As Kohonen pointed out in [11], LVQ instead of SOM should be used in decision and classification processes. This is the reason that LVQ is applied in case selection for building compact case base for CBR classifiers.

Basically, there are three different learning techniques of LVQ, i.e., LVQ1, LVQ2, and LVQ3[12]. In this work, we use the kind of LVQ algorithms which can be considered to be the supervised version of the SOMs. It defines a mapping from high dimensional input data space onto a regular two-dimensional array of nodes called competitive layer. Every node $i$ of the competitive layer is associated with an $m$-dimensional vector $v_i = [v_{i1}, v_{i2}, , v_{im}]$, where $m$ denotes the dimension of the cases called reference vectors. The basic assumption here is that the nodes near to the same input vector should locate near to each other.

Given an input vector, the most similar node in the competitive layer can be found as the winning node. Other nearby nodes for the input vector can be also found through similarity computation. Based on the mentioned assumption, the winning node and those nearby nodes should locate near to the input vector. The class information is also incorporated in the learning process. At each learning step, if the winning node and those nearby nodes are in the same class of input vector, the distances among these nodes are reduced; otherwise, these nodes are kept intact. This is different from the unsupervised learning process of SOM, where the winning node and those in its neighbourhood will move towards each other even they are not in the same class. The amount of decrease in distance is determined by the given learning rate. As a result, after the learning with the reference vectors, LVQ converges to a stable structure and the final weight vectors are the cluster centres. These weight vectors are considered as the generated prototypes which can represent the entire case base.

## 4.2   Rough LVQ Algorithm

Although LVQ has similar advantages of SOM, such as the robustness with noise and missing information, it does not mean that the data preprocessing is not required before the learning process. Since the basic assumption of LVQ is that similar feature values should lead to similar classification results, the similarity computation is critical in the learning process. Feature selection is one of the

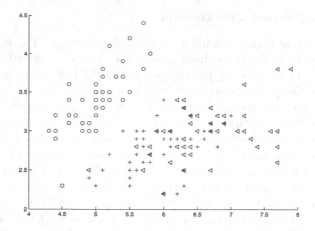

**Fig. 2.** Iris data on feature set {SL, SW}

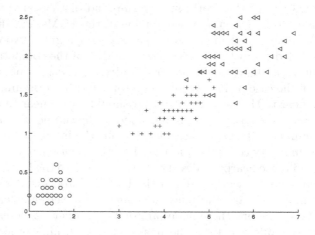

**Fig. 3.** Iris data on feature set {PL, PW}

most important preparations for LVQ which can achieve better clustering and similarity computation results.

Different subset of features will result different data distribution and clusters. Take the Iris data [17] for example. Fig. 2 and Fig. 3 show the two dimensional Iris data on two different subsets of features: {SL, SW} and {PL, PW}.

Based on the two subsets of features, LVQ is applied to learn three prototypes for the Iris data. The generated representative cases are shown in Tables 4-5. It shows that different subset of attributes can affect the LVQ learning process and different prototypes are generated. According to the classification accuracy, the feature set of {PL, PW} is better than {SL, SW}.

In this paper, the feature selection is handled using the approximate reduct-based method which given in the previous section. LVQ is then applied to generate representative cases for the entire case base. Here the learning rate $\alpha$ is

**Table 4.** Prototypes extracted using PL and PW

| Prototypes | SL | SW | PL | PW | Class label |
|:---:|:---:|:---:|:---:|:---:|:---:|
| P1 | 0.619 | 0.777 | 0.224 | 0.099 | 1 |
| P2 | 0.685 | 0.613 | 0.589 | 0.528 | 2 |
| P3 | 0.766 | 0.587 | 0.737 | 0.779 | 3 |

Classification accuracy using P1 P2 and P3: 0.98

**Table 5.** Prototypes extracted using SL and SW

| Prototypes | SL | SW | PL | PW | Class label |
|:---:|:---:|:---:|:---:|:---:|:---:|
| P1 | 0.649 | 0.842 | 0.211 | 0.094 | 1 |
| P2 | 0.712 | 0.550 | 0.572 | 0.212 | 2 |
| P3 | 0.980 | 0.840 | 1.096 | 1.566 | 3 |

Classification accuracy using P1, P2 and P3: 0.80

given in advance, and only the distance between the winning node and the given input vector is updated in each learning step. The number of weight vectors is determined as the number of classes in the given case base. The learning process is ended with a fixed number of iterations $T$, say, 5000 in this paper. Assume the given case base has $n$ cases which represented by $m$ features, and there are $c$ classes. $R$ is the approximate reduct computed by the feature selection process. The LVQ algorithm is given as follows:

Step 1. Initialize $c$ weight vectors $[v_1, v_2, ..., v_c]$ by randomly selecting one case from each class.

Step 2. Generate prototypes through LVQ.

$t \leftarrow 1$;
While $(t \leq T)$
  for $k = 1$ to $n$
    $x \in U$, $x_k \leftarrow x$, $U \leftarrow U - x_k$;
    1. Compute the distances $D = \{\|x_k - v_{i,t-1}\|_R : 1 \leq i \leq c\}$;
    2. Select $v_{win,t-1} = arg\{v_{i,t-1} : \|x_k - v_{i,t-1}\|_R = min\{d \in D\}\}$;
    3. If $Class(v_{win,t-1}) = Class(x_k)$
         Update $v_{win,t} = v_{i,t-1} + \alpha(x_k - v_{win,t-1})$;
  Output $V = [v_{1,T-1}, v_{2,T-1}, ..., v_{c,T-1}]$.

The output vectors are not the data points in the given case base, but modified during the learning process based on the provided information by the data. They are considered to be the generated prototypes which represent the entire case base. Each prototype can be used to describe the corresponding class and regarded as the cluster center.

# 5    Experimental Results

To illustrate the effectiveness of the developed rough LVQ case selection method, we describe here some results on three real life data from UCI (University of California, Irvine)Machine Learning Repository [17]. These databases are: (1)Iris, (2)Glass, and (3)Pima.

The characteristics are listed in Table 6. In all the experiments, 80% cases in each database are randomly selected for training and the remaining 20% cases are used for testing.

**Table 6.** The characteristics of three UCI databases

| Data set | Number of cases | Number of features | Category of features | Number of classes |
|----------|-----------------|--------------------|----------------------|-------------------|
| Iris     | 150             | 4                  | Numerical            | 3                 |
| Glass    | 214             | 10                 | Numerical            | 5                 |
| Pima     | 768             | 8                  | Numerical            | 2                 |

In this paper, four indices are used to evaluate the rough LVQ case generation method. The classification accuracy is one of the important factors to be considered for building classifiers. On the other hand, the efficiency of CBR classifiers in terms of case retrieval time should not be neglected. The storage space and clustering performance (in terms of intra-similarity and inter-similarity) are also tested in this section.

Based on these evaluation indices, comparisons are made between our developed method and others such as basic SOM, basic LVQ, Random case selection methods, and SVM for only binary data. Here, by "Random case selection", we mean randomly selecting some cases in the training data. In order to make it comparable, the number of randomly selected cases is equal to the number of generated prototypes by the rough LVQ method. These selected cases are then used for predicting the testing data. Notice that SVM is more preferred to be used on binary data than on multiple-class data, we therefore perform it on the Pima data.

As mentioned in Sections 3 and 4, the rough set-based feature selection is firstly used to find the approximate reduct of the given case bases. In the experiments of this section, the parameter $\beta$ is determined during the testing through populating the points in the interval [0.5, 1]. Initially, $\beta$ is set to be 0.5. In each step, the $\beta$ value increase at a constant rate 0.01 and this value is used in the feature selection process and being tested. The steps stop when attains 1. The value which can achieve the highest classification accuracy is selected as the suitable $\beta$. The learning rates for the three data sets are: $\alpha= 0.8$ (Iris data), $\alpha= 0.8$ (Glass data) and $\alpha= 0.5$ (Pima data).

## 5.1    Classification Accuracy

In this section, the results of classification accuracy for the three databases and four case selection methods are demonstrated and analyzed. The used accuracies here are defined as:

$$Accuracy_{Test} = \frac{|\{x, x \text{ can be correctly classified}, x \in Testdata\}|}{|Testdata|},$$

$$Accuracy_{All} = \frac{|\{x, x \text{ can be correctly classified}, x \in Entiredata\}|}{|Entiredata|},$$

where $| * |$ is the cardinality of set $*$; Testdata is the set of cases for testing; Entiredata is the set of cases in the whole data set. To be more specifically, "x can be correctly classified" means that "x can be correctly classified" by the extracted prototypes.

If the training cases are used for classify the testing cases, the classification accuracies on the three databases are: 0.980 (Iris), 0.977 (Glass), 0.662 (Pima). These accuracy values are called the original classification accuracies. The experimental results of using the generated prototypes are demonstrated in Table 7. Here we test the accuracy using both the testing cases and all cases, denoted by $Accu_{Test}$ and $Accu_{All}$, respectively. It is observed that after the case generation, the original accuracies are preserved and even improved. The rough LVQ method can achieve the highest classification accuracy in most of the testing. The basic LVQ method performs better than the other methods: Random and SOM. SVM has been used on Pima data, which obtains $Accu_{Test}$=0.334, and $Accu_{All}$=0.337.

**Table 7.** Classification accuracy using different case generation methods

| Methods | Iris data | | Glass data | | Pima data | |
|---|---|---|---|---|---|---|
| | $Accu_{Test}$ | $Accu_{All}$ | $Accu_{Test}$ | $Accu_{All}$ | $Accu_{Test}$ | $Accu_{All}$ |
| Random | 0.760 | 0.746 | 0.860 | 0.864 | 0.597 | 0.660 |
| SOM | 0.920 | 0.953 | 0.930 | 0.925 | 0.688 | 0.730 |
| LVQ | 0.980 | 0.953 | 0.930 | **0.935** | 0.708 | **0.743** |
| Rough LVQ | **1.000** | **0.960** | **0.930** | **0.935** | **0.714** | 0.740 |

## 5.2 Reduced Storage Space of Rough LVQ-Based Method

Due to both the feature selection and case selection processes, the storage space with respect to the features and cases is reduced substantially. Subsequently, the average case retrieval time will decrease. These results are shown in Table 8, where

$$Reduced\ features = (1 - \frac{|Selected\ features|}{|Original\ features|}) \times 100\%,$$

$$Reduced\ cases = (1 - \frac{|Prototypes|}{|Entiredata|}) \times 100\%,$$

$$Saved\ time\ of\ case\ retrieval = (t_{train} - t_p),$$

where $t_{train}$ is the case retrieval time using the training cases; $t_p$ is the case retrieval time using the extracted prototypes. The unit of time is second.

From Table 8, the storage requirements of features and cases are reduced dramatically. For example, the percentage of reduced features is 60% for Glass data, and the percentage of reduced cases is 99.6% for Pima data. The case retrieval time also decreases because that there are much fewer features and cases after applying the rough LVQ-based case selection method.

**Table 8.** Reduced storage and saved case retrieval time

| Data set | Reduced features | Reduced cases | Saved time of case retrieval |
|----------|------------------|---------------|------------------------------|
| Iris | 50% | 97.0% | 0.600 sec |
| Glass | 60% | 98.8% | 0.989 sec |
| Pima | 50% | 99.6% | 0.924 sec |

It should be noted that SVM can also be considered as a case generation method which can reduce the storage through the discovery of support vectors. In the experiments of using SVM on Pima data, about 395 support vectors has been generated, which is 64.5% of the original cases, i.e., 35.5% case storage has been reduced.

## 5.3 Intra-similarity and Inter-similarity

Intra-similarity and inter-similarity are two important indices to reflect the clustering performance. They are used in this section to prove that the developed rough LVQ-based approach can achieve better clustering than using random selected prototypes. Since the similarity between two cases is inverse proportional to the distance between them, we use inter-distance and intra-distance to describe the inter-similarity and intra-similarity. Assume there are $K$ classes for a given case base, $C_1, C_2, ..., C_K$. The intra-distance and inter-distance of the case base are defined as:

$$Intra\text{-}Distance = \sum_{x,y \in C_i} d(x,y),$$
$$Inter\text{-}Distance = \sum_{x \in C_i, y \in C_j} d(x,y), i,j = 1, 2, ..., K, i \neq j$$
$$Ratio = Inter\text{-}Distance / Intra\text{-}Distance.$$

The lower the intra-distance and the higher the inter-distance, the better is the clustering performance. Therefore, it is obvious that the higher the ration between the inter-distance and the intra-distance, the better is the clustering performance.

The results are shown in Table 9. Rough LVQ method demonstrates higher Ratio values and therefore achieves better clustering result.

**Table 9.** Inter-distance and inter-distance: Comparisons between the Random and Rough LVQ methods

| Data set | Methods | Inter-Distance | Intra-Distance | Ratio |
|----------|---------|----------------|----------------|-------|
| Iris | Random | 1284.52 | 102.13 | 12.577 |
| | Rough LVQ | 1155.39 | 51.99 | 22.223 |
| Glass | Random | 8640.20 | 4567.84 | 1.892 |
| | Rough LVQ | 7847.37 | 3238.99 | 2.423 |
| Pima | Random | 56462.83 | 54529.05 | 1.035 |
| | Rough LVQ | 28011.95 | 25163.45 | 1.113 |

# 6    Conclusions

In this paper, a rough LVQ approach is developed to address the case generation for building compact and competent CBR classifiers. Firstly, the rough set-based feature selection method is used to select features for LVQ learning. This method is built on the concept of approximate reduct instead of exact reuduct. It is a generalization of traditional discernibility matrix-based feature reduction. LVQ is then used to extract the prototypes to represent the entire case base. These prototypes are not the data points in the original case base, but are modified during the LVQ learning process. They are considered as the most representative cases for the given case base, and used to classify the unseen cases. Through the experimental results, using much fewer features (e.g., 40% of the original features for Glass data), the classification accuracies for the three real life data are higher using our method than those using methods of Random, basic SOM and LVQ. The case retrieval time for predicting class labels of unseen cases is also reduced. Furthermore, higher intra-similarity and lower inter-similarity are achieved using the rough LVQ approach than that using the random method.

# References

1. Kolodner, J.: Case-Based Reasoning. Morgan Kaufmann, San Francisco, (1993).
2. Pal S. K. and Shiu S. C. K.: Foundations of Soft Case-Based Reasoning. John Wiley, New York, (2004).
3. Kalapanidas E.and Avouris N.: Short-term air quality prediction using a case-based classifier. Environmental Modelling & Software, vol. 16, **3**, (2001), 263–272.
4. Emam K. E., Benlarbi S., Goel N. and Rai S. N.: Comparing case-based reasoning classifiers for predicting high risk software components. Journal of Systems and Software, vol. 55, **3**, (2001), 301–320.
5. Garrell J. M., Golobardes E., Bernad E., and Llor X.: Automatic diagnosis with genetic algorithms and case-based reasoning. Artificial Intelligence in Engineering, vol. 13, **4**, (1999), 367–372.
6. Pawlak, Z.: Rough sets: Theoretical aspects of reasoning about data. Kluwer Academic Publishers, Boston, (1991).
7. Nguyen, H. S. and Skowron, A.: Boolean reasoning for feature extraction problems. Proceedings of the 10th International Symposium on Methodologies for Intelligent Systems, (1997), 117–126.
8. Wang, J. and Wang, J.: Reduction algorithms based on discernibility matrix: The ordered attributes method. Journal of Computer Science & Technology, Vol. 16, **6**, (2001), 489–504.
9. Skowron, A. and Rauszer, C.: The discernibility matrices and functions in information systems. In: K. Slowinski (ed.): Intelligent Decision Support-Handbook of Applications and Advances of the Rough Sets Theory. Kluwer, Dordrecht, (1992), 331–362.
10. Shen, Q. and Chouchoulas, A.: A rough-fuzzy approach for generating classification rules. Pattern Recognition, Vol. 35, (2002), 2425–2438.
11. Kohonen, T.: Self-organization and associative memory. New York, Springer-verlag, (1988).
12. Kohonen, T.: Self-organizing maps. New York, Springer-verlag, (1997).

13. Mangiameli, P., Chen, S. K. and West, D.: A comparison of SOM neural network and hierarchical clustering methods. European Journal of Operational Research, Vol. 93, (1996), 402–417.
14. Pal, S. K., Dasgupta, B. and Mitra, P.: Rough-self organizing map. Applied Intelligence, Vol. 21, **3**, (2004), 289–299.
15. Han, J., Hu, X. and Lin, T. Y.: Feature subset selection based on relative dependency between attributes. Proceedings of the 4th International Conference of Rough Sets and Current Trends in Computing (RSCTC04), Springer-Verlag, Berlin, (2004), 176–185.
16. Bazan, J., Nguyen, H. S., Nguyen, S. H., Synak, P., Wróblewski, J.: Rough Set Algorithms in Classification Problem. In: Polkowski, L., Tsumoto, S. and Lin, T.Y. (eds.): Rough Set Methods and Applications. Physica-Verlag, Heidelberg, New York, (2000), 49–88.
17. UCI Machine Learning Data Repository:
    http://www.ics.uci.edu/ mlearn/MLRepository.html

# Mining Rough Association from Text Documents for Web Information Gathering

Yuefeng Li[1] and Ning Zhong[2]

[1] School of Software Engineering and Data Communications
Queensland University of Technology, Brisbane QLD 4001, Australia
y2.li@qut.edu.au
[2] Department of Information Engineering
Maebashi Institute of Technology, Maebashi 371-0816, Japan
zhong@maebashi-it.ac.jp

**Abstract.** It is a big challenge to guarantee the quality of association rules in some application areas (e.g., in Web information gathering) since duplications and ambiguities of data values (e.g., terms). Rough set based decision tables could be efficient tools for solving this challenge. This paper first illustrates the relationship between decision tables and association mining. It proves that a decision rule is a kind of closed pattern. It also presents an alternative concept of rough association rules to improve the quality of discovered knowledge in this area. The premise of a rough association rule consists of a set of terms (items) and a weight distribution of terms (items). The distinct advantage of rough association rules is that they contain more specific information than normal association rules. This paper also conducts some experiments to compare the proposed method with association rule mining and decision tables; and the experimental results verify that the proposed approach is promising.

**Keywords:** Association mining, Web information gathering, Rough sets.

## 1 Introduction

One of the important issues for Web information gathering is to apply data mining techniques within Web documents to discover some interesting patterns for user information needs. The motivation arises while we determine the interesting and useful Web pages or text documents to a specified topic. It is easier for users to answer which of Web pages or documents are relevant to the specified topic rather than describe what the specified topic they want. The challenging issue is how to discover satisfactory knowledge in order to response what users want effectively.

Data mining has been used in Web text mining, which refers to the process of searching through unstructured data on the Web and deriving meaning from it [7] [9] [14]. One of main purposes of text mining is association discovery [3], where the association between a set of terms and a category (e.g., a term or a set of terms) can be described as association rules. The current association discovery approaches include maximal patterns [8] [12], sequential patterns [27] and closed patterns [28] [13].

It is indubitable that the existing data mining techniques can return numerous discovered patterns (knowledge) from electronic data and information. However, it is a

J.F. Peters et al. (Eds.): Transactions on Rough Sets VII, LNCS 4400, pp. 103–119, 2007.

big challenge to use the discovered knowledge efficiently for making decisions due to the noise in the discovered knowledge. The concept of closed patterns is forward one more step for dealing with the noise, but there are still many meaningless patterns retained [28].

Another approach to improve the quality of association rules is to apply constraints to generate only those association rules that are interesting to users based on some constraints instead of all the association rules [13] [30]. Rough set based decision tables [23] [24] provide a promising structure for the representation of constraint-based association rules. Different to the association rule mining, decision tables do not attempt to represent all of interesting patterns, instead of; they only keep some sorts of larger patterns.

In terms of association mining, however, the puzzle for decision tables is that we do not understand what kinds of patterns used in the decision tables. In this research, we first present the concept of decision patterns to interpret this puzzle.

The association discovery and decision table based approaches only discuss relationship between terms in a broad-spectrum level. They pay no attention to the duplications of terms in a transaction (e.g., a document) and how to use labeled information in the training set. The consequential result is that the effectiveness of the systems is unsatisfactory. The cause of this problem is that the specified topic is an approximation concept. However, theses approaches clearly classify the document space into a set of equivalent classes (or granules); and only parts of some classes are what users want.

To improve the effectiveness of association discovery, in this paper we also present an alternative concept, rough association rules, to describe the approximation concept. The premise (precondition) of a rough association rule consists of not only a set of terms and a weight distribution of terms as well. In this way, the document space is classified based on two measures: term sets and term weight distributions. We also present a mining algorithm for discovery of the rough association rules in this paper. In addition, some experiments are also conducted to test the proposed approach.

The remainder of the paper is structured as follows. We begin by introducing the concept of association mining from text documents to summarize the main characteristic of normal association discovery methods in Section 2. In section 3, we present the concept of decision patterns to interpret decision tables in terms of association mining. In Section 4, we first discuss decision rules and text mining. We also describe Pawlak's method for discovery of decision rules as an algorithm. In Section 5, we present the concept of rough association rules. In Section 6, we propose an algorithm for discovery of rough association rules. We also compare this algorithm with Pawlak's method in this section. Section 7 evaluates the proposed approach for information gathering. Section 8 discusses related work and the last section is the conclusions.

## 2   Mining Association from Information Table

Formally the association discovery from text documents can be described as an information table $(D, V^D)$, where $D$ is a set of documents in which each document is a set of terms (the duplicate terms are removed here); and $V^D = \{t_1, t_2, \cdots, t_n\}$ is a set of selected terms for all documents in $D$. Usually $D$ consists of a set of positive documents $D^+$, and a set of negative documents $D^-$.

**Table 1.** An information table

| Documents | Terms | Positive |
|:---------:|:-----:|:--------:|
| $d_1$ | $t_1\ t_2$ | yes |
| $d_2$ | $t_3\ t_4\ t_6$ | yes |
| $d_3$ | $t_3\ t_4\ t_5\ t_6$ | yes |
| $d_4$ | $t_3\ t_4\ t_5\ t_6$ | yes |
| $d_5$ | $t_1\ t_2\ t_6\ t_7$ | yes |
| $d_6$ | $t_1\ t_2\ t_6\ t_7$ | yes |
| $d_7$ | $t_1\ t_2$ | no |
| $d_8$ | $t_3\ t_4$ | no |
| ... | | |

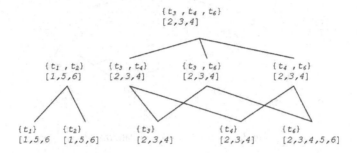

**Fig. 1.** Interesting patterns and their covering sets

**Definition 1.** A set of terms $X$ is referred to as a *termset* if $X \subseteq V^D$. Let $X$ be a termset, we use $[X]$ to denote the covering set of $X$, which includes all positive documents $d$ such that $X \subseteq d$, i.e., $[X] = \{d | d \in D^+, X \subseteq d\}$.

Given a termset $X$, its support is $|[X]|/|D^+|$. A termset $X$ is called frequent pattern if its *support* $\geq$ *min_sup*, a minimum support. The confidence of a frequent pattern is the fraction of the documents including the pattern that are positive. Given a frequent pattern $X$, its confidence is defined as $|[X]|/N$, where $N = \{d | d \in D, X \subseteq d\}$. The confidence shows the percentage of the pattern's occurrence frequency in the positive documents. A frequent pattern is called an interesting pattern if its *confidence* $\geq$ *min_conf*, a minimum confidence.

Table 1 lists a part of an information table, where $V^D = \{t_1, t_2, \cdots, t_7\}$, $D = \{d_1, \cdots, d_6, d_7, d_8\}$, $D^+ = \{d_1, d_2, \cdots, d_6\}$ and $D^- = \{d_7, d_8\}$. Let *min_sup* $=$ 50% and *min_conf* $=$ 75%, we can get 10 interesting patterns. Figure 1 shows these interesting patterns and their covering sets.

There are some noise patterns in Figure 1. For example, pattern $\{t_3, t_4\}$ always occurs with term $t_6$ in $D^+$. Therefore, we expect to keep only the larger patterns $\{t_3, t_4, t_6\}$ and prune the noise one, the shorter one $\{t_3, t_4\}$.

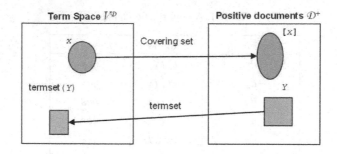

**Fig. 2.** Connection between Terms and Documents

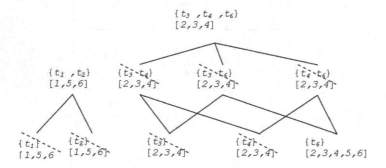

**Fig. 3.** Pruning Non-closed Patterns

**Definition 2.** Given a termset $X$, we know its covering set $[X]$ which is a subset of positive documents. Similarly, given a set of positive documents $Y$, we can define its termset, which satisfies

$$termset(Y) = \{t|t \in V^D, \forall d \in Y => t \in d\}.$$

Figure 2 shows the connection between terms and positive documents for these interesting patterns. Given an interesting pattern $X$, its closure $C(X) = termset([X])$. Different to the traditional definition about covering sets, here $[X]$ is defined on $D^+$, a subset of the information table. From the above definitions, we can prove the following theorem about the closure operation.

**Theorem 1.** Let $X$ and $Y$ be patterns. We have

(1) $C(X) \supseteq X$ for all patterns $X$;
(2) $X \subseteq Y => C(X) \subseteq C(Y)$.

**Definition 3.** An interesting pattern $X$ is closed if and only if $X = C(X)$.

Figure 3 illustrates the process of pruning the non-closed patterns, where only three patterns are closed patterns. They are $\{t_3, t_4, t_6\}$, $\{t_1, t_2\}$ and $\{t_6\}$.

Each closed pattern can be actually viewed as an association rule, e.g., closed pattern $\{t_1, t_2\}$ means

$$(t_1 \wedge t_2) \rightarrow (Positive = yes);$$

because that satisfies the definition about traditional association rules in data mining.

## 3   Mining Decision Patterns

Professor Z. Pawlak believed that the patterns hidden in databases can be characterized by decision tables, in which the premises of association rules (or called decision rules in [23] [24]) are interpreted as condition granules, and the post-conditions are interpreted as decision granules. The measure of uncertainties for decision rules is based on well-established statistical models. Decision tables are very efficient for dealing with very large sets of documents.

Table 2 illustrates a binary decision table about a training set of documents, where $N_g$ is the number of documents that are in the same granule; $t_1, t_2, \cdots, t_7$ are the condition attributes and $Positive$ is the decision attribute.

**Table 2.** A Binary Decision Table

| Granule | $t_1$ | $t_2$ | $t_3$ | $t_4$ | $t_5$ | $t_6$ | $t_7$ | Positive | $N_g$ |
|---------|-------|-------|-------|-------|-------|-------|-------|----------|-------|
| $g_1$ | 1 | 1 | 0 | 0 | 0 | 0 | 0 | yes | 80 |
| $g_2$ | 0 | 0 | 1 | 1 | 0 | 1 | 0 | yes | 140 |
| $g_3$ | 0 | 0 | 1 | 1 | 1 | 1 | 0 | yes | 490 |
| $g_4$ | 1 | 1 | 0 | 0 | 0 | 1 | 1 | yes | 220 |
| $g_5$ | 1 | 1 | 0 | 0 | 0 | 0 | 0 | no | 20 |
| $g_6$ | 0 | 0 | 1 | 1 | 0 | 0 | 0 | no | 50 |

Decision tables provide a straightforward way to represent discovered knowledge. For example, the first granule in Table 2 can be described as a decision rule

$$((t_1 = 1) \wedge (t_2 = 1) \wedge (t_3 = 0) \wedge \cdots \wedge (t_7 = 0)) \rightarrow (Positive = yes)$$

Different to association rule mining, decision tables cannot represent all of interesting patterns, instead of; they only keep some sorts of larger interesting patterns. In terms of association mining, however, the puzzle for decision tables is that we do not understand what kinds of interesting patterns used in the decision tables. In this section, we present the concept of decision patterns to interpret this puzzle. Let $X$ be an interesting pattern. We call it a decision pattern if $\exists d \in D^+$ such that $X = d$.

Different to an information table, zeros in a decision table make sense. Given a decision pattern $X$ (or later on granule), its *covering set* now is the set of documents $d$ that satisfy $X = d$. This definition is the generalization of the concept of covering sets in information tables if we consider all zeros in a decision table (see Table 2).

**Theorem 2.** Decision patterns are closed patterns.

**Proof:** Let $X$ be a decision pattern. From the definition of the decision patterns, we know there is a positive document $d$ such that $X = d$, that is $d \in [X]$.

Given a term $t \in termset([X])$, according to Definition 2 we have $t \in d_i$ for all $d_i \in [X]$, that is, $t \in d = X$. Therefore, $C(X) = termset([X]) \subseteq X$. We also have $X \subseteq C(X)$ from theorem 1, and hence we have $X = C(X)$.     □

Theorem 2 tells us that a decision rule $X \longmapsto (Positive = yes)$ is actually a kind of closed pattern in terms of association mining.

## 4   Multiple Dimensional Databases and Decision Tables

One important factor is missed in both closed patterns and decision patterns: the duplications of terms in a document. This factor is very important in terms of information retrieval. To consider this factor, we start to initialize this problem by using decision tables as multiple dimensional databases. Table 3 demonstrates the corresponding decision table $(D, A_C, A_D)$ if we consider the duplications of terms, where the numbers are frequencies of terms in their corresponding documents, the set of granules (objects) $D = \{d_1, d_2, d_3, d_4, d_5, d_6, d_7, d_8\}$; the set of condition attributes $A_C = \{t_1, t_2, t_3, t_4, t_5, t_6, t_7\}$, and the set of decision attributes $A_D = \{Positive\}$.

**Table 3.** A multiple dimensional databases

| Granule | $t_1$ | $t_2$ | $t_3$ | $t_4$ | $t_5$ | $t_6$ | $t_7$ | Positive | $N_d$ |
|---------|-------|-------|-------|-------|-------|-------|-------|----------|-------|
| $d_1$ | 2 | 1 | 0 | 0 | 0 | 0 | 0 | yes | 80 |
| $d_2$ | 0 | 0 | 2 | 1 | 0 | 1 | 0 | yes | 140 |
| $d_3$ | 0 | 0 | 3 | 1 | 1 | 1 | 0 | yes | 40 |
| $d_4$ | 0 | 0 | 1 | 1 | 1 | 1 | 0 | yes | 450 |
| $d_5$ | 1 | 1 | 0 | 0 | 0 | 1 | 1 | yes | 20 |
| $d_6$ | 2 | 1 | 0 | 0 | 0 | 1 | 1 | yes | 200 |
| $d_7$ | 2 | 2 | 0 | 0 | 0 | 0 | 0 | no | 20 |
| $d_8$ | 0 | 0 | 1 | 1 | 0 | 0 | 0 | no | 50 |

Every object in a decision table can be mapped into a decision rule [23]: either a positive decision rule or a negative decision rule. Therefore, we can obtain eight decision rules, e.g., $d_1$ in Table 3 can be read as the following rule: $(GERMAN, 2) \wedge (VW, 1) \rightarrow yes$ where $(term, frequency)$ denotes a term frequency pair in the corresponding object.

Let $termset(d) = \{t_1, \cdots, t_k\}$, formally every document $d$ determines a sequence: $(t_1, f(t_1, d)), \cdots, (t_k, f(t_k, d)), Positive(d)$.

---

**Algorithm 1.** Pawlak's Method

---

**input** : $D, A_C, A_D$ and $V^D$.
**output:** Decision rules.
$UN \longleftarrow 0;$   // $UN$ would be the number of all objects
**for** (*all* $d \in D$) **do**
$\quad | \quad UN \longleftarrow UN + N_d;$
**end**
**for** (*all* $d \in D$) **do**
$\quad$ $strength(d) \longleftarrow N_d/UN;$
$\quad$ $CN \longleftarrow N_d/UN;$   //$CN$, the number of objects that have the same condition
$\quad$ **for** (*all* $d' \in D, d' \neq d$) **do**
$\quad\quad$ **if** $(d(A_C) = d'(A_C))$ **then**
$\quad\quad | \quad CN \longleftarrow CN + N_{d'};$
$\quad\quad$ **end**
$\quad$ **end**
$\quad$ $certainty\_factor(d) \longleftarrow N_d/CN;$
**end**

---

The sequence can determine a decision rule:

$$(t_1, f(t_1, d)) \wedge \cdots \wedge (t_k, f(t_k, d)) \rightarrow Positive(d)$$

or in short $d(A_C) \rightarrow d(A_D)$.

Algorithm 1 describes Pawlak's idea for the discovery of decision rules (see [23] or [18]). If we assume the basic operation is the comparison between two objects (i.e., $d(A_C) = d'(A_C)$ ), then the time complexity is $(n-1) \times n = O(n^2)$, where $n$ is the number of granules in the decision table. It also needs a similar algorithm to determine interesting rules for Pawlak's method. We can obtain many decision rules as showed in the above example. However, there exists ambiguities whist we use the decision rules for determining other relevance information for a specified topic because there may be many rules' premises that have the same termset. For example, give an instance of a piece of information that it contains only four terms $t_3, t_4, t_5$ and $t_6$; but we can found two rules' premises ($d_3$ and $d_4$) that match this instance and have the same conclusion.

To remove such ambiguities, in next section we present the concept of rough association rules in order to compose some decision rules into a single granule if they have the same termset. We also use a weight distribution for the granule to specify the possible semantic meaning in it.

## 5   Rough Association Rules

For every attribute $a \in A_C$, its domain is denoted as $V_a$; especially in the above example, $V_a$ is the set of all natural numbers. Also $A_C$ determines a binary relation $I(A_C)$ on $D$ such that $(d_i, d_j) \in I(A_C)$ if and only if $a(d_i) > 0$ and $a(d_j) > 0$ for every $a \in A_C$, where $a(d_i)$ denotes the value of attribute a for object $d_i \in D$.

It is easy to prove that $I(A_C)$ is an equivalence relation, and the family of all equivalence classes of $I(A_C)$, that is a partition determined by $A_C$, is denoted by $D/I(A_C)$

or simply by $D/A_C$. The classes in $D/A_C$ are referred to $A_C - granules$ (or called the set of condition granules). The class which contains $d_i$ is called $A_C - granule$ induced by $d_i$, and is denoted by $A_C(d_i)$. We also can obtain an $A_D - granules(D/A_D)$ (or called the set of decision granules) in parallel.

For example, using Table 3, we can get the set of condition granules, $D/A_C = \{\{d_1, d_7\}, \{d_2\}, \{d_3, d_4\}, \{d_5, d_6\}, \{d_8\}\}$, and the set of decision granules, $D/A_D = \{Positive = yes, Positive = no\} = \{\{d_1, d_2, d_3, d_4, d_5, d_6\}, \{d_7, d_8\}\}$, respectively. In the following we let $D/A_C = \{cd_1, cd_2, cd_3, cd_4, cd_5\}$ and $D/A_D = \{dc_1, dc_2\}$.

We also need to consider the weight distribution of terms for the condition granules in order to consider the factor of the frequencies of terms in documents. Let $cd_i$ be $\{d_{i1}, d_{i2}, \cdots, d_{im}\}$, we can obtain a weigh distribution for terms $a_j$ in these documents using the following equation:

$$weight(a_j) = \frac{a_j(cd_i)}{\sum_{a \in A_c} a(cd_i)} \tag{1}$$

where we use the composition operation (see [18]) to assign a value to condition granules' attributes, which satisfies:

$$a(cd_i) = a(d_{i1}) + a(d_{i2}) + ... + a(d_{im}) \tag{2}$$

for all $a \in A_C$.

Table 4 illustrates granules and their covering sets we obtain from Table 3 according to previous definitions, where $\widehat{d_i}$ denotes all of documents that have the same representation (e.g., $|\widehat{d_1}| = N_{d_1} = 80$); and each condition granule consists of both a termset and a weight distribution. For example, $cd_1 = < \{t_1, t_2\}, (4/7, 3/7, 0, 0, 0, 0, 0) >$ or in short $cd_1 = \{(t_1, 4/7), (t_2, 3/7)\}$.

Table 4. Granules and their covering sets

| Condition granule | $t_1$ | $t_2$ | $t_3$ | $t_4$ | $t_5$ | $t_6$ | $t_7$ | Covering set |
|---|---|---|---|---|---|---|---|---|
| $cd_1$ | 4/7 | 3/7 | 0 | 0 | 0 | 0 | 0 | $\{\widehat{d_1}, \widehat{d_7}\}$ |
| $cd_2$ | 0 | 0 | 1/2 | 1/4 | 0 | 1/4 | 0 | $\{\widehat{d_2}\}$ |
| $cd_3$ | 0 | 0 | 2/5 | 1/5 | 1/5 | 1/5 | 0 | $\{\widehat{d_3}, \widehat{d_4}\}$ |
| $cd_4$ | 1/3 | 2/9 | 0 | 0 | 0 | 2/9 | 2/9 | $\{\widehat{d_5}, \widehat{d_6}\}$ |
| $cd_5$ | 0 | 0 | 1/2 | 1/2 | 0 | 0 | 0 | $\{\widehat{d_8}\}$ |

(a) Condition granules

| Decision granule | Positive | Covering set |
|---|---|---|
| $dc_1$ | yes | $\{\widehat{d_1}, \widehat{d_2}, \widehat{d_3}, \widehat{d_4}, \widehat{d_5}, \widehat{d_6}\}$ |
| $dc_2$ | no | $\{\widehat{d_7}, \widehat{d_8}\}$ |

(b) Decision granules

Using the associations between condition granules and decision granules, we can rewrite the six decision rules in Table 2 as follows:

$$cd_1 \rightarrow \{(dc_1, 80/100), (dc_2, 20/100)\}$$

$$cd_2 \rightarrow \{(dc_1, 140/140)\}$$

$$cd_3 \rightarrow \{(dc_1, 490/490)\}$$

$$cd_4 \rightarrow \{(dc_1, 220/220)\}$$

$$cd_5 \rightarrow \{(dc_2, 50/50)\}$$

Formally the association can be represented as the following mapping:

$$\Gamma : (D/A_C) \rightarrow 2^{(D/A_D) \times [0,1]}$$

where $\Gamma(cd_i)$ is the set of conclusions for premise $cd_i$ ($i = 1, \cdots, |D/A_C|$), which satisfies:

$$f_j = \frac{|[cd_i \wedge dc_j]|}{|[cd_i]|}$$

for all $(dc_j, f_j) \in \Gamma(cd_i)$ and

$$\sum_{(dc,f) \in \Gamma(cd_i)} f = 1$$

Now we consider the support degree for each condition granule. The obvious way is to use frequencies in the decision table, that is,

$$NC(cd_i) = |[cd_i]| = \sum_{d \in cd_i} N_d$$

for every condition granule $cd_i$. By normalizing, we can get a support function $sup$ on $(D/A_C)$ such that

$$sup(cd_i) = \frac{NC(cd_i)}{\sum_{cd \in (D/A_C)} NC(cd)}$$

for all $cd_i \in (D/A_C)$. It is obvious that $sup$ is a probability function on $D/A_C$. Therefore, the pair $(\Gamma, sup)$ is an association set defined in [19].

**Definition 4.** Let $\Gamma(cd_i) = \{(dc_{i,1}, f_{i,1}), \cdots (dc_{i,|\Gamma(cd_i)|}, f_{i,|\Gamma(cd_i)|})\}$ for all $cd_i$ $(D/A_C)$. We call "$cd_i \rightarrow dc_{i,j}$" a rough association rule, its strength is $sup(cd_i) \times f_{i,j}$ and its certainty factor is $f_{i,j}$, where $1 \leq j \leq |\Gamma(cd_i)|$.

**Theorem 3.** Let "$cd_i \rightarrow dc_{i,j}$" be a rough association rule and $(dc_{i,j}, f_{i,j}) \in \Gamma(cd_i)$. There exists a granule $g_k$ (see Table 2) such that

$$strength(cd_i \rightarrow dc_{i,j}) = N_{g_k}/UN.$$

**Proof:** Since each granule in a decision table is mapped into a rough association rule and vice versa, let $g_k$ be the corresponding granule of "$cd_i \rightarrow dc_{i,j}$" in the decision table, that is $\|[(cd_i \wedge dc_{i,j})]\| = N_{g_k}$.

According to the above definitions, we have

$$strength(cd_i \rightarrow dc_{i,j}) = \frac{NC(cd_i)}{\sum_{cd \in (D/A_C)} NC(cd)} \cdot \frac{\|[cd_i \wedge dc_{i,j}]\|}{\|[cd_i]\|}$$

$$= \frac{\|[cd_i]\|}{\sum_{cd \in (D/A_C)} \|[cd]\|} \cdot \frac{\|[cd_i \wedge dc_{i,j}]\|}{\|[cd_i]\|}$$

$$= \frac{\|[[cd_i \wedge dc_{i,j}]]\|}{\sum_{cd \in (D/A_C)} \|[cd]\|} = \frac{N_{g_k}}{UN}. \qquad \square$$

From Theorem 3, we can understand that the concepts of strengths and certainty factors defined in Definition 4 are the generalization of Pawlak's decision rules.

Figure 4 illustrates the data structure for the representation of rough associations between condition granules and decision granules.

| $Sup$ | $\mathcal{D}/A_C$ | | $\Gamma(cd_i)$ | |
|-------|-------------------|------|-----------------|-----------------|
| 0.10  | $cd_1$ | $\rightarrow$ | $(dc_1, 0.8)$ | $(dc_2, 0.2)$ |
| 0.14  | $cd_2$ | $\rightarrow$ | $(dc_1, 1)$ | |
| 0.49  | $cd_3$ | $\rightarrow$ | $(dc_1, 1)$ | |
| 0.22  | $cd_4$ | $\rightarrow$ | $(dc_1, 1)$ | |
| 0.05  | $cd_5$ | $\rightarrow$ | $(dc_2, 1)$ | |

**Fig. 4.** The data structure for associations between condition granules and decision granules

## 6   Mining Algorithm

In this section, we first present an algorithm (see Algorithm 2) to find the set of rough association rules. We also analyze the proposed algorithm and compare it with Pawlak's Method.

There are total 4 outer *for* loops in Algorithm 2. The first three *for* loops are used to create the data structure (see Fig. 4) for representing condition granules, decision granules and the associations between them. The last *for* loop generates all rough association rules.

The time complexity of Algorithm 2 is determined by the first outer *for* loop since other for loops all traverse pairs in $\Gamma(cd_i)$ ($i = 1, \cdots, |D/A_C|$) and the number of pairs in all $\Gamma(cd_i)$ ($i = 1, \cdots, |D/A_C|$) is just $n$, where $n$ is the number of granules in the decision table.

Checking if "($\exists cd \in (D/A_C)$ such that $termset(d) = termset(cd)$)" in the first for loop that takes $O(|D/A_C|)$, so the time complexity of the algorithm is $O(n \times |D/A_C|)$, where the basic operation is still the comparison between objects. Algorithm 2 is better than Algorithm 1 in time complexity since $|D/A_C| \leq n$.

**Definition 5.** A rough association rule "$cd_i \rightarrow dc_{i,j}$" is an interesting rule if $P(dc_{i,j}|cd_i) - P(dc_{i,j})$ is greater than a suitable constant.

Given a rough association rule $cd_i \rightarrow dc_{i,j}$, it is obvious that $P(dc_{i,j}|cd_i) = f_{i,j}$ if $(dc_{i,j}, f_{i,j}) \in \Gamma(cd_i)$. To decide the probability on the set of decision granules, we present the following theorem.

**Theorem 4.** Let $P : (D/A_D) \rightarrow [0,1]$ such that

$$P(dc) = \sum_{cd_i \in (D/A_C), (dc,f) \in \Gamma(cd_i)} sup(cd_i) \times f.$$

We have $P$ is a probability function on $(D/A_D)$.

**Proof.** It is obvious that the values of $P$ for all decision granules are not negative. We also have

$$\sum_{dc \in (D/A_D)} P(dc) = \sum_{dc \in (D/A_D)} \sum_{cd_i \in (D/A_C), (dc,f) \in \Gamma(cd_i)} sup(cd_i) \times f$$

$$= \sum_{cd \in (D/A_C)} \sum_{(dc,f) \in \Gamma(cd)} sup(cd) \times f$$

$$= \sum_{cd \in (D/A_C)} sup(cd) \times \sum_{(dc,f) \in \Gamma(cd)} f$$

$$= \sum_{cd \in (D/A_C)} sup(cd) \times 1 = 1. \qquad \square$$

## 7   Evaluations

We use Reuters Corpus Volume 1, also known as RCV1, to evaluate the proposed method. We also use the first 20 topics that TREC (Text REtrieval Conference, see $http : trec.nist.gov$) developed for filtering track in 2002. For each topic, a collection is divided into two sets: a training set of documents and a testing set of documents. We also use both precision and recall to measure the performance of systems, where the precision is the fraction of retrieved documents that are relevant to the topic, and the recall is the fraction of relevant documents that have been retrieved.

   We compare rough association mining model with two baseline models: closed pattern based association mining model (see Section 2 for details), and binary decision rule model (see Section 3). A common basic text processing is used for all models, which includes case folding, stemming, stop words removal and 150 term selection that uses $tf * idf$ (term frequency times inverse document frequency) technique.

---

**Algorithm 2.** Rough Association Mining Approach

---

**input** : $D, A_C, A_D$ and $V^D$.
**output**: Rough association rules.
$(D/A_C) \longleftarrow \emptyset$;
Let $UN$ be the number of all objects;
//create the data structure as shown in Figure 1.
**for** $(all\ d \in D)$ **do**
  **if** $(\exists cd \in (D/A_C)\ such\ that\ termset(d) = termset(cd))$ **then**
    $compose\ d(A_C)\ to\ cd$ using Eq. (2);
    $insert\ d(A_D)\ to\ \Gamma(cd)$;
  **else**
    $(D/A_C) \longleftarrow (D/A_C) \cup \{d(A_C)\}$;
    $\Gamma(d(A_C)) = d(A_D)$;
  **end**
**end**
**end**
**for** $(i = 1\ to\ |D/A_C|)$ **do**
  $sup(cd_i) = (1/UN) \times (\sum_{(dc,f) \in \Gamma(cd_i)} f)$;
  $calculate\ weights\ for\ cd_i\ using\ Eq.(1)$;
**end**
**for** $(i = 1\ to\ |D/A_C|)$ **do**
  $temp \longleftarrow 0$;
  **for** $(all\ (dc_{i,j}, f_{i,j}) \in \Gamma(cd_i))$ **do**
    $temp \longleftarrow temp + f_{i,j}$;
  **end**
  **for** $(all\ (dc_{i,j}, f_{i,j}) \in \Gamma(cd_i))$ **do**
    $f_{i,j} \longleftarrow f_{i,j} \div temp$;
  **end**
**end**
//generate all rough association rules.
**for** $(i = 1\ to\ |D/A_C|)$ **do**
  **for** $(all\ (dc_{i,j}, f_{i,j}) \in \Gamma(cd_i))$ **do**
    $strength(cd_i \to dc_{i,j}) \longleftarrow sup(cd_i) \times f_{i,j}$;
    $certainty\_factor(cd_i \to dc_{i,j}) \longleftarrow f_{i,j}$;
  **end**
**end**

---

The main theme of Web information gathering is to seek a suitable representation for a specified topic. To use rough association mining, we are interested in relevant information; therefore, we assume the training set only includes positive documents. Let $V^D$ include the 150 selected keywords and $D = D^+$, Algorithm 2 is used to find rough association rules. We also use the following equation to evaluate a weight for each term according to the discovered rough association rules:

$$weight(term) = \sum_{cd \in (D/A_C),(term,w) \in cd} sup(cd) \times w \qquad (3)$$

where $sup(cd)$ equals to the strength of rough association rule "$cd \rightarrow (Positive = yes)$" if $D = D^+$.

In addition, given a testing document $d$, we use the following equation to determine its relevance in rough association mining model:

$$rel(d) = \sum_{term \in V^d, term \in d} weight(term). \tag{4}$$

In the training phase, the closed pattern based model finds all closed patterns (where $min\_sup$ is the double of the average frequency of terms, the size of largest patterns is 5, and $min\_conf$ is half) and calculates their support and confidence values. In the testing phase, the relevance of document $d$ is the sum of the multiplications of support and confidence of all closed patterns that occur in $d$.

The binary decision rule model, first obtains a set of granules, $G$, in the training phase. It also uses the following equation to evaluate a weight for each term:

$$weight(term) = \sum_{g \in G, term \in g} \frac{strenth(g)}{|termset(g)|}.$$

Given a testing document, $d$, we also use Equation (3) to determine its relevance in binary decision rule model.

**Table 5.** Experimental results

| | Rough association rules | Binary decision rules | Closed association rules |
|---|---|---|---|
| Avg. of top25 precision | 53.60% | 50.60% | 49.20% |
| Avg. of breakeven points | 0.4943 | 0.4840 | 0.4922 |

Table 5 is the experimental results. We use two measures in the table: top 25 precision and breakeven points, where a breakeven point is a point in the precision and recall curve with the same x coordinate and y coordinate.

Figure 5 shows the differences between rough association mining and binary decision rule mining; and between rough association mining and closed pattern based association rule mining in top 25 precision for the 20 topics; and Figure 6 shows their differences in breakeven points for the 20 topics. The positive values (the bars above the horizontal axis) mean the rough association mining performed better than others. The negative values (the bars below the horizontal axis) mean others performed better than rough association mining.

It is no less impressed by the performance of the rough association rule mining since both top 25 precision and breakeven points gain a significant increase. As a result of the experiment we believe that the proposed method is significant since they can improve the effectiveness of the association discovery for Web text mining.

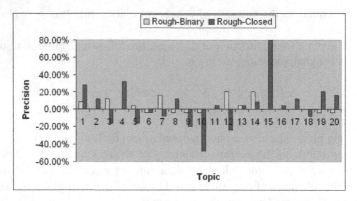

**Fig. 5.** The difference in top25 precision between models

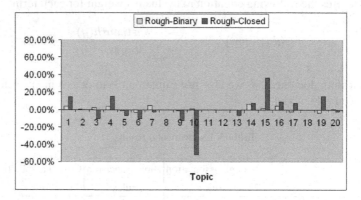

**Fig. 6.** The difference in breakeven points between models

## 8  Related Work

Web information gathering (WIG) systems tend to find useful information from the huge size of Web related data sources to meet their user information needs. The key issue regarding the effectiveness of WIG is automatic acquiring of knowledge from text documents for describing user profiles in order to response what users want efficiently and effectively [16] [19]. It is also a fundamental issue in Web personalization [4].

Traditional information retrieval (IR) techniques can be used to provide straight-forward solutions for this problem. We can classify the methods into two categories: single-vector models and multi-vector models. The former models produce one term-weight vector to represent the relevant information for the topic [2] [11] [5] [26]. The later models produce more than one vector [22] [15]. IR based techniques can be used to obtain efficient systems. This is the distinct merit of IR-based techniques. However, the main drawback of IR-based models is that it is hard to interpret the meaning of vectors, and hence the correlation between vectors cannot be explained using user acceptable concepts.

Text mining tries to derive meaning from documents. Association mining has been used in Web text mining for such purpose for association discovery, trends discovery, event discovery, and text classification [7] [10] [14].

The association between terms and categories (e.g., a term or a set of terms) can be described as association rules. The trends discovery means the discovery of phrases, a sort of sequence patterns. The event discovery is the identification of stories in continuous news streams [3]. Usually clustering based mining techniques can be used for such a purpose. It was also necessary to combine association rule mining with the existing taxonomies in order to determine useful patterns [18] [6].

To compare with IR-based models, data mining-based Web text mining models do not use term independent assumption [1] [20]. Also, Web mining models try to discover some unexpected useful data [3] [21]. The disadvantage of association rule mining is that there are too many discovered patterns that make the application of the discovered knowledge inefficient. Also there are many noise patterns that make the discovered knowledge contains much uncertainties. Although pruning non-closed patterns that can improve the quality of association mining in text mining in some extents [28], the performance of text mining systems are still ineffectively.

Decision rule mining [23] [18][24] can be an alternative solution to specify association rules. However, there exists ambiguities whist we use the decision rules for determining other relevance information for specified topics. We have demonstrated in the previous sections that rough association rule mining can be used to overcome these disadvantages.

## 9   Conclusions

In this paper, we discuss the convention of using rough set theory for text mining. We introduce the concept of decision patterns in order to interpret decision rules in terms of association mining. We have proved that the decision patterns are kinds of closed patterns. We also present a new concept of rough association rules to improve of the quality of association discovery for text mining. To compare with the traditional association mining, the rough association rules include more specific information and can be updated dynamically to produce more effective results. We have verified that the new algorithm is faster than Pawlak's decision rules mining algorithm. We also show that the proposed approach gains a better performance on both precision and recall. This research is significant since it takes one more step further to the development of association rule mining.

## Acknowledgments

This paper was partially supported by Grant DP0556455 from the Australian Research Council.

## References

1. M. L. Antonie and O. R. Zaiane, Text document categorization by term association, *2nd IEEE International Conference on Data Mining*, Japan, 2002, 19-26.
2. R. Baeza-Yates and B.Ribeiro-Neto, *Modern Information Retrieval*, Addison Wesley, 1999.

3. G. Chang, M.J. Healey, J. A. M. McHugh, and J. T. L. Wang, *Mining the World Wide Web: an information search approach*, Kluwer Academic Publishers, 2001.
4. M. Eirinaki and M. Vazirgiannis, Web mining for web personalization, *ACM Transactions on Internet Technology*, 2003, **3(1):** 1-27.
5. D. A. Evans, et al., CLARIT experiments in batch filtering: term selection and threshold optimization in IR and SVM Filters, *TREC02*, 2002.
6. U. Fayyad, G. Piatetsky-Shapiro, P. Smyth, and R. Uthrusamy (eds.), *Advances in knowledge discovery and data mining*, Menlo Park, California: AAAI Press/ The MIT Press, 1996.
7. R. Feldman and H. Hirsh, Mining associations in text in presence of background knowledge, *2nd ACM SIGKDD International Conference on Knowledge Discovery and Data Mining*, 1996, 343-346.
8. R. Feldman, et. al., Maximal association rules: a new tool for mining for keyword co-occurrences in document collection, *3rd International conference on knowledge discovery (KDD)*, 1997, 167-170.
9. R. Feldman, et. al., Text mining at the term level, *Lecture Notes in Artificial Intelligence 1510*, Springer, 65-73.
10. R. Feldman, I. Dagen, and H. Hirsh, Mining text using keywords distributions, *Journal of Intelligent Information Systems*, 1998, **10(3):** 281-300.
11. D. A. Grossman and O. Frieder, *Information retrieval algorithms and heuristics*, Kluwer Academic Publishers, Boston, 1998.
12. J. W. Guan, D. A. Bell, D. Y. Liu, The rough set approach to association rules, *3rd IEEE International Conference on Data Mining*, 2003, Melbourne, Florida, USA, 529-532.
13. J. Han, Y. Fu, Minino Multiple-level Association Rules in Large Databases, *IEEE Trans. On Knowledge and Data Engineering*, 1999, **11(5):** 798-805.
14. J. D. Holt and S. M. Chung, Multipass algorithms for mining association rules in text databases, *Knowledge and Information Systems*, 2001, **3:** 168-183.
15. X. Li and B. Liu, Learning to classify texts using positive and unlabeled data, *IJCAI*, 2003, 587-592.
16. Y. Li and N. Zhong, Web mining model and its applications on information gathering, *Knowledge-Based Systems*, 2004, **17:** 207-217.
17. Y. Li and N. Zhong, Capturing evolving patterns for ontology-based, *IEEE/WIC/ACM International Conference on Web Intelligence*, 2004, Beijing, China, 256-263.
18. Y. Li and N. Zhong, Interpretations of association rules by granular computing, *3rd IEEE International Conference on Data Mining*, 2003, Melbourne, Florida, USA, 593-596.
19. Y. Li and N. Zhong, Mining ontology for automatically acquiring Web user information needs, *IEEE Transactions on Knowledge and Data Engineering*, 2006, **18(4):** 554-568.
20. B. Liu, Y. Dai, X. Li, W. S. Lee, and P. S. Yu, Building text classifiers using positive and un-labeled examples, *3rd IEEE International Conference on Data Mining*, Melbourne, Florida, USA, 2003, pp.179-186.
21. B. Liu, Y. Ma, and Philip S. Yu, Discovering business intelligence information by comparing company Web sites, in: N. Zhong, J. Liu and Y. Y. Yao (eds.) *Web Intelligence*, Springer-Verlag, 2003, 105-127.
22. J. Mostafa, W. Lam and M. Palakal, A multilevel approach to intelligent information filtering: model, system, and evaluation, *ACM Transactions on Information Systems*, 1997, **15(4):** 368-399.
23. Z. Pawlak, In pursuit of patterns in data reasoning from data, the rough set way, *3rd International Conference on Rough Sets and Current Trends in Computing*, USA, 2002, 1-9.
24. Z. Pawlak, Flow graphs and decision algorithms, *9th International Conference on Rough Set, Fuzzy Sets, Data Mining and Granular Computing*, Chongqing, China, 2003, 1-10.
25. S. Robertson, and D. A. Hull, The TREC-9 filtering track final report, *TREC-9*, 2000.

26. F. Sebastiani, Machine learning in automated text categorization, *ACM Computing Surveys*, 2002, **34(1):** 1-47.
27. P. Tzvetkov, X. Yan and J. Han, TSP: Mining top-K closed sequential patterns, *3rd IEEE International Conference on Data Mining*, 2003, Melbourne, Florida, USA, 347-354.
28. S.-T. Wu, Y. Li, Y. Xu, B. Pham and P. Chen, Automatic pattern taxonomy exatraction for Web mining, *IEEE/WIC/ACM International Conference on Web Intelligence*, 2004, Beijing, China, 242-248.
29. H. Yu, J. Han, and K. Chang, PEBL: positive example based learning for Web page classification using SVM, *KDD02*, 2002, 239-248.
30. G. I. Webb and S. Zhang, K-optimal rule discovery, *Data Mining and Knowledge Discovery*, 2004, **10:** 39-79.

# Applications of Rough Set Based K-Means, Kohonen SOM, GA Clustering

Pawan Lingras

Department of Mathematics and Computing Science, Saint Mary's University
Halifax, Nova Scotia, B3H 3C3, Canada
Pawan.Lingras@StMarys.CA

**Abstract.** Rough set theory provides an alternative way of representing sets whose exact boundary cannot be described due to incomplete information. Rough sets have been widely used for classification and can be equally beneficial in clustering. The clusters in practical data mining do not necessarily have crisp boundaries. An object may belong to more than one cluster. This paper describes modifications of clustering based on Genetic Algorithms, K-means algorithm, and Kohonen Self-Organizing Maps (SOM). These modifications make it possible to represent clusters as rough sets. Rough clusters are shown to be useful for representing groups of highway sections, Web users, and supermarket customers. The rough clusters are also compared with conventional and fuzzy clusters.

## 1 Introduction

Pawlak [15] proposed the concept of rough sets to describe the fact that exact membership of certain objects in a set may not be precisely defined. The usefulness of the original proposal was demonstrated for learning rules from a database [16]. Rough set theory has since made substantial progress as a classification tool in data mining [17]. The basic concept of representing a set as lower and upper bounds can be used in broader context. Clustering in relation to rough set theory is attracting increasing interest among researchers [3,18,22]. Lingras [7] described how a rough set theoretic classification scheme can be represented using a rough set genome. In subesquent publications [10,12], modifications of K-means and Kohonen Self-Organizing Maps (SOM) were proposed to create intervals of clusters based on rough set theory. Recently, Wojna [24] showed how K nearest neighbour classification can be augmented with rule induction.

This paper describes the evolutionary, neural, and statistical approaches for creating rough clusters. Rough clustering of highway sections, Web users, and supermarket customers is used to demonstrate the range of applications. A comparison of rough clusters with conventional and fuzzy clusters is used to show that rough clustering may provide a happy medium between conventional and fuzzy clusters.

J.F. Peters et al. (Eds.): Transactions on Rough Sets VII, LNCS 4400, pp. 120–139, 2007.

## 2    Adaptation of Rough Set Theory for Clustering

Due to space limitations familiarity with rough set theory is assumed [17]. Rough sets were originally proposed using equivalence relations. However, it is possible to define a pair of upper and lower bounds $(\underline{A}(X), \overline{A}(X))$ or a rough set for every set $X \subseteq U$ as long as the properties specified by Pawlak [15,17] are satisfied. Yao et al. [26,25] described various generalizations of rough sets by relaxing the assumptions of an underlying equivalence relation. Such a trend towards generalization is also evident in rough mereology proposed by Polkowski and Skowron [19] and the use of information granules in a distributed environment by Skowron and Stepaniuk [21]. The present study uses such a generalized view of rough sets. If one adopts a more restrictive view of rough set theory, the rough sets developed in this paper may have to be looked upon as interval sets.

Let us consider a hypothetical classification scheme

$$U/P = \{X_1, X_2, \ldots, X_k\} \tag{1}$$

that partitions the set $U$ based on an equivalence relation $P$. Let us assume that due to insufficient knowledge it is not possible to precisely describe the sets $X_i, 1 \le i \le k$, in the partition. However, it is possible to define each set $X_i \in U/P$ using its lower $\underline{A}(X_i)$ and upper $\overline{A}(X_i)$ bounds based on the available information. We will use vector representations, $\mathbf{u}, \mathbf{v}$ for objects and $\mathbf{x_i}$ for cluster $X_i$.

We are considering the upper and lower bounds of only a few subsets of $U$. Therefore, it is not possible to verify all the properties of the rough sets [15,17]. However, the family of upper and lower bounds of $\mathbf{x_i} \in \mathbf{U/P}$ are required to follow some of the basic rough set properties such as:

> (C1) An object $\mathbf{v}$ can be part of at most one lower bound
>
> (C2) $\mathbf{v} \in \underline{A}(\mathbf{x}_i) \Longrightarrow \mathbf{v} \in \overline{A}(\mathbf{x}_i)$
>
> (C3) An object $\mathbf{v}$ is not part of any lower bound
>
> $\Updownarrow$
>
> $\mathbf{v}$ belongs to two or more upper bounds.

Property (C1) emphasizes the fact that a lower bound is included in a set. If two sets are mutually exclusive, their lower bounds should not overlap. Property (C2) confirms the fact that the lower bound is contained in the upper bound. Property (C3) is applicable to the objects in the boundary regions, which are defined as the differences between upper and lower bounds. The exact membership of objects in the boundary region is ambiguous. Therefore, property (C3) states that an object cannot belong to only a single boundary region. Note that (C1)-(C3) are not necessarily independent or complete. However, enumerating them will be helpful in understanding the rough set adaptation of evolutionary, neural, and statistical clustering methods.

## 3  Rough Set Genome and Its Evaluation

The origin of Genetic Algorithms (GAs) is attributed to Holland's [4] work on cellular automata. There has been significant interest in GAs over the last two decades. The range of applications of GAs includes such diverse areas as job shop scheduling, training neural nets, image feature extraction, and image feature identification [1]. This section contains some of the basic concepts of genetic algorithms as described in [1].

A genetic algorithm is a search process that follows the principles of evolution through natural selection. The domain knowledge is represented using a candidate solution called an *organism*. Typically, an organism is a single *genome* represented as a vector of length $n$:

$$c = (c_i \mid 1 \leq i \leq n),  \tag{2}$$

where $c_i$ is called a *gene*.

An abstract view of a generational GA is given in Fig. 1. A group of organisms is called a *population*. Successive populations are called *generations*. A generational GA starts from initial generation $G(0)$, and for each generation $G(t)$ generates a new generation $G(t+1)$ using genetic operators such as *mutation* and *crossover*. The mutation operator creates new genomes by changing values of one or more genes at random. The crossover operator joins segments of two or more genomes to generate a new genome.

A rough set genome consists of $n$ genes, one gene per object in $U$. A gene for an object is a string of bits that describes which lower and upper approximations the object belongs to. Properties (C1)-(C3) provide certain restrictions on the memberships. An object $\mathbf{u} \in U$ can belong to the lower approximation of at most one class $\mathbf{x}_i$. If an object belongs to the lower approximation of $\mathbf{x}_i$ then it also belongs to the upper approximation of $\mathbf{x}_i$. If an object does not belong to the lower approximation of any $\mathbf{x}_i$, then it belongs to the upper approximation of at least two (possibly more) $\mathbf{x}_i$.

Based on these observations the string for a gene can be partitioned into two parts, *lower* and *upper*. Both the lower and upper parts of the string consist of $k$ bits each. The $i^{th}$ bit in lower/upper string tells whether the object is in the lower/upper approximation of $\mathbf{x}_i$.

If $\mathbf{u} \in \underline{A}(\mathbf{x}_i)$, then based on the property (C2), $\mathbf{u} \in \overline{A}(\mathbf{x}_i)$. Therefore, the $i^{th}$ bit in both the lower and upper strings will be turned on. Based on the property (C1) all the other bits must be turned off.

Genetic Algorithm:
         generate initial population, $G(0)$;
         evaluate $G(0)$;
         for($t = 1$; solution is not found, t++)
                  generate $G(t)$ using $G(t-1)$;
                  evaluate $G(t)$;

**Fig. 1.** Abstract view of a generational genetic algorithm

**Valid genes**

| | Lower | | | Upper | | |
|---|---|---|---|---|---|---|
| | $\underline{A}(\mathbf{x}_3)$ | $\underline{A}(\mathbf{x}_2)$ | $\underline{A}(\mathbf{x}_1)$ | $\overline{A}(\mathbf{x}_3)$ | $\overline{A}(\mathbf{x}_2)$ | $\overline{A}(\mathbf{x}_1)$ |
| $gene_1$ | 0 | 0 | 0 | 0 | 1 | 1 |
| $gene_2$ | 0 | 0 | 0 | 1 | 0 | 1 |
| $gene_3$ | 0 | 0 | 0 | 1 | 1 | 0 |
| $gene_4$ | 0 | 0 | 0 | 1 | 1 | 1 |
| $gene_5$ | 0 | 0 | 1 | 0 | 0 | 1 |
| $gene_6$ | 0 | 1 | 0 | 0 | 1 | 0 |
| $gene_7$ | 1 | 0 | 0 | 1 | 0 | 0 |

**Some examples of invalid genes**

| | Lower | | | Upper | | |
|---|---|---|---|---|---|---|
| | $\underline{A}(\mathbf{x}_3)$ | $\underline{A}(\mathbf{x}_2)$ | $\underline{A}(\mathbf{x}_1)$ | $\overline{A}(\mathbf{x}_3)$ | $\overline{A}(\mathbf{x}_2)$ | $\overline{A}(\mathbf{x}_1)$ |
| $invalidGene_1$ | 0 | 0 | 1 | 0 | 0 | 0 |
| $invalidGene_2$ | 0 | 1 | 0 | 1 | 1 | 0 |
| $invalidGene_3$ | 1 | 0 | 1 | 0 | 0 | 0 |
| $invalidGene_4$ | 0 | 0 | 0 | 0 | 0 | 1 |

**Fig. 2.** Genes in a rough set genome

If $\mathbf{u}$ is not in any of the lower approximations, then according to property (C3), it must be in two or more upper approximations of $\mathbf{x}_i, 1 \leq i \leq k$, and corresponding $i^{th}$ bits in the upper string will be turned on.

Fig. 2 shows examples of all the valid and some of the invalid genes for $k = 3$. Genes $gene_1$ to $gene_7$ are all the acceptable values of genes for $k = 3$. An object represented by $gene_1$ belongs to $\overline{A}(\mathbf{x}_1)$ and $\overline{A}(\mathbf{x}_2)$. An object represented by $gene_6$ belongs to $\underline{A}(\mathbf{x}_2)$, and by property (C2) to $\overline{A}(\mathbf{x}_2)$.

Any other value not given by $gene_1$ to $gene_7$ is not valid. Fig. 2 also shows four of the 57 invalid values. The $invalidGene_1$ is invalid because an object cannot be in $\underline{A}(\mathbf{x}_1)$ and not be in $\overline{A}(\mathbf{x}_1)$. The $invalidGene_2$ is invalid because an object cannot be in $\underline{A}(\mathbf{x}_2)$ and in $\overline{A}(\mathbf{x}_3)$ at the same time. The $invalidGene_3$

is invalid because an object cannot be in $\underline{A}(\mathbf{x}_1)$ and in $\underline{A}(\mathbf{x}_3)$ at the same time. Since the object represented by $invalidGene_4$ only belongs to $\overline{A}(\mathbf{x}_1)$, according to property (C3) it is invalid.

A genetic algorithm package such as the one used in the study [23] makes it possible to describe a set of valid gene values or alleles. All the standard genetic operations will then only create genomes that have these values. Therefore, the conventional genetic operations can be used with rough set genomes in such a package.

The quality of a conventional classification scheme is determined by using the within-group-error [20] $\triangle$ given by:

$$\triangle = \sum_{i\,=\,1}^{k} \sum_{\mathbf{u},\mathbf{v}\in\mathbf{x}_i} d(\mathbf{u},\mathbf{v}), \tag{3}$$

where $\mathbf{u}$ and $\mathbf{v}$ are objects from the same class $\mathbf{x}_i$.

The function $d$ provides the distance between two objects. The distance $d(\mathbf{u},\mathbf{v})$ is given by:

$$d(\mathbf{u},\mathbf{v}) = \sqrt{\frac{\sum_{j=1}^{m}(u_j - v_j)^2}{m}} \tag{4}$$

For a rough set classification scheme, the exact values of classes $\mathbf{x}_i \in U/P$ are not known. Given two objects $\mathbf{u},\mathbf{v}\in U$ we have three distinct possibilities:

1. Both $\mathbf{u}$ and $\mathbf{v}$ are in the same lower approximation $\underline{A}(\mathbf{x}_i)$.
2. Object $\mathbf{u}$ is in a lower approximation $\underline{A}(\mathbf{x}_i)$ and $\mathbf{v}$ is in the corresponding upper approximation $\overline{A}(\mathbf{x}_i)$, and case 1 is not applicable.
3. Both $\mathbf{u}$ and $\mathbf{v}$ are in the same upper approximation $\overline{A}(\mathbf{x}_i)$, and cases 1 and 2 are not applicable.

For these possibilities, one can define three corresponding types of within-group-errors, $\triangle_1, \triangle_2,$ and $\triangle_3$ as:

$$\triangle_1 = \sum_{i\,=\,1}^{k} \sum_{\mathbf{u},\mathbf{v}\in\underline{A}(\mathbf{x}_i)} d(\mathbf{u},\mathbf{v}),$$

$$\triangle_2 = \sum_{i\,=\,1}^{k} \sum_{\mathbf{u}\in\underline{A}(\mathbf{x}_i),\mathbf{v}\in\overline{A}(\mathbf{x}_i);\mathbf{v}\notin\underline{A}(\mathbf{x}_i)} d(\mathbf{u},\mathbf{v}),$$

$$\triangle_3 = \sum_{i\,=\,1}^{k} \sum_{\mathbf{u},\mathbf{v}\in\overline{A}(\mathbf{x}_i);\mathbf{u},\mathbf{v}\notin\underline{A}(\mathbf{x}_i)} d(\mathbf{u},\mathbf{v}).$$

The total error of rough set classification will then be a weighted sum of these errors:

$$\triangle_{total} = w_1 \times \triangle_1 + w_2 \times \triangle_2 + w_3 \times \triangle_3. \tag{5}$$

Since $\triangle_1$ corresponds to situations where both objects definitely belong to the same class, the weight $w_1$ should have the highest value. On the other hand, $\triangle_3$ corresponds to a situation where both objects may or may not belong to the same class. Hence, $w_3$ should have the lowest value. In other words, $w_1 > w_2 > w_3$. There are many possible ways of developing an error measure for rough set classifications. The measure $\triangle_{total}$ is perhaps one of the simplest. More sophisticated alternatives may be used, depending upon the application.

If we used genetic algorithms to minimize $\triangle_{total}$, the genetic algorithms would try to classify all the objects in upper approximations by taking advantage of the fact that $w_3 < w_1$. This may not necessarily be the best classification scheme. We want the rough set classification to be as precise as possible. Therefore, a precision measure needs to be used in conjunction with $\triangle_{total}$ for evaluating the quality of a rough set genome. A possible precision measure can be defined [15] as:

$$precision = \frac{\text{Number of objects classified in lower approximations}}{\text{Total number of objects}} \qquad (6)$$

The objective of the genetic algorithms will then be to maximize the quantity:

$$objective = p \times precision + \frac{e}{\triangle_{total}}, \qquad (7)$$

where $p$ and $e$ are additional parameters. The parameter $p$ describes the importance of the precision measure in determining the quality of a rough set genome. Higher values of $p$ will result in smaller boundary region. Similarly, $e$ indicates the importance of within-group-errors relative to the size of the boundary region.

## 4   Rough Clustering Highway Sections Using GAs

Seasonal and permanent traffic counters scattered across a highway network are the major sources of traffic data. These traffic counters measure the traffic volume – the number of vehicles that have passed through a particular section of a lane or highway in a given time period. Traffic volumes can be expressed in terms of hourly or daily traffic. More sophisticated traffic counters record additional information such as the speed, length and weight of the vehicle. Highway agencies generally have records from traffic counters collected over a number of years. In addition to obtaining data from traffic counters, traffic engineers also conduct occasional surveys of road users to get more information.

The permanent traffic counter (PTC) sites are grouped to form various road classes. These classes are used to develop guidelines for the construction, maintenance and upgrading of highway sections. In one commonly used system, roads are classified on the basis of trip purpose and trip length characteristics [20]. Examples of resulting classes are commuter, business, long distance, and recreational highways. Tthe trip purpose provides information about the road users, an important criterion in a variety of traffic engineering analyses. Trip purpose information can be obtained directly from the road users, but since all users cannot be surveyed, traffic engineers study various traffic patterns obtained from seasonal and permanent traffic counters and sample surveys of a few road users.

The present study is based on a sample of 264 monthly traffic patterns - variation of monthly average daily traffic volume in a given year - recorded between 1987 and 1991 on Alberta highways. The distribution of PTCs in various regions are determined based on the traffic flow through the the the provincial highway networks. The patterns obtained from these PTCs represent traffic from all major regions in the province.

The rough set genomes used in the experiment consisted of 264 genes, one gene per pattern. The hypothetical classification scheme consisted of three classes:

1. Commuter/business,
2. Long distance, and
3. Recreational.

The rough set classification scheme was expected to specify lower and upper bounds of these classes.

Each gene was allowed to take the valid values shown in Fig. 2. Each rough set genome corresponded to a rough set classification scheme. Genetic algorithms attempted to evolve a genome such that the value of an objective function was maximal. The objective function given by eq. 7 was used during the evolution process. Since the precision was in the range $[0, 1]$ and the total error $\Delta_{total}$ could be as high as 15000, it was decided to modify the parameters $e$ as:

$$e = e' \times \Delta_{max}. \tag{8}$$

Various values of $e'$ and $p$ as well other genetic parameters such as number of generations, probabilities of crossover and mutation, and population sizes were used.

The resulting rough set classification schemes were subjectively compared with the conventional classification scheme. The upper and lower approximations of the commuter/business, long distance, and recreational classes were also

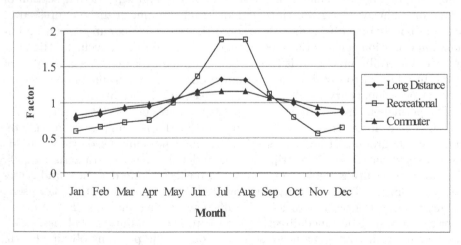

**Fig. 3.** Monthly patterns for the lower approximations

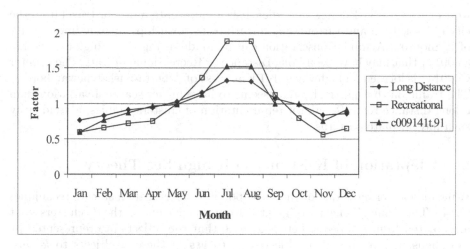

**Fig. 4.** Monthly pattern that may be long distance or recreational

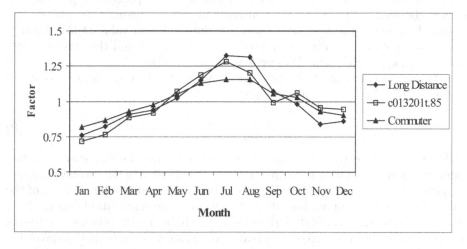

**Fig. 5.** Monthly pattern that may be commuter/business or long distance

checked against the geography of Alberta highway networks. More details of the experiment can be found in [7].

Fig. 3 shows the monthly patterns for the lower approximations of the three groups: commuter/business, long distance, and recreational. The average pattern for the lower approximation of commuter/business class has the least variation over the year. The recreational class, conversely, has the most variation. The variation for long distance class is less than the recreational but more than the commuter/business class. Fig. 4 shows one of the highway sections near counter number C013201 that may have been Commuter/Business or Long Distance in 1985. It is clear that the monthly pattern for the highway section falls in between the two classes. The counter C013201 is located on highway 13, 20 Km.

west of Alberta-Saskatchewan border. It is an alternate route for travel from the city of Saskatoon and surrounding townships to townships surrounding the city of Edmonton. A similar observation can be made in Fig. 5 for highway section C009141 that may have been Long Distance or Recreational in 1991. The counter C009141 is located on highway 9, 141 Km. west of Alberta-Saskatchewan border. The traffic on that particular road seems to have higher seasonal variation than a long distance road. Rough set representation of clusters enables us to identify such intermediate patterns.

## 5   Adaptation of K-Means to Rough Set Theory

K-means clustering is one of the most popular statistical clustering techniques [2,14]. The name K-means originates from the means of the $k$ clusters that are created from $n$ objects. Let us assume that the objects are represented by $m$-dimensional vectors. The objective is to assign these $n$ objects to $k$ clusters. Each of the clusters is also represented by an $m$-dimensional vector, which is the centroid or mean vector for that cluster. The process begins by randomly choosing $k$ objects as the centroids of the $k$ clusters. The objects are assigned to one of the $k$ clusters based on the minimum value of the distance $d(\mathbf{v}, \mathbf{x})$ between the object vector $\mathbf{v} = (v_1, ..., v_j, ..., v_m)$ and the cluster vector $\mathbf{x} = (x_1, ..., x_j, ..., x_m)$. The distance $d(\mathbf{v}, \mathbf{x})$ is given by:

After the assignment of all the objects to various clusters, the new centroid vectors of the clusters are calculated as:

$$x_j = \frac{\sum_{\mathbf{v} \in \mathbf{x}} v_j}{\text{Size of cluster } \mathbf{x}}, \text{where } 1 \le j \le m. \tag{9}$$

The process stops when the centroids of clusters stabilize, i.e. the centroid vectors from the previous iteration are identical to those generated in the current iteration.

Incorporating rough sets into K-means clustering requires the addition of the concept of lower and upper bounds. Calculation of the centroids of clusters from conventional K-Means needs to be modified to include the effects of lower as well as upper bounds. The modified centroid calculations for rough sets are then given by:

if     $\underline{A}(\mathbf{x}) \ne \emptyset$ and $\overline{A}(\mathbf{x}) - \underline{A}(\mathbf{x}) = \emptyset$

$$x_j = \frac{\sum_{\mathbf{v} \in \underline{A}(\mathbf{x})} v_j}{|\underline{A}(\mathbf{x})|}$$

else if   $\underline{A}(\mathbf{x}) = \emptyset$ and $\overline{A}(\mathbf{x}) - \underline{A}(\mathbf{x}) \ne \emptyset$

$$x_j = \frac{\sum_{\mathbf{v} \in (\overline{A}(\mathbf{x}) - \underline{A}(\mathbf{x}))} v_j}{|\overline{A}(\mathbf{x}) - \underline{A}(\mathbf{x})|} \tag{10}$$

else

$$x_j = w_{lower} \times \frac{\sum_{\mathbf{v} \in \underline{A}(\mathbf{x})} v_j}{|\underline{A}(\mathbf{x})|} + w_{upper} \times \frac{\sum_{\mathbf{v} \in (\overline{A}(\mathbf{x}) - \underline{A}(\mathbf{x}))} v_j}{|\overline{A}(\mathbf{x}) - \underline{A}(\mathbf{x})|}$$

where $1 \leq j \leq m$. The parameters $w_{lower}$ and $w_{upper}$ correspond to the relative importance of lower and upper bounds, and $w_{lower} + w_{upper} = 1$. If the upper bound of each cluster were equal to its lower bound, the clusters would be conventional clusters. Therefore, the boundary region $\overline{A}(\mathbf{x}) - \underline{A}(\mathbf{x})$ will be empty, and the second term in the equation will be ignored. Thus, eq. 10 will reduce to conventional centroid calculations.

The next step in the modification of the K-means algorithms for rough sets is to design criteria to determine whether an object belongs to the upper or lower bound of a cluster given as follows For each object vector, $\mathbf{v}$, let $d(\mathbf{v}, \mathbf{x}_j)$ be the distance between itself and the centroid of cluster $\mathbf{x}_j$. Let $d(\mathbf{v}, \mathbf{x}_i) = \min_{1 \leq j \leq k} d(\mathbf{v}, \mathbf{x}_j)$. The ratio $d(\mathbf{v}, \mathbf{x}_i)/d(\mathbf{v}, \mathbf{x}_j)$, $1 \leq i, j \leq k$, are used to determine the membership of $\mathbf{v}$. Let $T = \{j : d(\mathbf{v}, \mathbf{x}_i)/d(\mathbf{v}, \mathbf{x}_j) \leq threshold$ and $i \neq j\}$.

1. If $T \neq \emptyset$, $\mathbf{v} \in \overline{A}(\mathbf{x}_i)$ and $\mathbf{v} \in \overline{A}(\mathbf{x}_j), \forall j \in T$. Furthermore, $\mathbf{v}$ is not part of any lower bound. The above criterion gurantees that property (C3) is satisfied.
2. Otherwise, if $T = \emptyset$, $\mathbf{v} \in \underline{A}(\mathbf{x}_i)$. In addition, by property (C2), $\mathbf{v} \in \overline{A}(\mathbf{x}_i)$.

It should be emphasized that the approximation space $A$ is not defined based on any predefined relation on the set of objects. The upper and lower bounds are constructed based on the criteria described above.

## 6    Rough Clustering Web Users Using K-Means

The study data was obtained from the Web access logs of the first three courses in computing science at Saint Mary's University over a sixteen-week period. Students' attitudes towards the course vary a great deal. It was hoped that the profile of visits would reflect some of the distinctions between the students. For the initial analysis, it was assumed that the visitors could fall into one of the following three categories:

1. Studious: These visitors download the current set of notes. Since they download a limited/current set of notes, they probably study classnotes on a regular basis.
2. Crammers: These visitors download a large set of notes. This indicates that they have stayed away from the classnotes for a long period of time. They are planning for pre-test cramming.
3. Workers: These visitors are mostly working on class or lab assignments or accessing the discussion board.

The rough set classification scheme was expected to specify lower and upper bounds for these classes.

It was hoped that the variety of user behaviours mentioned above would be identifiable based on the number of Web accesses, types of documents downloaded, and time of day. Certain areas of the Web site were protected and the users could only access them using their IDs and passwords. The activities in the restricted parts of the Web site consisted of submitting a user profile, changing a password, submission of assignments, viewing the submissions, accessing the discussion board, and viewing current class marks. The rest of the Web site was public. The public portion consisted of viewing course information, the lab manual, classnotes, class assignments, and lab assignments.

If the users only accessed the public Web site, their IDs would be unknown. Therefore, the Web users were identified based on their IP address. This also assured that the user privacy was protected. A visit from an IP address started when the first request was made from the IP address. The visit continued as long as the consecutive requests from the IP address had sufficiently small delay.

The Web logs were preprocessed to create an appropriate representation of each user corresponding to a visit. The abstract representation of a Web user is a critical step that requires a good knowledge of the application domain. Previous personal experience with the students in the course suggested that some of the students print preliminary notes just before a class and an updated copy after the class. Some students view the notes on-line on a regular basis. Some students print all the notes around important days such as midterm and final examinations. In addition, there are many accesses on Tuesdays and Thursdays, when the in-laboratory assignments are due. On and off campus points of access can also provide some indication of a user's objectives for the visit. Based on some of these observations, it was decided to use the following attributes for representing each visitor:

1. On campus/Off campus access.
2. Day time/Night time access: 8 a.m. to 8 p.m. was considered to be the day time.
3. Access during lab/class days or non-lab/class days: All the labs and classes were held on Tuesday and Thursday. The visitors on these days are more likely to be workers.
4. Number of hits.
5. Number of classnotes downloaded.

The first three attributes had binary values of 0 or 1. The last two values were normalized. Since the classnotes were the focus of the clustering, the last variable was assigned higher importance.

The modified K-means algorithm was run for various values of *threshold* and initial centroid vectors. Similarly, various pairs of $(w_{lower}, w_{upper})$ ranging from $(0.95, 0.05)$ to $(0.55, 0.45)$ were tried. It was found that when the value of $w_{lower}$ was set at 0.75 and $w_{upper}$ was equal to 0.25, the resulting intervals provided good representations of clusters of Web visitors. The resulting rough set classification schemes were subjectively analyzed. The results were compared with conventional and fuzzy C-means clustering. In fuzzy C-means clustering objects are assinged membership values for each of the three clusters. The details of the fuzzy C-means can be found in [5]. More details about the experiments can be found in [13].

Table 1 shows the cardinalities of conventional clusters, the modified K-means based on rough set theory, and the sets with fuzzy memberships greater than 0.6. The actual numbers in each cluster vary based on the characteristics of each

**Table 1.** Cardinalities of the clusters for three techniques

| Course | Cluster | FCM >0.6 | Lower bound | Conventional Clusters |
|--------|---------|----------|-------------|-----------------------|
| First  | Studious | 1382 | 1412 | 1814 |
|        | Crammers | 414  | 288  | 406  |
|        | Workers  | 4354 | 5350 | 5399 |
| Second | Studious | 1750 | 1197 | 1699 |
|        | Crammers | 397  | 443  | 634  |
|        | Workers  | 1322 | 1677 | 3697 |
| Third  | Studious | 265  | 223  | 318  |
|        | Crammers | 84   | 69   | 89   |
|        | Workers  | 717  | 906  | 867  |

**Table 2.** The conventional K-means cluster center vectors

| Course | Cluster | Campus Access | Day Night | Lab Day | Hits | Doc Req |
|--------|---------|---------------|-----------|---------|------|---------|
| First  | Studious | 0.67 | 0.76 | 0.44 | 2.97 | 2.78 |
|        | Crammers | 0.62 | 0.72 | 0.32 | 4.06 | 8.57 |
|        | Workers  | 0.67 | 0.74 | 0.49 | 0.98 | 0.85 |
| Second | Studious | 0.00 | 0.68 | 0.28 | 0.67 | 0.55 |
|        | Crammers | 0.66 | 0.72 | 0.36 | 2.43 | 2.92 |
|        | Workers  | 1.00 | 0.82 | 0.46 | 0.66 | 0.51 |
| Third  | Studious | 0.69 | 0.75 | 0.50 | 3.87 | 3.15 |
|        | Crammers | 0.60 | 0.71 | 0.44 | 5.30 | 10.20 |
|        | Workers  | 0.62 | 0.74 | 0.50 | 1.41 | 1.10 |

course. For example, in the fuzzy C-means clustering results, the first term course had significantly more workers than studious visitors, while the second term course had more studious visitors than workers. The increase in the percentage of studious visitors in the second term seems to be a natural progression. It should be noted that the progression from workers to studious visitors was more obvious with fuzzy clusters than the conventional clusters and the rough K-means clusters. Interestingly, the second year course had significantly large number of workers than studious visitors. This seems to be counter-intuitive. However, it can be explained based on the structure of the Websites. Unlike the two first year courses, the second year course did not post the classnotes on the Web. The notes downloaded by these students were usually sample programs that were essential during their laboratory work.

Table 2 shows cluster center vectors from the conventional K-means. It was possible to classify the three clusters as studious, workers, and crammers, from the results obtained using the conventional K-means algorithm. The crammers had the highest number of hits and classnotes in every data set. The average number of notes downloaded by crammers varied from one set to another. The studious visitors downloaded the second highest number of notes. The distinction between

**Table 3.** The modified K-means cluster center vectors

| Course | Cluster | Campus Access | Day Night | Lab Day | Hits | Doc Req |
|--------|---------|--------|-----|-----|------|-----|
| First | Studious | 0.67 | 0.75 | 0.43 | 3.16 | 3.17 |
| | Crammers | 0.61 | 0.72 | 0.33 | 4.28 | 9.45 |
| | Workers | 0.67 | 0.75 | 0.49 | 1.00 | 0.86 |
| Second | Studious | 0.14 | 0.69 | 0.03 | 0.64 | 0.55 |
| | Crammers | 0.64 | 0.72 | 0.34 | 2.58 | 3.29 |
| | Workers | 0.97 | 0.88 | 0.88 | 0.66 | 0.49 |
| Third | Studious | 0.70 | 0.74 | 0.48 | 4.09 | 3.91 |
| | Crammers | 0.55 | 0.72 | 0.43 | 5.48 | 10.99 |
| | Workers | 0.62 | 0.75 | 0.51 | 1.53 | 1.13 |

**Table 4.** Fuzzy C-Means cluster center vectors

| Course | Cluster | Campus Access | Day Night | Lab Day | Hits | Doc Req |
|--------|---------|--------|-----|-----|------|-----|
| First | Studious | 0.68 | 0.76 | 0.44 | 2.30 | 2.21 |
| | Crammers | 0.64 | 0.72 | 0.34 | 3.76 | 7.24 |
| | Workers | 0.69 | 0.77 | 0.51 | 0.91 | 0.75 |
| Second | Studious | 0.60 | 0.75 | 0.13 | 0.63 | 0.52 |
| | Crammers | 0.64 | 0.73 | 0.33 | 2.09 | 2.54 |
| | Workers | 0.83 | 0.87 | 0.75 | 0.62 | 0.47 |
| Third | Studious | 0.69 | 0.75 | 0.50 | 3.36 | 2.42 |
| | Crammers | 0.59 | 0.72 | 0.43 | 5.14 | 9.36 |
| | Workers | 0.62 | 0.77 | 0.52 | 1.28 | 1.06 |

**Table 5.** Average vectors for fuzzy C-means with membership>0.6

| Course | Cluster | Campus Access | Day Night | Lab Day | Hits | Doc Req |
|--------|---------|--------|-----|-----|------|-----|
| First | Studious | 0.70 | 0.78 | 0.45 | 2.37 | 2.41 |
| | Crammers | 0.65 | 0.72 | 0.33 | 3.74 | 7.92 |
| | Workers | 0.67 | 0.75 | 0.50 | 0.82 | 0.67 |
| Second | Studious | 0.52 | 0.89 | 0.00 | 0.49 | 0.40 |
| | Crammers | 0.65 | 0.75 | 0.34 | 2.18 | 0.96 |
| | Workers | 1.00 | 1.00 | 1.00 | 0.52 | 0.36 |
| Third | Studious | 0.69 | 0.75 | 0.51 | 3.69 | 2.28 |
| | Crammers | 0.58 | 0.70 | 0.43 | 5.38 | 10.39 |
| | Workers | 0.60 | 0.75 | 0.52 | 1.19 | 1.00 |

workers and studious visitors for the second course was also based on other attributes. For example, in the second data set, the workers were more prone to come on lab days, access Websites from on-campus locations during the daytime.

It is also interesting to note that the crammers had higher ratios of document requests to hits. The workers, on the other hand, had the lowest ratios of document requests to hits. Table 3 shows the modified K-means center vectors. The fuzzy center vectors are shown in Table 4. These center vectors are comparable to the conventional centroid vectors. In order to compare fuzzy and conventional clustering, visits with fuzzy membership greater than 0.6 were grouped together. Similar characteristics can be found in these tables. For the second data set, the modified K-means is more sensitive to the differences between studious and crammers in the first three attributes than the other two techniques.

## 7    Modifications of Kohonen Algorithm

Fig. 6 illustrates the conventional Kohonen network architecture for the one-dimensional case. The unsupervised learning with the Kohonen rule [6] uses competitive learning approach. In competitive learning, the output neurons compete with each other. The winner output neuron has the output of 1, the rest of the output neurons have outputs of 0. The competitive learning is suitable for classifying a given pattern into exactly one of the mutually exclusive clusters. The network is used to group patterns represented by $m$-dimensional vectors into $k$ groups. The network consists of two layers. The first layer is called the input layer and the second layer is called the Kohonen layer. The network receives the input vector for a given pattern. If the pattern belongs to the $i$th group, then $i$th neuron in the Kohonen layer has an output value of one and other Kohonen layer neurons have output values of zero. Each connection is assigned a weight $x_i$. Weights of all the connections to a Kohonen layer neuron make up an $m$-dimensional weight vector $\mathbf{x}$. The weight vector $\mathbf{x}$ for a Kohonen layer neuron is the vector representation of the group corresponding to that neuron. For any input vector $\mathbf{v}$, the network compares the input with the weight vector for a group using the measure such as Eq. 4.

The pattern $\mathbf{v}$ belongs to the group with minimum value for $d(\mathbf{x}, \mathbf{v})$. The Kohonen neural network generates the clusters through a learning process as follows: Initially, the network connections are assigned somewhat arbitrary weights. The training set of input vectors is presented to the network several times. For

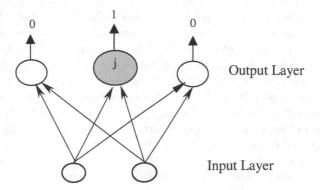

**Fig. 6.** Kohonen Neural Network

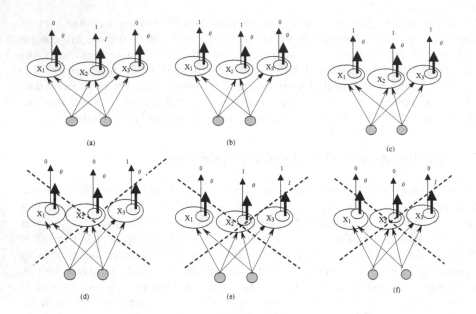

**Fig. 7.** Modified Kohonen Neural Network

each iteration, the weight vector **x** for a group that is closest to the pattern **v** is modified using the equation:

$$\mathbf{x}_{new} = \mathbf{x}_{old} + \alpha(t) \times (\mathbf{v} - \mathbf{x}_{old}), \qquad (11)$$

where $\alpha(t)$ is a learning factor which starts with a high value at the beginning of the training process and is gradually reduced as a function of time.

The rough set based Kohonen algorithm uses the concept of lower and upper bounds in the equations for updating the weights of winners. The Kohonen rough set architecture is similar to the conventional Kohonen architecture. It consists of two layers, an input layer and the Kohonen rough set layer (rough set output layer). These two layers are fully connected. Each input layer neuron has a feed forward connection to each output layer neuron. Fig. 7 illustrates the Kohonen rough set neural network architecture for a one-dimensional case. A neuron in the Kohonen layer consists of two parts, a lower neuron and an upper neuron. The lower neuron has an output of 1 if an object belongs to the lower bound of the cluster. Similarly, a membership in the upper bound of the cluster will result in an output of 1 from the upper neuron. Since an object belonging to the lower bound of a cluster also belongs to its upper bound, when the lower neuron has an output of 1, the upper neuron also has an output of 1. However, a membership in the upper bound of a cluster does not necessarily imply the membership in its lower bound. Therefore, the upper neuron contains the lower neuron. Fig. 7 provides some cases to explain outputs from the Kohonen rough set neural network works based on properties (C1), (C2), and (C3). Fig. 7(a-c) shows some of the possible outputs, while Fig. 7(d-f) shows some of the invalid outputs from the network. Fig. 7(a) shows a case where an object belongs to lower bound of cluster $x_2$. Based on the property (C2), it also belongs to the upper bound of $x_2$. Fig. 7(b) shows a situation where an object belongs to the upper bounds of clusters $x_1$ and $x_2$. The object in Fig. 7(c) belongs to the upper bounds of clusters $x_1$, $x_2$ and $x_3$. Fig. 7(d) shows an invalid situation where an object belongs only to the upper bound of the cluster $x_3$. This is a violation of the property (C3). Fig. 7(e) shows a violation of property (C1), where an object belongs to lower bound of $x_3$ as well as the upper bound of $x_2$. Similarly, a violation of property (C2) can be seen in an invalid case shown in Fig. 7(f). Here the object only belongs to the lower bound of cluster $x_3$ and not its upper bound. The modification of the Kohonen algorithm must make sure that the properties (C1)-(C3) are obeyed by avoiding cases such as the ones shown in Fig. 7(d-f). The interval clustering provides good results, if initial weights are obtained by running the conventional Kohonen learning. The next step in the modification of the Kohonen algorithm for obtaining rough sets is to design criteria to determine whether an object belongs to the upper or lower bounds of a cluster. The assignment criteria for the modified Kohonen algorithm is the same as the modified K-means algorithm discussed in the previous section. For each object vector, $v$, let $d(v, x_j)$ be the distance between itself and the weight vector $x_j$ of cluster $X_i$. Let $d(v, x_i) = \min_{1 \le j \le k} d(v, x_j)$ The ratios $\frac{d(v, x_i)}{d(v, x_j)}$, were used to determine the membership of $v$ as follows. Let $T = \{j : d(v, x_i)/d(v, x_j) \le threshold$ and $i \ne j\}$.

1. If $T \ne \emptyset$, $v \in \overline{A}(x_i)$ and $v \in \overline{A}(x_j), \forall j \in T$. Furthermore, $v$ is not part of any lower bound. The above criterion guarantees that property (C3) is satisfied. The weight vectors $x_i$ and $x_j$ are modified as:
$x_i{}^{new} = x_i{}^{old} + \alpha_{upper}(t) \times (v - x_i{}^{old})$, and
$x_j{}^{new} = x_j{}^{old} + \alpha_{upper}(t) \times (v - x_j{}^{old})$.
2. Otherwise, if $T = \emptyset$, $v \in \underline{A}(x_i)$. In addition, by property (C2), $v \in \overline{A}(x_i)$. The weight vectors $x_i$ is modified as:
$x_i{}^{new} = x_i{}^{old} + \alpha_{lower}(t) \times (v - x_i{}^{old})$.

Usually, $\alpha_{lower} > \alpha_{upper}$. It can be easily verified that the above algorithm preserves properties (C1)-(C3). It should be emphasized that the approximation space $A$ is not defined based on any predefined relation on the set of objects. The upper and lower bounds are constructed based on the criteria described above. Lingras *et al.* [12] conducted experiments with Web logs on three Web sites, which suggests that the modification of the Kohonen neural networks provide reasonable interval set representations of clusters. The following section describes another experiment reported in [11] for clustering supermarket customers.

## 8    Rough Clustering Supermarket Customers Using Kohonen SOM

The data used in the study was supplied by a supermarket chain. The data consisted of transactional records from three regions. The first region, S1,

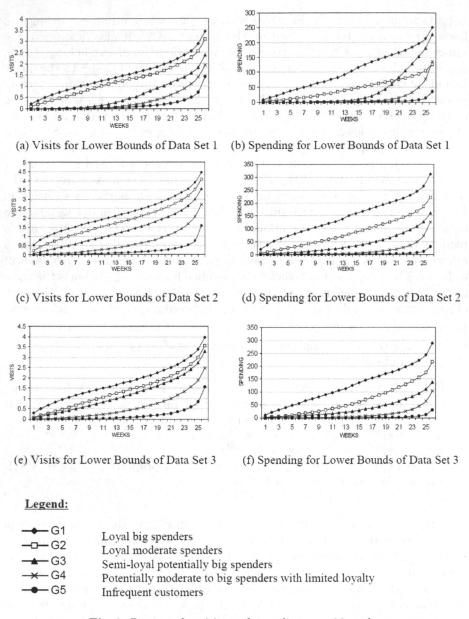

(a) Visits for Lower Bounds of Data Set 1    (b) Spending for Lower Bounds of Data Set 1

(c) Visits for Lower Bounds of Data Set 2    (d) Spending for Lower Bounds of Data Set 2

(e) Visits for Lower Bounds of Data Set 3    (f) Spending for Lower Bounds of Data Set 3

**Legend:**

| | |
|---|---|
| ◆— G1 | Loyal big spenders |
| ▫— G2 | Loyal moderate spenders |
| ▲— G3 | Semi-loyal potentially big spenders |
| ✱— G4 | Potentially moderate to big spenders with limited loyalty |
| ●— G5 | Infrequent customers |

**Fig. 8.** Patterns for visits and spending over 26 weeks

consisted of one store in a rural setting. The second rural region (S2) was served by five stores, while the third region was an urban area with six stores. The data was collected over a twenty-six week period: October 22, 2000 - April 21, 2001. Lingras and Adams [9] used data on the spending and visits of supermarket customers for clustering those customers. The use of average values of these variables may hide some of the important information present in the temporal

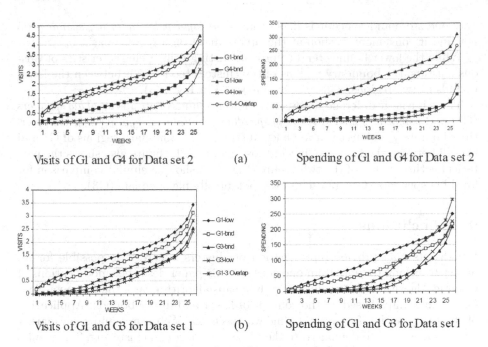

Visits of G1 and G4 for Data set 2    (a)    Spending of G1 and G4 for Data set 2

Visits of G1 and G3 for Data set 1    (b)    Spending of G1 and G3 for Data set 1

**Fig. 9.** Comparison of two interval clusters

patterns. Therefore, Lingras and Adams [9] used the weekly time series values. It is possible that customers with similar profiles may spend different amounts in a given week. However, if the values were sorted, the differences between these customers may vanish. For example, three weeks spending of customer A may be $10, $30, and $20. Customer B may spend $20, $10, and $30 in those three weeks. If the two time-series were compared with each other, the two customers may seem to have completely different profiles. However, if the time-series values were sorted, the two customers would have identical patterns. Therefore, the values of these variables for 26 weeks were sorted, resulting in a total of 52 variables. A variety of values for K (number of clusters) were used in the initial experiments. After experimenting with a range of values, the threshold was set at 0.7, $\alpha_{lower}$ was chosen to be 0.01, 0.005 was used as the value of $\alpha_{upper}$, and 1000 iterations were used for the training phase of each data set.

Fig. 8 shows the average spending and visit patterns for the lower bounds of the five clusters. The patterns enable us to distinguish between the five types of customers as: loyal big spenders (G1), loyal moderate spenders (G2), semi-loyal potentially big spenders (G3), potentially moderate to big spenders with limited loyalty (G4), and infrequent customers (G5).

The patterns of these classes for the three regions were mostly similar. However, there was an interesting difference in S1 region. Even though for most weeks *loyal moderate spenders* (G2) had higher spending than *semi-loyal potentially big spenders* (G3), the highest spending of G3 was higher than G2. The region has only one store and hence it is likely that *semi-loyal potentially big spenders* do not find it convenient to shop at the supermarket on a regular basis.

While the lower bounds tend to provide distinguishing characteristics of various clusters, the boundary regions of the clusters tend to fall between the lower bounds of two regions. This fact is illustrated in Fig. 9(a).

There is a large difference between the lower bounds of groups *loyal big spenders* (G1) and *potentially moderate to big spenders with limited loyalty* (G4). However, their boundary regions seem to be less distinct. The boundary regions of G1 and G4 fall between the lower bounds of those groups. The figure also shows the patterns for the overlap of the two groups. Fig. 9(b) shows a similar comparison for loyal big spenders (G1) and semi-loyal potentially big spenders (G3).

## 9    Conclusions

Rough sets may provide represenation of clusters, where it is possible for an object to belong to more than one cluster. This paper describes modifications of Genetic Algorithms based clustering, K-means algorithm, and the Kohonen Self-Organizing Maps based on the concept of lower and upper bounds. Application of these approaches for grouping highway sections, Web users, and supermarket customers is used to demonstrate the versatility of rough clustering. The results are also compared with conventional and fuzzy clustering.

## Acknowledgment

The author would like to thank the Natural Sciences and Engineering Research Council of Canada and the Faculty of Graduate Studies and Research, Saint Mary's University for funding. Data from Alberta Highways, Saint Mary's University, and the supermarket is also appreciated.

## References

1. Buckles BP and Petry FE (1994) Genetic Algorithms. IEEE Computer Press, Los Alamitos, California
2. Hartigan JA and Wong MA (1979) Algorithm AS136: A K-Means Clustering Algorithm. Applied Statistics 28:100–108.
3. Hirano S and Tsumoto S (2000) Rough Clustering and Its Application to Medicine. Journal of Information Science 124:125–137
4. Holland JH (1975) Adaptation in Natural and Artificial Systems. University of Michigan Press, Ann Arbor
5. Joshi A, Krishnapuram R (1998) Robust Fuzzy Clustering Methods to Support Web Mining. Proceedings of the workshop on Data Mining and Knowledge Discovery, SIGMOD '98, 15/1-15/8
6. Kohonen T (1988) Self-Organization and Associative Memory. Springer Verlag, Berlin
7. Lingras P (2001) Unsupervised Rough Set Classification using GAs. Journal Of Intelligent Information Systems 16(3):215–228.

8. Lingras P (2002) Rough set clustering for Web mining. Proceedings of 2002 IEEE International Conference on Fuzzy Systems.
9. Lingras P, Adams G. (2002) Selection of Time-Series for Clustering Supermarket Customers. Technical Report 2002_006, Department of Mathematics and Computing Science, Saint Mary's University, Halifax, N.S., Canada. http://cs.stmarys.ca/tech_reports/
10. Lingras P, West C (2004) Interval Set Clustering of Web Users with Rough K-means. Journal of Intelligent Information Systems 23(1):5–16
11. Lingras P, Hogo M, Snorek M, Leonard B (2003) Clustering Supermarket Customers using Rough Set Based Kohonen Networks. Proceedings of Fourteenth International Symposium on Methodologies for Intelligent Systems, Lecture Notes in Artificial Intelligence Series, 2871, Springer, 169–173
12. Lingras P, Hogo M, Snorek M (2004) Interval Set Clustering of Web Users using Modified Kohonen Self-Organizing Maps based on the Properties of Rough Sets. Web Intelligence and Agent Systems: An International Journal 2(3)
13. Lingras P, Yan R, West C (2003) Fuzzy C-Means Clustering of Web Users for Educational Sites. Proceedings of Sixteenth Conference of the Canadian Society of Computational Studies of Intelligence Advances in Artificial Intelligence Series 2671 Toronto, Springer, 557–562
14. MacQueen J (1967) Some Methods fir Classification and Analysis of Multivariate Observations. Proceedings of Fifth Berkeley Symposium on Mathematical Statistics and Probability 1:281–297
15. Pawlak Z (1982) Rough Sets. International Journal of Information and Computer Sciences 11:(145-172)
16. Pawlak Z (1984) Rough classification. International Journal of Man-Machine Studies 20:469–483
17. Pawlak Z (1992) Rough Sets: Theoretical Aspects of Reasoning about Data. Kluwer Academic Publishers
18. Peters JF, Skowron A, Suraj Z, Rzasa W, Borkowski M (2002) Clustering: A rough set approach to constructing information granules. Soft Computing and Distributed Processing, Proceedings of 6th International Conference, SCDP 2002, 57–61
19. Polkowski L, Skowron A (1996) Rough Mereology: A New Paradigm for Approximate Reasoning. International Journal of Approximate Reasoning 15(4):333-365
20. Sharma SC, Werner A (1981) Improved method of grouping provincewide permanent traffic counters. Transportation Research Record 815:13–18
21. Skowron A, Stepaniuk J (1999) Information granules in distributed environment. In: Zhong N, Skowron A, Ohsuga S (eds) New Directions in Rough Sets, Data Mining, and Granular-Soft Computing. Lecture notes in Artificial Intelligence. Springer Verlag, Tokyo, 1711:357–365
22. Voges KE, Pope NKLl, Brown MR (2002) Cluster Analysis of Marketing Data: A Comparison of K-Means, Rough Set, and Rough Genetic Approaches. In: Abbas HA, Sarker RA, Newton CS (eds) Heuristics and Optimization for Knowledge Discovery, Idea Group Publishing 208–216
23. Wall M (1993) Galib, A C++ Library of Genetic Components. http://lancet.mit.edu/ga/
24. Wojna A (2005) Analogy-Based Reasoning in Classifier Construction. Transations of Rough Sets 4:277–374
25. Yao YY (1998) Constructive and algebraic methods of the theory of rough sets. Information Sciences 109:21–47
26. Yao YY, Lin TY (1996) Generalization of rough sets using modal logic. Intelligent Automation and Soft Computing 2(2):103-120

# Characterizing Pawlak's Approximation Operators

Victor W. Marek

Department of Computer Science
University of Kentucky
Lexington, KY 40506-0046, USA

To the memory of Zdzisław Pawlak, in
recognition of his friendship and guidance

**Abstract.** We investigate the operators associated with approximations
in the rough set theory introduced by Pawlak in his [14,11] and extensively studied by the Rough Set community [16]. We use universal algebra
techniques to establish a natural characterization of operators associated
with rough sets.

## 1   Introduction

The concept of *rough set determined by an equivalence relation R* has been introduced by Pawlak [14,11] in his studies of data mining. It is a natural extension of
a model of database introduced in [10] that treats records as objects which may
be indiscernible in the language (i.e. the tables are bags, not sets, of records).
Rough sets and a set of associated numerical measures allow for capturing various
degrees of similarity of objects such as records, documents, or other data units.
The point of departure of Pawlak was the realization that the descriptive languages are often inadequate to correctly describe concepts (i.e. – in set-theoretic
terms – subsets of the domain). The express goal of rough sets was to operate
in the following situation: we have a collection of objects $X$ and some description language $L$. We have some collection of objects $Y \subseteq X$. We would like to
describe $Y$ in the language $L$. That is we would like to find a formula $\varphi$ of $L$
so that

$$Y = \{x \in X : \varphi[x]\}.$$

We call such sets $Y$ *definable*. While usually the number of available definitions
is infinite, even in the situation when $X$ is finite, not every subset has to be
definable. Yet another point, made in [12], is that a set $Y$ may be definable in the
language $L$ but all the definitions are prohibitively large. In such circumstances
we may want to find a smaller language $L'$ where $Y$ is *not* definable $L'$, but the
approximations are definable in $L'$ by short formulas. This is certainly the case
in various medical applications.

In his analysis, Pawlak observed that in the case of finite set $X$, there is a
*largest* subset of $Y$ that is definable, and a *least* superset of $Y$ that is definable.

J.F. Peters et al. (Eds.): Transactions on Rough Sets VII, LNCS 4400, pp. 140–150, 2007.
© Springer-Verlag Berlin Heidelberg 2007

There is a way to compute these largest and least definable subsets of $X$ that are associated with $Y$. Specifically, this is done with the *indiscernibility relation* associated with the language $L$. In finite case, there is always a formula defining a *least definable set* containing a given object $x$. Let us call these sets *monads* (for the lack of better name, and for the fact that they resemble Leibniz monads). Then, it turns out that the largest definable set included in $Y$ is the union of monads that are entirely included in $Y$, while the least definable set containing $Y$ consists of those monads that have a nonempty intersection with $Y$. Abstracting from the existence of a specific language and its logical operations, Pawlak introduced the notion of *indiscernibility relation* in the set $X$. This is the equivalence relation $R$ so that the monads are its cosets.

We believe that the guiding examples motivating Pawlak were standard medical terminologies such as SNOMED ([19]) or ICD-9 ([5]) and their inadequacies for description of classes of medical cases. It is worth mentioning that for many years Pawlak collaborated with physicians interested in Medical Informatics (needless to say, this started long before the term *Medical Informatics* were even coined). Pawlak was concerned with the fact that medical reasoning approximates the available data, often disregarding values of some attributes. As a result, it is often difficult, for a variety of reasons, to classify medical cases. If one treats a medical condition as an ideal set of cases and attempts to describe it within a concrete language of some terminology then all a physician can do is to produce a differential diagnosis. This leads, naturally, to lower and upper approximations of the classes of medical cases. While Pawlak's intuitions were motivated by his collaborations with practicing physicians, it turned out that the methodology of approximations and indiscernibility relations are a common phenomenon. We refer the reader to monographs and journals devoted to rough set theory ([16]) for further motivations.

Let us assume that the underlying set $X$ is finite. Denoting by $\underline{R}(Y)$ and $\overline{R}(Y)$, respectively, the largest definable subset of $Y$ and the least definable superset of $Y$, we get the desired approximation relationships

$$\underline{R}(Y) \subseteq Y \subseteq \overline{R}(Y).$$

The sets $\underline{R}(Y)$ and $\overline{R}(Y)$ provide collectively measure of adequacy of the underlying language to the task of describing $Y$. Moreover, by various statistical operations on those sets, and on other sets derived by set-theoretic means, we can analyze the properties of the set $Y$ itself. For that reason we would like to know more about the sets $\overline{R}(Y)$ and $\underline{R}(Y)$. We would like to know what are possible operators of the form $\overline{R}(\cdot)$ and $\underline{R}(\cdot)$, and how those behave when $R$ vary (i.e. when the language changes). These issues, to some extent were addressed in recent [9], but the review of the literature indicates that the Rough Sets community investigated a number of possible explanations for the rough set formalism by immersing it into various well-known mathematical areas. Those areas are all related to a variety of ways in which one can describe databases. We will list several different areas which were explored, although more could be mentioned. The references are, by necessity, incomplete. The first one is the idea of topological interpretation of approximations (already explored in [11]). This was for instance studied in [20]. Another approach was to look at modal logic interpretations of rough sets (see for instance [21] and a series of papers by Orłowska and collaborators [3], and more generally [13]). One could think about approximations using Kleene [8] three-valued logic as it was done in [12]. It is also possible to look at abstractions of rough sets via techniques of universal algebra. This last area explores Boolean algebra with operators [7,6]. The rough set community investigated these connections, with a varying degree of generality, in [3,13,18]. Our contribution belong to this direction of research. We attempt to apply the techniques of universal algebra and in particular of [7,18] to find an *esthetically appealing* characterization of Pawlak's operators. In this quest our results find, indeed, a clean and interesting characterization of these operators. By necessity, some of the results discussed in this paper are known. After all operators over Boolean algebras have been introduced by Tarski and his collaborators over 50 years ago. For instance at least points (1)-(4) of Proposition 5 are known. The terms in which we characterize the Pawlak's operators are mostly known in the literature. The ones that we introduce and which (in conjunction with other properties) appear to be new are the following:

$$Y_1 \cap f(Y_2) \neq \emptyset \quad \text{if and only if} \quad Y_2 \cap f(Y_1) \neq \emptyset. \qquad \textit{Exchange}$$

and

$$Y_1 \cup f(Y_2) \neq X \quad \text{if and only if} \quad Y_2 \cup f(Y_1) \neq X \qquad \textit{Dual exchange}$$

As we will see, in addition to the well known properties of operators, these properties characterize the lower and upper Pawlak approximations, respectively.

Thus, in this paper, we prove four results that pertain to the explanation of Pawlak's approximation operators. First, we show a simple and elegant characterization of upper approximation. Much later we state but not prove the dual characterization (an indirect proof of this other property follow from duality considerations, and the point (5) of Proposition 5). We also prove the duality of exchange and dual exchange properties, and we show how one can introduce a structure of a complete lattice in upper approximation operators.

We believe that Pawlak, who believed in elegance of mathematical formulation of tools that are useful in practice, would enjoy the simplicity of our description of his operators.

## 2    Preliminaries

Given a set $X$ and an equivalence (indiscernibility) relation $R$ in $X$, we write $[x]_R$ for the $R$-*coset* of the element $x$ in $X$, that is $\{y : xRy\}$. Given an equivalence relation $R$, the cosets of elements of $X$ form a partition of $X$ into nonempty blocks. We may drop the subscript $R$ when $R$ is determined by the context.

Let $R$ be an equivalence relation in the set $X$. The relation $R$ determines, for every set $Y \subseteq X$, two sets: the lower and upper $R$-bounds (also known as approximations) of $Y$. Specifically, following Pawlak [14,11,15] we define

$$\overline{R}(Y) = \{x \in X : [x] \cap X \neq \emptyset\}$$

and

$$\underline{R}(Y) = \{x \in X : [x] \subseteq X\}.$$

It is a simple consequence of the properties of equivalence relations and of De Morgan laws that for every subset $Y$ of $X$, the complement of $Y$, $-Y$, has the following properties:

$$-\underline{R}(-Y) = \overline{R}(Y)$$

and

$$-\overline{R}(-Y) = \underline{R}(Y)$$

We now introduce the notion of an operator in a set $X$ and introduce various classes of operators. Let $X$ be a set. The set $\mathcal{P}(X)$ is the powerset of $X$, the collection of all subsets of $X$. Given a set $X$, by an *operator* in $X$ we mean any function $f : \mathcal{P}(X) \to \mathcal{P}(X)$. An operator $f$ in the set $X$ is *additive* if for all $Y_1, Y_2 \subseteq X$, $f(Y_1 \cup Y_2) = f(Y_1) \cup f(Y_2)$. An operator $f$ in the set $X$ is *multiplicative* if for all $Y_1, Y_2 \subseteq X$, $f(Y_1 \cap Y_2) = f(Y_1) \cap f(Y_2)$. An operator $f$ in $X$ is *progressive* if for all $Y \subseteq X$, $Y \subseteq f(Y)$. An operator $f$ in $X$ is *regressive* if for all $Y \subseteq X$, $f(Y) \subseteq Y$. An operator $f$ in $X$ is *idempotent* if for all $Y \subseteq X$, $f(f(Y)) = f(Y)$. An operator $f$ in $X$ *preserves empty set* if $f(\emptyset) = \emptyset$ (Operators preserving empty set are called *normal* in [7].) Finally, we say that an operator $f$ in $X$ *preserves unit* if $f(X) = X$.

All the properties of operators introduced above are pretty standard. Here are two properties (characteristic for our intended application) which are non-standard. Let $X$ be a set and let $f$ be an operator in $X$. We say that $f$ has an *exchange property* if for all $Y_1, Y_2 \subseteq X$,

$$Y_1 \cap f(Y_2) \neq \emptyset \quad \text{if and only if} \quad Y_2 \cap f(Y_1) \neq \emptyset.$$

This property of the operator will turn out to be crucial in our characterization of the upper approximation in Pawlak's rough sets.

Likewise, we say that that $f$ has a *dual exchange property* if for all $Y_1, Y_2 \subseteq X$

$$Y_1 \cup f(Y_2) \neq X \quad \text{if and only if} \quad Y_2 \cup f(Y_1) \neq X.$$

The dual exchange property will be used to characterize lower approximations of rough sets.

## 3  Characterizing  $\overline{R}$

We now show the principal result of this note, the characterization of operations $\overline{R}$ for equivalence relations $R$ (The characterization of lower approximations will follow from this result and the general facts regarding duality properties of operators.) We have the following result.

**Proposition 1.** *Let $X$ be a finite set and let $f$ be an operator in $X$. Then there exists an equivalence relation $R$ such that $f = \overline{R}$ if and only if: $f$ preserves empty set; $f$ is additive; $f$ is progressive; $f$ is idempotent; and $f$ has the exchange property.*

Proof: First, we need to show that whenever $R$ is an equivalence relation in $X$ then the operator $\overline{R}$ has the five properties listed above. The first four of these are pretty obvious; $\overline{R}$ preserves emptyset because when there is no element, then there is no coset. The additivity follows from the distributivity of existential quantifier with respect to disjunction, progressiveness follows from the fact that for all $x \in X$, $x \in [x]_R$, and the idempotence follows from the transitivity of the relation $R$. We will now show that the operator $\overline{R}$ possesses the exchange property. We observe that the exchange property is symmetric with respect to $Y_1$ and $Y_2$. Therefore all we need to prove is that whenever $Y_1 \cap \overline{R}(Y_2) \neq \emptyset$ then also $Y_2 \cap \overline{R}(Y_1) \neq \emptyset$. Let us reformulate slightly the statement $Y_1 \cap \overline{R}(Y_2) \neq \emptyset$. This statement is equivalent to the fact that there is an $x \in Y_1$, and an $y \in Y_2$ so that $xRy$. We now proceed as follows. Since $Y_1 \cap \overline{R}(Y_2) \neq \emptyset$, there is an element $x$ that belongs to $Y_1$ and an element $y \in Y_2$ such that $xRy$. But then $[x]_R = [y]_R$, and so $y$ is an element of $Y_2$ for which there is an element $x' \in Y_1$ so that $yRx'$. Namely $x$ is that element $x'$. Therefore $Y_2 \cap \overline{R}(Y_1)$ is nonempty.

Now, let us assume that $f$ is an operator in $X$, and that $f$ has the five properties mentioned above, that is $f$ preserves empty set, $f$ is additive, $f$ is progressive, $f$ is idempotent, and that $f$ has the exchange property. Then we need to construct an equivalence relation $R_f$ so that $f$ coincides with $\overline{R}$. Here is how we define relation $R_f$:

$$x R_f y \quad \text{if} \quad x \in f(\{y\}).$$

Our first task is to prove that, indeed, $R_f$ is an equivalence relation in $X$. To see reflexiveness, let us observe that since $f$ is progressive, for every $x$,

$$\{x\} \subseteq f(\{x\})$$

that is, $x \in f(\{x\})$. But this means that $x R_f x$, for every $x \in X$.

For the symmetry of $R_f$, let us assume $x R_f y$, that is $x \in f(\{y\})$. This means that

$$\{x\} \cap f(\{y\}) \neq \emptyset.$$

By the exchange property of $f$,

$$\{y\} \cap f(\{x\}) \neq \emptyset.$$

That is $y \in f(\{x\})$. In other words, $y R_f x$.

Finally, let us assume that $x, y, z$ have the property that $x R_f y$ and $y R_f z$. That is:

$$x \in f(\{y\}) \quad \text{and} \quad y \in f(\{z\}).$$

That is

$$\{x\} \subseteq f(\{y\}) \quad \text{and} \quad \{y\} \subseteq f(\{z\}).$$

From the second equality we have

$$\{y\} \cup f(\{z\}) = f(\{z\}).$$

By the additivity of $f$ we have

$$f(\{y\}) \cup f(f(\{z\})) = f(f(\{z\})).$$

By idempotence of $f$ we have, then

$$f(\{y\}) \cup f(\{z\}) = f(\{z\}).$$

This means that

$$f(\{y\}) \subseteq f(\{z\}).$$

But $x \in f(\{y\})$ and so $x \in f(\{z\})$, that is $xR_f z$, as desired.

To complete the proof of our assertion we need to prove that for all $Y \subseteq X$, $f(Y) = \overline{R_f}(Y)$. Our proof will use the fact that we deal with a finite set. We will comment on the dependence on this assumption later.

First, let us assume that $Y \subseteq X$, and that $x \in f(Y)$. Since $X$ is finite, so is $Y$. Then

$$Y = \bigcup_{x \in Y} \{x\}.$$

Now, let us observe that since the operator $f$ is additive, it is finitely additive that is it distributes with respect to finite unions. Thus:

$$f(Y) = \bigcup_{x \in Y} f(\{x\}).$$

This means that, since our assumption was that $x$ belongs to $f(Y)$, for some $y \in Y$, $x \in f(\{y\})$. But then $xR_f y$ for some $y \in Y$, that is $x \in \overline{R_f}(Y)$. In other words, for an arbitrary $Y \subseteq X$, $f(Y) \subseteq \overline{R_f}(Y)$.

Conversely, let us assume that $x \in \overline{R_f}(Y)$. Then, since we proved that $R_f$ is an equivalence relation, for some $y \in Y$, $xR_f y$. That is, according to the definition of the relation $R_f$, $x \in f(\{y\})$. Next, we observe that $f$ is monotone, that is $Y_1 \subseteq Y_2$ implies that $f(Y_1) \subseteq f(Y_2)$. Indeed, if $Y_1 \subseteq Y_2$ then $Y_1 \cup Y_2 = Y_2$, thus $f(Y_1 \cup Y_2) = f(Y_2)$ and by additivity $f(Y_1) \cup f(Y_2) = f(Y_2)$, that is $f(Y_1) \subseteq f(Y_2)$. Returning to the argument, since $y \in Y$, $\{y\} \subseteq Y$, and by our remark on monotonicity:

$$f(\{y\}) \subseteq f(Y).$$

This implies that $x \in f(Y)$ and since $x$ was an arbitrary element of $\overline{R_f}(Y)$, $\overline{R_f}(Y) \subseteq f(Y)$. Thus we proved the other inclusion and since $Y$ was the arbitrary subset of $X$, we proved that $f$ and $\overline{R_f}$ coincide.     □

In the proof of our Proposition 1 we computed, out of the operator $f$, a relation $R_f$ so that $f = \overline{R_f}$. But this relation is unintuitive (at least for non-specialists). We will now provide a more intuitive description of the same relation. Given an operator $f$, we define a relation $S_f$ as follows:

$$xS_f y \quad \text{if} \quad \forall_{Y \subseteq X}(x \in f(Y) \Leftrightarrow y \in f(Y)).$$

We now have the following result.

**Proposition 2.** *If the operator $f$ satisfies the conditions of Proposition 1, then $R_f = S_f$.*

Proof: We need to prove two implications:

(a) $\forall_{x,y}(xR_fy \Rightarrow xS_fy)$, and
(b) $\forall_{x,y}(xS_fy \Rightarrow xR_fy)$

To show (a) let $x, y$ be arbitrary elements of $X$, and let us assume $xR_fy$. Then, since $R_f$ is symmetric, $yR_fx$, that is $y \in f(\{x\})$. It is sufficient to prove that for all subsets $Y$ of $X$, if $x \in f(Y)$ then $y \in f(Y)$ (the proof of the converse is very similar, except that we use the fact that $x \in f(\{y\})$). So, let $x \in f(Y)$. Then $\{x\} \subseteq f(Y)$, so, by monotonicity, $f(\{x\}) \subseteq f(f(Y)) = f(Y)$ (last equality uses idempotence of $f$). Thus, $f(\{x\}) \subseteq f(Y)$, and since $y \in f(\{x\})$, $y \in f(Y)$. Thus, taking into account the other implication, proved as discussed above, we proved that $xR_fy$ implies $xS_fy$.

Next, let us assume that $xS_fy$. That is,

$$\forall_{Y \subseteq X}(x \in f(Y) \Leftrightarrow y \in f(Y)).$$

We need to prove that $x \in f(\{y\})$. But $y \in f(\{y\})$, since for $Y = \{y\}$, $y \in Y$, and for every $Y$, $Y \subseteq f(Y)$ ($f$ is progressive). But now specializing the above equivalence to $Y = \{y\}$, we find that $x \in f(\{y\})$, as desired.          □

In the proof of Proposition 1 we used the assumption that $X$ was a finite space. In fact, we could relax this assumption, but at a price. Recall that we assumed that the operator $f$ was additive (i.e. $f(Y_1 \cup Y_2) = f(Y_1) \cup f(Y_2)$, for all subsets $Y_1, Y_2$ of $X$). In the case when $X$ is finite we have for any family $\mathcal{X}$ of subsets of $X$

$$f(\bigcup \mathcal{X}) = \bigcup_{Y \in \mathcal{X}} f(Y).$$

This is easily proved by induction on the size of $\mathcal{X}$. Let us call an operator $f$ *completely additive* if the equality

$$f(\bigcup \mathcal{X}) = \bigcup_{Y \in \mathcal{X}} f(Y).$$

holds for *every* family $\mathcal{X}$ of subsets of $X$. Under the assumption of complete additivity the assumption of finiteness can be eliminated.

## 4    Structure of the Family of Upper Closure Operators

We will now look at the situation when the set $X$ has several different equivalence relations, that is several corresponding notions of rough sets. This is, actually, quite common situation; for instance we may have different medical nomenclature systems that are used to describe medical cases. In fact it is a well-known fact that the medical nomenclatures of various nations are not translatable.

We now face the question of the relationship between the different upper closure operators. Specifically, we may want to check the relationship between $\overline{R_1}$ and $\overline{R_2}$ given relations $R_1$ and $R_2$.

**Proposition 3.** *Let $R_1, R_2$ be two equivalence relations. Then*

$$R_1 \subseteq R_2 \quad \text{if and only if} \quad \forall_{Y \subseteq X}(\overline{R_1}(Y) \subseteq \overline{R_2}(Y)).$$

Proof: First, let us assume that $R_1 \subseteq R_2$, and let $Y$ be an arbitrary subset of $X$. We need to prove $\overline{R_1}(Y) \subseteq \overline{R_2}(Y)$. To this end, let $x \in \overline{R_1}(Y)$. Then there is an element $y \in Y$ such that $xR_1y$. But then $xR_2y$ and so $x \in \overline{R_2}(Y)$.

Conversely, let us assume that for every $Y$, $\overline{R_1}(Y) \subseteq \overline{R_2}(Y)$. We want to prove that $R_1 \subseteq R_2$. Let us assume that $xR_1y$. Then $x \in \overline{R_1}(\{y\})$, thus $x \in \overline{R_2}(\{y\})$. In other words, there is some $y' \in \{y\}$ such that $xR_2y'$. But $\{y\}$ has unique element, $y$. Thus $xR_2y$, as desired. $\qquad\square$

The structure of the family of all equivalence relations in a set $X$ is well-known. Let $\langle EqR_X, \subseteq \rangle$ be the relational structure with $EqR_X$ equal to the set of all equivalence relations in $X$, ordered by inclusion. Then $\langle EqR_X, \subseteq \rangle$ is a complete lattice (regardless whether $X$ is finite or not) but $\langle EqR_X, \subseteq \rangle$ is not a distributive lattice, in general ([4, p. 227]).

Proposition 3 allows us to transfer the properties of equivalence relations to operators. Let us define an *upper rough set operator* in the set $X$ as any operator that preserves empty set, is completely additive (thus we no longer assume $X$ to be finite), progressive, idempotent, and has the exchange property. We denote by $\mathcal{R}_X$ the set of all upper rough set operators in $X$, and $\preceq_X$ the *dominance relation* in $\mathcal{R}_X$ defined by

$$f \preceq g \quad \text{if} \quad \forall_{Y \subseteq X}(f(Y) \subseteq g(Y)).$$

Then applying our discussion of the lattice of equivalence relations in $X$ to Proposition 3 we get the following fact.

**Proposition 4.** *The structure $\langle \mathcal{R}_X, \preceq \rangle$ is a poset. In fact $\langle \mathcal{R}_X, \preceq \rangle$ is a complete lattice, but in general not a distributive one.*

## 5    Duality

Let $f$ be an operator in a set $X$. The *dual* of the operator $f$, $f_d$, is an operator defined by the following equality:

$$f_d(Y) = -f(-Y).$$

Here $Y$ ranges over arbitrary subsets of $X$, $-Y = X \setminus Y$ is the complement operation. The dual operators are used in various places in mathematics and computer science. One example is the operator dual to van Emden-Kowalski operator $T_P$ ([2, p. 83]).

While we defined the notion of dual operator in the Boolean lattice, $\langle \mathcal{P}(X), \subseteq \rangle$, as long as the lattice has a complement operation $-$, the notion of a dual operator can be defined. Moreover, if for all $x$, $--x = x$, then $(f_d)_d = f$. This is certainly the case in our application.

Now, let us assume that we are dealing with operators in a set $X$. We have the following fact.

**Proposition 5.** *Let $X$ be a set and $f$ an operator in $X$. Then:*

1. *The operator $f$ preserves the empty set (unit) if and only if the operator $f_d$ preserves the unit (empty set)*
2. *The operator $f$ is progressive (regressive) if and only if the operator $f_d$ is regressive (progressive)*
3. *The operator $f$ is additive (multiplicative) if and only if the operator $f_d$ is multiplicative (additive)*
4. *The operator $f$ is idempotent if and only if the operator $f_d$ is idempotent*
5. *The operator $f$ possesses the exchange property (dual exchange property) if and only if the operator $f_d$ possesses the dual exchange property (exchange property).*

Proof: The points (1)-(3) are entirely routine. To see the point (4), let us assume that the operator $f$ is idempotent. Then for an arbitrary $Y$,

$$f_d(f_d(Y)) = -f(-f_d(Y)) = -f(--f(-Y)) = -f(f(-Y)) = -f(-Y) = f_d(Y).$$

The penultimate equality uses the idempotence of $f$. The other direction of (4) follows from the fact that $(f_d)_d = f$, and the argument above.

To see (5), we first assume that $f$ has the exchange property. We prove that $f_d$ has the dual exchange property. To this end we need to prove that for arbitrary $Y_1, Y_2 \subseteq X$,

$$Y_1 \cup f_d(Y_2) \neq X \quad \text{if and only if} \quad Y_2 \cup f_d(Y_1) \neq X.$$

Since this formula is symmetric with respect to $Y_1$ and $Y_2$, it is enough to prove the implication:
$$Y_1 \cup f_d(Y_2) \neq X \;\Rightarrow\; Y_2 \cup f_d(Y_1) \neq X.$$

So let us assume that $Y_1 \cup f_d(Y_2) \neq X$. Then, substituting $--Y_1$ for $Y_1$, and expanding the definition of $f_d$, we get:

$$(--Y_1) \cup -f(-Y_2) \neq X$$

that is:
$$-(-Y_1 \cap f(-Y_2)) \neq X.$$

This is, of course, equivalent to:

$$-Y_1 \cap f(-Y_2) \neq \emptyset.$$

Since $f$ has the exchange property,

$$-Y_2 \cap f(-Y_1) \neq \emptyset.$$

Thus we get:
$$-(-Y_2 \cap f(-Y_1)) \neq X,$$

which reduces to
$$Y_2 \cup -f(-Y_1) \neq X.$$

that is
$$Y_2 \cup f_d(Y_1) \neq X,$$
as desired. The proof of the other part of (5) namely that whenever $f$ has the dual exchange property then $f_d$ has the exchange property, is similar.    $\square$

Now, let us look at the familiar equality $\underline{R}(Y) = -\overline{R}(-Y)$. This, in the language of operators, says that for every equivalence relation $R$, the operator $\underline{R}$ is simply $\overline{R}_d$. So now we compare the characterization of the upper approximation by five conditions (Proposition 1) and the duality result above (Proposition 5). We get the following result.

**Proposition 6.** *Let $X$ be a finite set and let $f$ be an operator in $X$. Then there exists an equivalence relation $R$ such that $f = \underline{R}$ if and only if: $f$ preserves unit; $f$ is multiplicative; $f$ is regressive; $f$ is idempotent; and $f$ has the dual exchange property.*

Again, we can also study the family of all operators that have the five properties of operators characterizing lower approximation and introduce a complete, but non-distributive lattice structure in that set. That is, we can prove the result analogous to the Proposition 4.

# 6    Conclusions

Algebraic methods, whenever applicable, provide a clean foundations for an underlying subject. They abstract from unnecessary details, showing the properties that really matter. This is certainly the case in the area of rough sets. Our results confirm that, as observed by numerous authors [13,18] the theory of rough sets relates to the operators in lattices, a theory well-developed ([1, p. 86, ff.]) and with many deep results. Rough sets approximate elements of one lattice (Boolean lattice of all sets) with elements of a sublattice (of definable sets). Abstract approach to this idea of approximation and characterization of approximations in algebraic terms will only improve our understanding of the concept of rough set. We find it amazing that the ideas of Tarski (who certainly shied from applications) found its expression in Pawlak's, very applied, research.

# Acknowledgments

This research was supported by the National Science Foundation under Grant 0325063. We are grateful to A. Skowron for directing us to a number of references. An anonymous referee pointed relations to several cited papers. We express to her/him gratitude for indicating the connections to the universal algebra research.

# References

1. Davey, B.A., and Priestley, H.A., *Introduction to Lattices and Order*, Cambridge University Press, 1992.
2. K. Doets, *From Logic to Logic Programming*, MIT Press, 1994.

3. I. Düntsch and E. Orłowska Beyond Modalities: Sufficiency and Mixed Algebras. Chapter 16 of [OS01], 2001.
4. W. Hodges, *Model Theory*. Cambridge University Press, 1993.
5. System for medical diagnoses ICD-9, http://pmiconline.com.
6. B. Jonsson A Survey of Boolean Algebras with Oprators. In: *Algebras and Order*, pages 239–284. Kluwer, 1991.
7. B. Jonsson and A. Tarski. Boolean Algebras with Operators. *American Journal of Mathematics* 73:891–939, 1951.
8. S. C. Kleene. *Introduction to Metamathematics*. North-Holland, 1967. Fifth reprint.
9. Liu, G-L., The axiomatization of the Rough Set Upper Approximation Operations, *Fundamenta Informaticae* 69:331–342, 2006.
10. W. Marek and Z. Pawlak. Information storage and retrieval systems, mathematical foundations, *Theoretical Computer Science* 1(4):331–354, 1976.
11. W. Marek and Z. Pawlak. Rough sets and information systems, *Fundamenta Informaticae* 7(1):105–115, 1984.
12. V.W. Marek and M. Truszczyński. Contributions to the Theory of Rough Sets, *Fundamenta Informaticae* 39(4):389–409, 1999.
13. Ewa Orłowska and Andrzej Szałas. *Relational Methods for Computer Science Applications. Selected Papers from 4th International Seminar on Relational Methods in Logic, Algebra and Computer Science (RelMiCS'98)*, Studies in Fuzziness and Soft Computing 65, Physica-Verlag/Springer, 2001.
14. Z. Pawlak. Rough Sets, *International Journal of Computer and Information Sciences* 11:341–356, 1982.
15. Pawlak, Z., *Rough Sets – theoretical aspects of reasoning about data*, Kluwer, 1991.
16. Rough Sets, a journal series of *Springer Lecture Notes in Computer Science*.
17. E. SanJuan and L. Iturrioz. Duality and informational representability of some information algebras. pages 233-247, in: *Rough Sets in Knowledge Discovery, Methodology and Applications*, L. Polkowski and A. Skowron, (eds). Physica-Verlag, 1998.
18. E. SanJuan and L. Iturrioz. An application of standard BAO theory to some abstract Information algebras. Chapter 12 of [OS01], 2001.
19. System of medical terminology SNOMED, http://www.snomed.org.
20. Y.-Y. Yao. Two views of the theory of rough sets in finite universes. *International Journal of Approximate Reasoning* 15:291–317, 1996.
21. Y.-Y. Yao and T.Y. Lin. Generalization of rough sets using modal logic, *Intelligent Automation and Soft Computing* 2:103–120, 1996.

# Application of Rough Sets
# in Pattern Recognition

Sushmita Mitra and Haider Banka

Center for Soft Computing Research: A National Facility,
Indian Statistical Institute, Kolkata 700 108, India
{sushmita,hbanka_r}@isical.ac.in

**Abstract.** This article provides an overview of recent literature on some tasks of pattern recognition using rough sets and its hybridization with other soft computing paradigms. Rough set theory is an established tool for dealing with imprecision, noise, and uncertainty in data. In this article we will focus on two recent applications using rough sets; *viz.*, feature selection in high dimensional gene expression data, and collaborative clustering. The experimental results demonstrate that the incorporation of rough set improves the performance of the system.

## 1 Introduction

Rough set theory [1] provides an important and mathematically established tool, for dimensionality reduction in large data. Rough set theory (RST) was developed by Pawlak as a tool to deal with inexact and incomplete data. Over the years, RST has become a topic of great interest to researchers and has been applied to many domains, in particular to knowledge databases. This success is due in part to the following aspects of the theory, *viz.*, only the facts hidden in data are analyzed, no additional information about the data is required, and a minimal knowledge representation is generated.

One of the important problems in extracting and analyzing information from large databases is the associated high complexity. Feature selection is helpful as a preprocessing step for reducing dimensionality, removing irrelevant data, improving learning accuracy and enhancing output comprehensibility. Microarray data is a typical example presenting an overwhelmingly large number of features (genes), the majority of which are not relevant to the description of the problem and could potentially degrade the classification performance by masking the contribution of the relevant features. The key informative features represent a base of reduced cardinality, for subsequent analysis aimed at determining their possible role in the analyzed phenotype. This highlights the importance of feature selection, with particular emphasis on microarray data.

In microarray data, many of the attributes may be redundant, and we may find *minimal* subsets of attributes which give the same classification as the whole set *A*. These subsets are called *reducts* in RST, and correspond to the *minimal*

J.F. Peters et al. (Eds.): Transactions on Rough Sets VII, LNCS 4400, pp. 151–169, 2007.

*feature sets* that are *necessary* and *sufficient* to represent a *correct* decision about classification. Thus RST provides a methodology for addressing the problem of relevant feature selection.

The task of finding reducts is reported to be NP-hard [2]. The high complexity of this problem has motivated investigators to apply various approximation techniques to find near-optimal solutions. There are some studies reported in literature, e.g. [3,4], where genetic algorithms have been applied to find reducts. Genetic algorithms (GAs) provide an efficient search technique in a large solution space, based on the theory of evolution. When there are two or more conflicting characteristics to be optimized, often the single-objective GA requires an appropriate formulation of the single fitness function in terms of an additive combination of the different criteria involved. In such cases *multi-objective* GAs (MOGAs) [5] provide an alternative, more efficient, approach to search for optimal solutions. In this article, we present various attempts of using GA's (both single and multi-objective) in order to obtain reducts, and hence provide some solution to the challenging task of feature selection.

The use of soft computing in clustering has been reported in literature [6,7]. Fuzzy sets and rough sets have been incorporated in the $c$-means framework to develop the fuzzy $c$-means (FCM) [8] and rough $c$-means (RCM) [9] algorithms. While membership in FCM enables efficient handling of overlapping partitions, the rough sets [1] deal with uncertainty, vagueness and incompleteness in data. Collaborative clustering was first investigated by Pedrycz [10], using standard FCM algorithm. This helps reveal a structure that is common or similar to a number of subsets of the data. One may proceed by clustering each sub-population locally as a module, considering small random samples, thereby enabling faster convergence of clustering. Subsequently there is collaboration between these modules by intercommunicating the individual cluster centroids. These representatives from the other sub-populations serve to globally influence and refine the clustering result of each module. Eventually, since the sub-populations are derived from the same large population, we converge to a stable global clustering after effective collaboration between the modules.

The concept of collaborative clustering can be further extended to the rough domain using collaborative rough or collaborative rough-fuzzy clustering. The use of rough sets helps in automatically controlling the effect of uncertainty among patterns lying between the upper and lower approximations, during collaboration between the modules. Thereby patterns within the lower approximation play a more pivotal role during clustering. Incorporation of membership, in the RCM framework, is seen to enhance the robustness of clustering as well as collaboration.

The rest of the sections is organized as follows. Section 2 introduces the relevant preliminaries on rough set theory and multi-objective genetic algorithms. Sections 3 and 4 describe feature selection and clustering in the rough set framework. Section 5 concludes the article.

# 2    Preliminaries

In this section we briefly discuss the basic concepts of RST and Multi-Objective Genetic Algorithms (MOGA), while assuming that readers are familiar with the preliminaries of GAs.

## 2.1    Rough Sets

Rough sets [1] constitute a major mathematical tool for managing uncertainty that arises from granularity in the domain of discourse – due to incomplete information about the objects of the domain. The granularity is represented formally in terms of an *indiscernibility* relation that partitions the domain. The intention is to approximate a *rough* (imprecise) concept in the domain, by a pair of *exact* concepts. These exact concepts are called the lower and upper approximations, and are determined by the indiscernibility relation. The lower approximation is the set of objects definitely belonging to the rough concept, whereas the upper approximation is the set of objects possibly belonging to the same. Fig. 1 provides an illustration of a rough set with its approximations. The formal definitions of the above notions and others required for the present work are given below.

**Definition 1.** *An **Information System** $\mathcal{A} = (U, A)$ consists of a non-empty, finite set $U$ of objects (cases, observations, etc.) and a non-empty, finite set $A$ of attributes a (features, variables), such that $a : U \to V_a$, where $V_a$ is a value set. We shall deal with information systems called **decision tables**, in which the attribute set has two parts $(A = C \cup D)$ consisting of the condition and decision attributes (in the subsets C, D of A respectively). In particular, the decision tables we take will have a single decision attribute d, and will be consistent, i.e., whenever objects $x, y$ are such that for each condition attribute $a$, $a(x) = a(y)$, then $d(x) = d(y)$.*

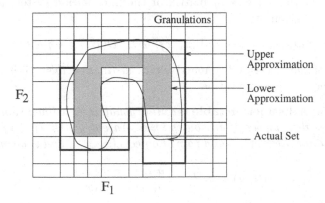

**Fig. 1.** Lower and upper approximations in a rough set

**Definition 2.** *Let* $B \subset A$. *Then a* **$B$-indiscernibility relation** $IND(B)$ *is defined as*

$$IND(B) = \{(x, y) \in U : a(x) = a(y), \forall a \in B\}. \tag{1}$$

It is clear that $IND(B)$ partitions the universe $U$ into equivalence classes

$$[x_i]_B = \{x_j \in U : (x_i, x_j) \in IND(B)\}, \ x_i \in U. \tag{2}$$

**Definition 3.** *The* **$B$-lower** *and* **$B$-upper approximations** *of a given set* $X(\subseteq U)$ *are defined, respectively, as follows:*

$\underline{B}X = \{x \in U : [x]_B \subseteq X\},$
$\overline{B}X = \{x \in U : [x]_B \cap X \neq \phi\}.$

*The* **$B$-boundary** *region is given by* $BN_B(X) = \overline{B}X \setminus \underline{B}X$.

**Reducts.** In a decision table $\mathcal{A} = (U, C \cup D)$, one is interested in eliminating redundant *condition* attributes, and actually *relative* $(D)$-reducts are computed. Let $B \subseteq C$, and consider the *$B$-positive region* of $D$, viz., $POS_B(D) = \bigcup_{[x]_D} \underline{B}[x]_D$. An attribute $b \in B(\subseteq C)$ is $D$-**dispensable** in $B$ if $POS_B(D) = POS_{B\setminus\{b\}}(D)$, otherwise $b$ is $D$-**indispensable** in $B$. Here $B$ is said to be $D$-**independent** in $\mathcal{A}$, if every attribute from $B$ is $D$-indispensable in $B$.

**Definition 4.** $B(\subseteq C)$ *is called a* **$D$-reduct** *in* $\mathcal{A}$, *if* $B$ *is* $D$-*independent in* $\mathcal{A}$ *and* $POS_C(D) = POS_B(D)$.

Notice that, as decision tables with a single decision attribute $d$ are taken to be consistent, $U = POS_C(d) = POS_B(D)$, for any $d$-reduct $B$.

The *core* is the set of essential attributes of any information system. Mathematically, $core(A) = \bigcap reduct(A)$, *i.e.*, the set consists of those attributes, which are members of all reducts.

**Discernibility Matrix.** $D$-reducts can be computed with the help of $D$-discernibility matrices [2]. Let $U = \{x_1, \cdots, x_m\}$. A $D$-**discernibility matrix** $M_D(\mathcal{A})$ is defined as an $m \times m$ matrix of the information system $\mathcal{A}$ with the $(i, j)$th entry $c_{ij}$ given by

$$c_{ij} = \{a \in C : a(x_i) \neq a(x_j), \text{and } (x_i, x_j) \notin IND(D)\}, \ i, j \in \{1, \cdots, m\}. \tag{3}$$

A variant of the discernibility matrix, *viz.*, *distinction table* [3] is used in our work to enable faster computation.

**Definition 5.** *A* **distinction table** *is a binary matrix with dimensions* $\frac{(m^2 - m)}{2} \times N$, *where* $N$ *is the number of attributes in* $A$. *An entry* $b((k, j), i)$ *of the matrix corresponds to the attribute* $a_i$ *and pair of objects* $(x_k, x_j)$, *and is given by*

$$b((k, j), i) = \begin{cases} 1 \text{ if } a_i(x_k) \neq a_i(x_j), \\ 0 \text{ if } a_i(x_k) = a_i(x_j). \end{cases} \tag{4}$$

The presence of a '1' signifies the ability of the attribute $a_i$ to discern (or distinguish) between the pair of objects $(x_k, x_j)$.

## 2.2 Multi-Objective Genetic Algorithms (MOGA)

Most real-world search and optimization problems typically involve multiple objectives. A solution that is better with respect to one objective requires a compromise in other objectives. Unlike single-objective optimization problems, the multiple-objective GA tries to optimize two or more conflicting characteristics represented by fitness functions. Modeling this situation with a single-objective GA would amount to a heuristic determination of a number of parameters involved in expressing such a scalar-combination-type fitness function. MOGA, on the other hand, generates a set of *Pareto-optimal* solutions [5] which simultaneously optimize the conflicting requirements of the multiple fitness functions.

Among the different multi-objective algorithms, it is observed that Non-dominated Sorting Genetic Algorithm (NSGA-II) [11] possesses all the features required for a good MOGA. It has been shown that this can converge to the global Pareto front, while simultaneously maintaining the diversity of the population. We describe here the characteristics of NSGA-II, like non-domination, crowding distance and the crowding selection operator.

**Non-domination.** The concept of optimality, behind the multi-objective optimization, deals with a set of solutions. The conditions for a solution to be *dominated* with respect to the other solutions are given below.

**Definition 6.** *If there are $M$ objective functions, a solution $x^{(1)}$ is said to dominate another solution $x^{(2)}$, if both conditions (a) and (b) are true:*

(a) *The solution $x^{(1)}$ is no worse than $x^{(2)}$ in all the $M$ objective functions.*
(b) *The solution $x^{(1)}$ is strictly better than $x^{(2)}$ in at least one of the $M$ objective functions.*

Otherwise the two solutions are *non-dominating* to each other. When a solution $i$ dominates solution $j$, then **rank** $r_i < r_j$.

The major steps for finding the non-dominated set in a population $P$ of size $|P|$ are outlined below.

1. Set solution counter $i = 1$ and create an empty non-dominated set $P'$.
2. **For** a solution $j \in P$ $(j \neq i)$, check if solution $j$ dominates solution $i$.
   **If** yes **then go to** Step 4.
3. **If** more solutions are left in $P$, increment $j$ by one and **go to** Step 2
   **Else** set $P' = P' \cup \{i\}$.
4. Increment $i$ by one.
   **If** $i \leq |P|$ **then go to** Step 2 **else** declare $P'$ as the non-dominated set.

After all the solutions of $P$ are checked, the members of $P'$ constitute the non-dominated set at the first level (front with rank $= 1$). In order to generate solutions for the next higher level (dominated by the first level), the above procedure is repeated on the reduced population $P = P - P'$. This is iteratively continued until $P = \emptyset$.

**Crowding distance.** In order to maintain diversity in the population, a measure called **crowding distance** is used. This assigns the highest value to the boundary solutions and the average distance of two solutions $[(i + 1)$th and $(i - 1)$th] on either side of solution $i$ along each of the objectives. The following algorithm computes the crowding distance $d_i$ of each point in the front $\mathcal{F}$.

1. Let the number of solutions in $\mathcal{F}$ be $l = |\mathcal{F}|$ and assign $d_i = 0$ for $i = 1, 2, \ldots, l$.
2. **For** each objective function $f_k, k = 1, 2, \ldots, M$, sort the set in its worse order.
3. Set $d_1 = d_l = \infty$.
4. **For** $j = 2$ to $(l - 1)$ increment $d_j$ by $f_{k_{j+1}} - f_{k_{j-1}}$.

**Crowding selection operator. Crowded tournament selection** operator is defined as follows. A solution $i$ wins tournament with another solution $j$ if any one of the following is true:

– Solution $i$ has better rank, *i.e.*, $r_i < r_j$.
– Both the solutions are in the same front, *i.e.*, $r_i = r_j$, but solution $i$ is less densely located in the search space, *i.e.*, $d_i > d_j$.

**The NSGA-II.** The NSGA-II algorithm in its modified form for handling high dimensional data is presented in Section 3.4.

# 3    Application to Feature Selection

We describe here the reduct generation *i.e.*, *feature selection* procedure, incorporating initial redundancy reduction, in single and multi-objective framework using rough sets. We focus our analysis to two-class problems.

## 3.1    Feature Selection Using Rough Sets

It is a process that selects a minimum subset of $M$ features from an original set of $N$ features $(M \leq N)$, so that the feature space is optimally reduced according to an evaluation criterion. Finding the best feature subset is often intractable or NP-hard. Feature selection typically involves the following steps:

– Subset generation: For $N$ features, the total number of candidate subsets is $2^N$. This makes an exhaustive search through the feature space infeasible, even with moderate value of $N$. Often heuristic and non-deterministic strategies are found to be more practical.
– Subset evaluation: Each generated subset needs to be evaluated by a criterion, and compared with the previous best subset.
– Stopping criterion: The algorithm may stop when either of the following holds.
  • A predefined number of features are selected,
  • a predefined number of iterations are completed,

- when addition or deletion of any feature does not produce a better subset, or
- an optimal subset is obtained according to the evaluation criterion.
- Validation: The selected best feature subset needs to be validated with different tests.

Search is a key issue in feature selection, involving search starting point, search direction, and search strategy. One also needs to measure the goodness of the generated feature subset. Feature selection can be supervised as well as unsupervised, depending on class information availability in data. The algorithms are typically categorized under filter and wrapper models [12], with different emphasis on dimensionality reduction or accuracy enhancement.

The essential properties of reduct are

- to classify among all elements of the universe with the same accuracy as the starting attribute (feature) set, and
- to be of small cardinality.

A close observation reveals that these two characteristics are of a conflicting nature. Hence the determination of reducts is better represented as a bi-objective problem and [13] investigates the multi-objective feature selection criteria for classification of cancer microarray data.

## 3.2 Single Objective Feature Selection Approach

Here we will discuss algorithms proposed by Wroblewski [3]. Wroblewski has proposed three heuristic based approaches for finding minimal reducts based on classical GAs, and permutation-based greedy approaches.

**Method 1.** This method uses greedy algorithms to generate the reducts. Here the aim is to find the proper order of attributes as:

**Step 1 :** Generate an initial set of random permutations of attributes $\tau(a_1, \ldots, a_N)$, each of them representing an ordered list of attributes, i.e., $(b_1, \ldots, b_N) = \tau(a_1, \ldots, a_N)$.
**Step 2 :** For each ordered list, start with empty reduct $R = \phi$ and set $i \leftarrow 0$.
**Step 3 :** Check whether $R$ is reduct. If $R$ is reduct, Stop.
**Step 4 :** Else, add one more element from the ordered list of attributes, i.e. define $R := R \cup b_{i+1}$.
**Step 5 :** Go to step 3.

The result of this algorithm will be either a reduct or a set of attributes containing a reduct as a subset. GAs help to find the reducts of different order.

**Method 2.** This method again uses greedy algorithms to generate the reducts. We can describe this method as follows:

**Step 1 :** Generate an initial set of random permutations of attributes $\tau(a_1, \ldots, a_N)$, each of them representing an ordered list of attributes, i.e., $(b_1, \ldots, b_N) = \tau(a_1, \ldots, a_N)$.

**Step 2 :** For each ordered list, define reduct $R$ as whole set of attributes.

**Step 3 :** Set $i \leftarrow 1$ and let $R := R - b_i$.

**Step 4 :** Check whether R is reduct. If it is not, then undo step 3 and $i \leftarrow i + 1$. Go back to step 3.

All genetic operators are chosen as in method 2. The result of this algorithm will always be a reduct, the proof of which is discussed in [3]. However, a disadvantage of this method is its high complexity.

**Method 3.** Solutions are represented by binary strings of length $N$, where $N$ is number of attributes (features). In the bit representation '1' means that the attribute is present and '0' means that it is not. The following fitness function is considered for each individual:

$$F_1(\nu) = \frac{N - L_\nu}{N} + \frac{C_\nu}{(m^2 - m)/2}, \tag{5}$$

where $\nu$ is reduct candidate, $N$ is number of available attributes, $L_\nu$ is number of 1's in $\nu$, $C_\nu$ is number of object combinations that $\nu$ can discern and $m$ is number of objects.

First part of the fitness function gives the candidate credit for containing less attributes (few 1's) and the second part of function determines the extent to which the candidate can discern among objects.

## 3.3   Using MOGA

The algorithms reported in literature, e.g. in [3,4], vary more or less in defining the fitness function, and typically use combined single objective functions. In order to optimize the pair of conflicting requirements, the fitness function of [3] was split in a two-objective GA setting. We use these two objective functions in the present work in a modified form.

Accordingly, two **fitness functions** $f_1$ and $f_2$ are considered for each individual. We have

$$f_1(\nu) = \frac{N - L_\nu}{N} \tag{6}$$

and

$$f_2(\nu) = \frac{C_\nu}{(m^2 - m)/2}, \tag{7}$$

where $\nu$ is the reduct candidate, $L_\nu$ represents the number of 1's in $\nu$, $m$ is the number of objects, and $C_\nu$ indicates the number of object combinations $\nu$ can discern between. The fitness function $f_1$ gives the candidate credit for containing less attributes (fewer 1's), while the function $f_2$ determines the extent to which the candidate can discern among objects.

Thus, by generating a reduct we are focusing on that minimal set of attributes which can essentially distinguish between all patterns in the given set. In this manner, a reduct is mathematically more meaningful as the most appropriate set of non-redundant features selected from a high-dimensional data.

## 3.4   The Algorithm

NSGA-II is modified in Ref. [13] to effectively handle large datasets. Since we are interested in inter-class distinction, the fitness function of eqn. (7) is modified as

$$f_2(\boldsymbol{\nu}) = \frac{C_\nu}{m_1 * m_2},\tag{8}$$

where $m_1$ and $m_2$ are the number of objects in the two classes. The basic steps of the proposed algorithm are summarized as follows.

1. Preprocessing for redundancy reduction is made for the high-dimensional microarray data [13], to get the reduced attribute value table $\mathcal{A}_r$.
2. $d$-distinction table is generated from $\mathcal{A}_r$ for the two classes being discerned.
3. A random population of size $n$ is generated.
4. The two fitness values $f_1$, $f_2$, for each individual, are calculated using eqns. (6), (8).
5. Non-Domination Sorting is done as discussed in Section 2.2, to identify different fronts.
6. Crowding sort based on crowding distance is performed to get a wide spread of solutions.
7. Offspring solution of size $n$ is created using *fitness* tournament selection, crossover and mutation operators. This is a modification of crowded tournament selection of Section 2.2, with $f_1$ being accorded a higher priority over $f_2$ during solution selection from the same front. Specifically, for $r_i = r_j$ we favour solution $i$ if $f_{1_i} > f_{1_j}$ (instead of $d_i > d_j$).
8. Select the best populations of size $\frac{n}{2}$ each from both the parent and offspring solutions, based on non-dominated sorting, to generate a combined population of size $n$. This modification enables effective handling of larger population sizes in case of large datasets, along with computational gain.
9. *Steps 4 to 7* are **repeated** for a pre-specified number of generations.

## 3.5   Experimental Results

Results for feature selection provided algorithm on benchmark microarray data, like *Colon*[1], *Lymphoma*[2], and *Leukemia*[3] cancer samples. Availability of literature about performance of other related algorithms on these datasets, summarized in Table 1, prompted us to select them for our study.

---

[1] http://microarray.princeton.edu/oncology
[2] http://llmpp.nih.gov/lymphoma/data/figure1/figure1.cdt
[3] http://www.genome.wi.mit.edu/MPR

**Table 1.** Usage details of the two-class microarray data

| Data used | # attributes | Classes | # samples | No. of attributes by | | |
|---|---|---|---|---|---|---|
| | | | | preprocessing | GA | MOGA |
| Colon | 2000 | Colon cancer | 40 | | | |
| | | Normal | 22 | 1102 | 15 | 9 |
| Lymphoma | 4026 | Other type | 54 | | | |
| | | B-cell lymphoma | 42 | 1867 | 18 | 2 |
| Leukemia | 7129 | ALL | 47 | | | |
| | | AML | 25 | 3783 | 19 | 2 |

Reduct generation with a single-objective (classical) GA [3] was investigated for different population sizes. The fitness function

$$F_t = \alpha_1 f_1(\boldsymbol{\nu}) + \alpha_2 f_2(\boldsymbol{\nu}) \tag{9}$$

was used, in terms of eqns. (6) and (8), with the parameters $\alpha_1 = \alpha_2 = 1$. Additionally, we investigated with $0 < \alpha_1, \alpha_2 < 1$ for $\alpha_1 = 1 - \alpha_2$. Sample results are provided in Table 1 for population size of 100, with optimal values being generated for $\alpha_1 = 0.9$. It is observed that, for different choices of $\alpha_1$ and $\alpha_2$, the size of the minimal reduct was 15, 18, 19 for *Colon, Lymphoma, Leukemia* data respectively.

The MOGA of Section 3.4 is run on the $d$-distinction table by using the fitness functions of eqns. (6)-(8), with different population sizes, to generate reducts upon convergence. Here the two fitness functions $f_1(\boldsymbol{\nu})$ and $f_2(\boldsymbol{\nu})$ offset each other, such that the priority accorded to $f_1(\boldsymbol{\nu})$ in Step 7 of the proposed algorithm of Section 3.4 allows weaker reducts (with less than 100% discrimination on object pairs from the $d$-distinction table) to appear in the best front. Some results are also provided in Table 1 on the three sets of two-class microarray gene expression data, after 15,000 generations. The corresponding recognition scores (%) (on test set) by the powerful $k$-nearest neighbors ($k$-NN) classifier [14], for different values of $k$, are presented in Fig. 2.

## 4  Application to Clustering

In this section we describe the different partitive algorithms used for clustering, like $c$-means, fuzzy $c$-means, rough $c$-means, and rough-fuzzy $c$-means. Our objective is to contrast these algorithms while underlining the commonalities existing between them. Let us consider $N$ samples, $c$ clusters, with cardinality $|c_i|$ for $i$th cluster $U_i$. Here $\mathbf{v}_i$ is the $i$th prototype, $m$ the fuzzifier, $\mathbf{x}_k$ the $k$th sample or pattern at a distance $d_{ik}$ from $\mathbf{v}_i$ with membership $u_{ik}$ in $U_i$. $\underline{B}U_i$, $\overline{B}U_i$ indicate lower and upper approximations of $U_i$, with $w_{low}$, $w_{up}$ being the importance or weight of lower and upper approximations. $S(U_i)$ is the within-cluster distance of $U_i$, and $d(U_i, U_j)$ is the between-cluster separation among $U_i$ and $U_j$.

### Recognition score (%)
### MOGA vs SGA (Colon data)

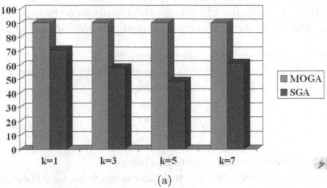

(a)

## Recognition score (%)
## MOGA vs SGA (Lymphoma data)

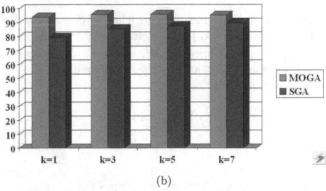

(b)

## Recognition score (%)
## MOGA vs SGA (Leukemia data)

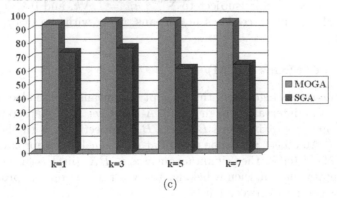

(c)

**Fig. 2.** Comparative performance of MOGA and GA using $k$−NN classifier on (a) *Colon*, (b) *Lymphoma*, and (c) *Leukemia* datasets

## 4.1    C-Means Clustering

The algorithm proceeds by partitioning $N$ objects into $c$ nonempty subsets. During each iteration of clustering algorithm, the centroids or means of the clusters are computed. The main steps of the $c$-means algorithm [14] are as follows:

1. Assign initial means $\mathbf{v}_i$ (also called centroids).
2. Assign each data object (pattern) $\mathbf{x}_k$ to the cluster $U_i$ for the closest mean.
3. Compute new mean for each cluster using

$$\mathbf{v}_i = \frac{\sum_{\mathbf{x}_k \in U_i} \mathbf{x}_k}{|c_i|}, \tag{10}$$

   where $|c_i|$ is the number of objects in cluster $U_i$.
4. Iterate Steps 2–3 **until** criterion function converges, *i.e.*, there are no more new assignments of objects.

## 4.2    Fuzzy $C$-Means (FCM)

This is a fuzzification of the $c$-means algorithm, proposed by Bezdek [8]. It partitions a set of $N$ patterns $\{\mathbf{x}_k\}$ into $c$ clusters by minimizing the objective function $J = \sum_{k=1}^{N} \sum_{i=1}^{c} (u_{ik})^m ||\mathbf{x}_k - \mathbf{v}_i||^2$, where $1 \leq m < \infty$ is the fuzzifier, $\mathbf{v}_i$ is the $i$th cluster center, $u_{ik} \in [0,1]$ is the membership of the $k$th pattern to it, and $||.||$ is the distance, such that

$$\mathbf{v}_i = \frac{\sum_{k=1}^{N} (u_{ik})^m \mathbf{x}_k}{\sum_{k=1}^{N} (u_{ik})^m} \tag{11}$$

and

$$u_{ik} = \frac{1}{\sum_{j=1}^{c} \left(\frac{d_{ik}}{d_{jk}}\right)^{\frac{2}{m-1}}}, \tag{12}$$

$\forall i$, with $d_{ik} = ||\mathbf{x}_k - \mathbf{v}_i||^2$, subject to $\sum_{i=1}^{c} u_{ik} = 1$, $\forall k$, and $0 < \sum_{k=1}^{N} u_{ik} < N$, $\forall i$. The algorithm proceeds as in $c$-means, along with the incorporation of membership.

## 4.3    Rough $C$-Means (RCM)

In the rough $c$-means algorithm, the concept of $c$-means is extended by viewing each cluster as an interval or rough set [9]. A rough set $X$ is characterized by its lower and upper approximations $\underline{B}X$ and $\overline{B}X$ respectively, with the following properties. (i) An object $\mathbf{x}_k$ can be part of at most *one* lower approximation. (ii) If $\mathbf{x}_k \in \underline{B}X$ of cluster $X$, then simultaneously $\mathbf{x}_k \in \overline{B}X$. (iii) If $\mathbf{x}_k$ is not a part of any lower approximation, then it belongs to two or more upper approximations. This permits overlaps between clusters.

Computation of the cluster prototypes is modified in the rough framework, by incorporating the concepts of upper and lower approximations. The right hand side of eqn. (10) is split into two parts. The centroid $\mathbf{v}_i$ of cluster $U_i$ is evaluated as $\mathbf{v}_i =$

$$
\begin{cases}
w_{low}\dfrac{\sum_{\mathbf{x}_k \in \underline{B}U_i}\mathbf{x}_k}{|\underline{B}U_i|} \\
\quad + w_{up}\dfrac{\sum_{\mathbf{x}_k \in (\overline{B}U_i - \underline{B}U_i)}\mathbf{x}_k}{|\overline{B}U_i - \underline{B}U_i|} & \text{if } \underline{B}U_i \neq \emptyset \wedge \overline{B}U_i - \underline{B}U_i \neq \emptyset, \\
\dfrac{\sum_{\mathbf{x}_k \in (\overline{B}U_i - \underline{B}U_i)}\mathbf{x}_k}{|\overline{B}U_i - \underline{B}U_i|} & \text{if } \underline{B}U_i = \emptyset \wedge \overline{B}U_i - \underline{B}U_i \neq \emptyset, \\
\dfrac{\sum_{\mathbf{x}_k \in \underline{B}U_i}\mathbf{x}_k}{|\underline{B}U_i|} & \text{otherwise,}
\end{cases}
\tag{13}
$$

where the parameters $w_{low}$ and $w_{up}$ correspond to the relative importance of the lower and upper approximations respectively. Here $|\overline{B}U_i - \underline{B}U_i|$ is the number of patterns in the rough boundary lying between the two approximations. RCM is found to generate three types of clusters, such as those having objects (i) in both the lower and upper approximations, (ii) only in lower approximation, and (iii) only in upper approximation. Thereby the three cases of eqn. (13) need to be considered while computing the cluster prototypes.

We now explain the condition under which an object may belong to the lower or upper bound of a cluster. Let $\mathbf{x}_k$ be an object at distance $d_{ik}$ from centroid $\mathbf{v}_i$ of cluster $U_i$. The actual algorithm is outlined as follows.

1. Assign initial means $\mathbf{v}_i$ for the $c$ clusters.
2. Assign each data object (pattern) $\mathbf{x}_k$ to the lower approximation $\underline{B}U_i$ or upper approximation $\overline{B}U_i$, $\overline{B}U_j$ of cluster pairs $U_i$, $U_j$ by computing the difference in its distance $d_{ik} - d_{jk}$ from cluster centroid pairs $\mathbf{v}_i$ and $\mathbf{v}_j$.
3. Let $d_{ik}$ be minimum and $d_{jk}$ be the next to minimum.
   **If** $d_{jk} - d_{ik}$ is less than some *threshold*
   **then** $\mathbf{x}_k \in \overline{B}U_i$ and $\mathbf{x}_k \in \overline{B}U_j$ and $\mathbf{x}_k$ cannot be a member of any lower approximation [*Property (iii)*],
   **else** $\mathbf{x}_k \in \underline{B}U_i$ such that distance $d_{ik}$ is minimum over the $c$ clusters [*Property (ii)*].
4. Compute new mean for each cluster $U_i$ using eqn. (13).
5. **Repeat** Steps 2-4 **until** convergence, *i.e.*, there are no more new assignments of objects.

The expression in eqn. (13) boils down to eqn. (10) when the lower approximation is equal to the upper approximation, implying an empty boundary region. It is observed that the performance of the algorithm is dependent on the choice of $w_{low}$, $w_{up}$ and *threshold*. We allowed $w_{up} = 1 - w_{low}$, $0.5 < w_{low} < 1$ and $0 < threshold < 0.5$.

## 4.4   Rough-Fuzzy $C$-Means (RFCM)

A new rough-fuzzy $c$-means algorithm has been designed [15]. This allows one to incorporate fuzzy membership value $u_{ik}$ of a sample $\mathbf{x}_k$ to a cluster mean $\mathbf{v}_i$, relative to all other means $\mathbf{v}_j \ \forall \ j \neq i$, instead of the absolute individual distance $d_{ik}$

from the centroid. This sort of relativistic measure, in terms of eqns. (11)-(12), enhances the robustness of the clustering with respect to different choices of parameters. The major steps of the algorithm are provided below.

1. Assign initial means $\mathbf{v}_i$ for the $c$ clusters.
2. Compute $u_{ik}$ by eqn. (12) for $c$ clusters and $N$ data objects.
3. Assign each data object (pattern) $\mathbf{x}_k$ to the lower approximation $\underline{B}U_i$ or upper approximation $\overline{B}U_i$, $\overline{B}U_j$ of cluster pairs $U_i$, $U_j$ by computing the difference in its membership $u_{ik} - u_{jk}$ to cluster centroid pairs $\mathbf{v}_i$ and $\mathbf{v}_j$.
4. Let $u_{ik}$ be maximum and $u_{jk}$ be the next to maximum.
   **If** $u_{ik} - u_{jk}$ is less than some *threshold*
      **then** $\mathbf{x}_k \in \overline{B}U_i$ and $\mathbf{x}_k \in \overline{B}U_j$ and $\mathbf{x}_k$ cannot be a member of any lower approximation,
      **else** $\mathbf{x}_k \in \underline{B}U_i$ such that membership $u_{ik}$ is maximum over the $c$ clusters.
5. Compute new mean for each cluster $U_i$, incorporating eqns. (11)-(12) into eqn. (13), as $\mathbf{v}_i =$

$$
\begin{cases}
w_{low} \dfrac{\sum_{\mathbf{x}_k \in \underline{B}U_i} u_{ik}^m \mathbf{x}_k}{\sum_{\mathbf{x}_k \in \underline{B}U_i} u_{ik}^m} + w_{up} \dfrac{\sum_{\mathbf{x}_k \in (\overline{B}U_i - \underline{B}U_i)} u_{ik}^m \mathbf{x}_k}{\sum_{\mathbf{x}_k \in (\overline{B}U_i - \underline{B}U_i)} u_{ik}^m} \\
\qquad\qquad\qquad \text{if } \underline{B}U_i \neq \emptyset \wedge \overline{B}U_i - \underline{B}U_i \neq \emptyset, \\[4pt]
\dfrac{\sum_{\mathbf{x}_k \in (\overline{B}U_i - \underline{B}U_i)} u_{ik}^m \mathbf{x}_k}{\sum_{\mathbf{x}_k \in (\overline{B}U_i - \underline{B}U_i)} u_{ik}^m} \quad \text{if } \underline{B}U_i = \emptyset \wedge \overline{B}U_i - \underline{B}U_i \neq \emptyset, \\[4pt]
\dfrac{\sum_{\mathbf{x}_k \in \underline{B}U_i} u_{ik}^m \mathbf{x}_k}{\sum_{\mathbf{x}_k \in \underline{B}U_i} u_{ik}^m} \qquad\qquad\qquad\quad \text{otherwise.}
\end{cases}
\tag{14}
$$

6. **Repeat** Steps 2-5 **until** convergence, *i.e.*, there are no more new assignments.

As indicated earlier, we use $w_{up} = 1 - w_{low}$, $0.5 < w_{low} < 1$, $m = 2$, and $0 < threshold < 0.5$.

An optimal selection of these parameters is an issue of reasonable interest. Typically experimental investigation is done for different combinations. GAs have been used for tuning the parameters *threshold* and $w_{low}$, while minimizing a fitness function based on clustering validity index, for generating an optimal number of clusters [16].

## 4.5    Collaborative RCM and RFCM Clustering

In this section we introduce a collaborative rough and rough-fuzzy $c$-means clustering by incorporating collaboration between different partitions or sub-populations.

Let a dataset be divided into $P$ sub-populations or modules. Each sub-population is independently clustered to reveal its structure. Collaboration is incorporated by exchanging information between the modules regarding the local partitions, in terms of the collection of prototypes computed within the individual modules. This sort of divide-and-conquer strategy enables efficient mining of large databases. The required communication links are hence at a higher level of abstraction, thereby representing information granules (rough or rough-fuzzy clusters) in terms of their prototypes.

The higher the value of the *threshold*, the larger is the number of samples in the boundary regions of the rough-fuzzy clusters. This leads to a stronger collaboration between the prototypes of different modules, resulting in the movement of the prototypes of corresponding clusters (from different modules) towards each other. Often this is eventually followed by a merger of the corresponding prototypes, and hence clusters. This implies that the cluster prototypes from different modules influence and approach each other, due to the collaboration existing mainly in the overlapping (or boundary) regions of the corresponding clusters. The impact of the collaboration on the ensemble of modules is expressed in terms of the changes occurring in the prototypes of the individual clusters. Since the modules correspond to partitions from the same large dataset, this sort of collaborative clustering stabilizes the ensemble towards efficient determination of a globally existent structure.

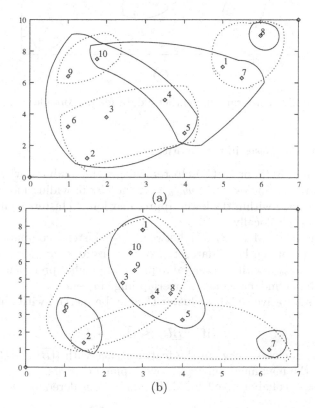

**Fig. 3.** Collaborative clustering on synthetic data for (a) Module A and (b) Module B, with RFCM

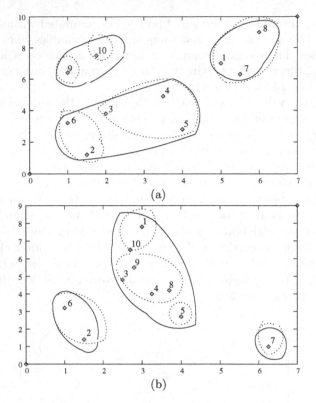

**Fig. 4.** Collaborative clustering on synthetic data for (a) Module A and (b) Module B, with RCM

There exist two phases in the algorithm.

- Generation of RCM or RFCM clusters within the modules, without collaboration. Here we employ $0.5 < w_{low} < 1$, thereby providing more importance to samples lying within the lower approximation of clusters while computing their prototypes locally.
- Collaborative RCM or RFCM between the clusters, computed locally for each module of the large dataset. Now we use $0 < w_{low} < 0.5$ (we chose $w_{low} = 1 - w_{low}$), with a lower value providing higher precedence to samples lying in the boundary region of overlapping clusters.
  - In collaborative RCM, a cluster $U_i$ may be merged with an overlapping cluster $U_j$

$$\text{if} \quad |\underline{B}U_i| \leq |\overline{B}U_i - \underline{B}U_i| \tag{15}$$

and $\mathbf{v}_i$ is closest to $\mathbf{v}_j$ in the feature space with $(|\overline{B}U_i - \underline{B}U_i| - |\underline{B}U_i|)$ being the maximum among all overlapping clusters.
  - In case of collaborative RFCM, $U_i$ can be considered for merging with $U_j$

$$\text{if} \quad \sum_{\mathbf{x}_k \in \underline{B}U_i} u_{ik} \leq \sum_{\mathbf{x}_k \in (\overline{B}U_i - \underline{B}U_i)} u_{ik} \tag{16}$$

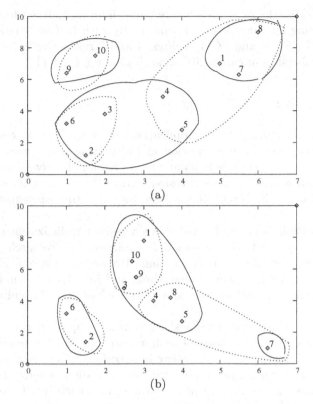

**Fig. 5.** Collaborative clustering on synthetic data for (a) Module A and (b) Module B, with FCM

and $\mathbf{v}_i$ is closest to $\mathbf{v}_j$ in the feature space with $(\sum_{\mathbf{x}_k \in (\overline{B}U_i - \underline{B}U_i)} u_{ik} - \sum_{\mathbf{x}_k \in \underline{B}U_i} u_{ik})$ being the maximum among all overlapping clusters.

Collaboration is done by exchanging cluster prototypes between modules, leading to a global determination of the overall structure within the data.

Let there be $c_1$ and $c_2$ clusters, generated by RCM or RFCM, in a pair of modules $(P = 2)$ under consideration. During collaboration, we begin with $c_1 + c_2$ cluster prototypes and merge using eqns. (15) or (16) respectively.

### 4.6   Experimental Results

Results are provided on a small synthetic dataset [15], [10], with the distance being represented in terms of the traditional Euclidean metric for numeric features.

The two-dimensional synthetic dataset [10], [15], before and after collaboration, are shown in Figs. 3–5. There are two modules (A and B) corresponding to ten samples each, partitioned into three clusters. Sample results using $threshold = 0.2$, $w_{low} = 0.9$ for RFCM and RCM, and $w_{low} = 1 - 0.9 = 0.1$ for collaborative RFCM and RCM, are provided.

Fig. 3 indicates the clustering before (solid line) and after (dotted line) collaboration, using modules A and B respectively with RFCM. Analogous results are depicted for RCM and FCM, in Figs. 4 and 5 respectively. Note that there exists no membership in case of RCM, such that $u_{ik} \in \{0, 1\}$.

## 5    Conclusion

We have described some recent applications of rough sets to certain pattern recognition tasks like feature selection and clustering. The evolutionary-rough feature selection [13], using redundancy reduction for effective handling of high-dimensional microarray gene expression data, serves as an interesting study in Bioinformatics. The NSGA-II has been modified to more effectively handle large data.

Microarray Bioinformatics has aided in a massive parallelization of experimental biology [17], and the associated explosion of research has led to astonishing progress in our understanding of molecular biology. Future hybrid approaches, combining powerful algorithms and interactive visualization tools with the strengths of fast processors, hold further promise for enhanced performance.

Collaborative clustering [10,15] is a promising approach towards modeling agent-based systems. A multi-agent system is one in which a number of agents cooperate and interact with each other in a complex and distributed environment, thereby achieving a global objective based on distributed data and control [18]. While handling large data in this framework, each intelligent agent may concentrate on information discovery (or clustering) within a module. Subsequently these agents can communicate with each other at the cluster interface, using appropriate protocol, their cluster profiles represented in terms of the centroids. Thereby, an agent can refine the partitioning within its own module by collaborating with the other agents.

## References

1. Pawlak, Z.: Rough Sets, Theoretical Aspects of Reasoning about Data. Kluwer Academic, Dordrecht (1991)
2. Skowron, A., Rauszer, C.: The discernibility matrices and functions in information systems. In Slowiński, R., ed.: Intelligent Decision Support, Handbook of Applications and Advances of the Rough Sets Theory. Kluwer Academic, Dordrecht (1992) 331–362
3. Wroblewski, J.: Finding minimal reducts using genetic algorithms. Technical Report 16/95, Warsaw Institute of Technology - Institute of Computer Science, Poland (1995)
4. Bjorvand, A.T.: 'Rough Enough' – A system supporting the rough sets approach. In: Proceedings of the Sixth Scandinavian Conference on Artificial Intelligence, Helsinki, Finland (1997) 290–291
5. Deb, K.: Multi-Objective Optimization using Evolutionary Algorithms. John Wiley, London (2001)

6. Zadeh, L.A.: Fuzzy logic, neural networks, and soft computing. Communications of the ACM **37** (1994) 77–84
7. Mitra, S., Acharya, T.: Data Mining: Multimedia, Soft Computing, and Bioinformatics. John Wiley, New York (2003)
8. Bezdek, J.C.: Pattern Recognition with Fuzzy Objective Function Algorithms. Plenum Press, New York (1981)
9. Lingras, P., West, C.: Interval set clustering of Web users with rough $k$-means. Technical Report No. 2002-002, Dept. of Mathematics and Computer Science, St. Mary's University, Halifax, Canada (2002)
10. Pedrycz, W.: Collaborative fuzzy clustering. Pattern Recognition Letters **23** (2002) 1675–1686
11. Deb, K., Agarwal, S., Pratap, A., Meyarivan, T.: A fast and elitist multi-objective genetic algorithm : NSGA-II. IEEE Transactions on Evolutionary Computation **6** (2002) 182–197
12. Yu, L., Liu, H.: Efficient feature selection via analysis of relevance and redundancy. Journal of Machine Learning Research **5** (2004) 1205–1224
13. Banerjee, M., Mitra, S., Banka, H.: Evolutionary-rough feature selection in gene expression data. IEEE Transactions on Systems, Man, and Cybernetics, Part C: Applications and Reviews (to appear)
14. Tou, J.T., Gonzalez, R.C.: Pattern Recognition Principles. Addison-Wesley, London (1974)
15. Mitra, S., Banka, H., Pedrycz, W.: Rough-fuzzy collaborative clustering. IEEE Transactions on Systems, Man, and Cybernetics, Part B: Cybernetics **36** (2006) 795–805
16. Mitra, S.: An evolutionary rough partitive clustering. Pattern Recognition Letters **25** (2004) 1439–1449
17. : Special Issue on Bioinformatics. IEEE Computer **35** (2002)
18. Ferber, J.: Multi-Agent Systems: An Introduction to Distributed Artificial Intelligence. Addison-Wesley Longman, New York (1999)

# Lower and Upper Approximations in Data Tables Containing Possibilistic Information

Michinori Nakata[1] and Hiroshi Sakai[2]

[1] Faculty of Management and Information Science,
Josai International University
1 Gumyo, Togane, Chiba, 283-8555, Japan
nakatam@ieee.org
[2] Department of Mathematics and Computer Aided Sciences,
Faculty of Engineering, Kyushu Institute of Technology,
Tobata, Kitakyushu, 804-8550, Japan
sakai@mns.kyutech.ac.jp

**Abstract.** An extended method of rough sets, called a method of weighted equivalence classes, is applied to a data table containing imprecise values expressed in a possibility distribution. An indiscerniblity degree between objects is calculated. A family of weighted equivalence classes is obtained via indiscernible classes from a binary relation for indiscernibility between objects. Each equivalence class in the family is accompanied by a possibilistic degree to which it is an actual one. By using the family of weighted equivalence classes we derive a lower approximation and an upper approximation. These approximations coincide with those obtained from methods of possible worlds. Therefore, the method of weighted equivalence classes is justified.

**Keywords:** Rough sets, Imprecise value, Correctness criterion, Weighted equivalence class, Lower and upper approximations.

## 1 Introduction

Rough sets proposed by Pawlak [27,29,30,31] play a significant role in the field of knowledge discovery and data mining. The framework of rough sets has the premise that data tables consisting of perfect information are obtained. However, there ubiquitously exists imperfect information containing imprecision and uncertainty in the real world [26]. Under these circumstances, it has been investigated to apply rough sets to data tables containing imprecise values represented by a missing value, an or-set, a possibility distribution, etc [5,6,7,8,9,10,11,14,15,16,17,18,19,20,21,22,23,24,33,34,35,36,37,38,39,40].

The methods are broadly separated into three ways. The first method is one based on possible worlds, which is called a method of possible worlds [25,33,34], [35,36]. In the method, a data table is divided into possible tables that consist of crisp values. Each possible table is dealt with by the conventional method of

J.F. Peters et al. (Eds.): Transactions on Rough Sets VII, LNCS 4400, pp. 170–189, 2007.

rough sets and then the results from possible tables are aggregated. The second method is to use assumptions on indiscernibility of missing values [5,6,9,11,14,15], [16,17,18,19,38,40]. Under the assumptions, we can obtain a binary relation for indiscernibility between objects. To the binary relation the methods of rough sets are applied using indiscernible classes. The third method directly deals with imprecise values under extending the conventional method of rough sets [20,21,22,23,24,38,39,40]. In the method, imprecise values are handled probabilistically[1] or possibilistically and the conventional method is extended probabilistically or possibilistically [20,21,22,23,24,38,39,40]. A degree for indiscernibility between any objects is calculated.

For the first method, the conventional method that is already established is applied to each possible table. Therefore, there is no doubt for correctness of the treatment. However, the method has difficulties for knowledge discovery at the level of a set of possible values, although it is suitable for finding knowledge at the level of possible values. This is because the number of possible tables exponentially increases as the number of imprecise attribute values increases.

For the second method, assumptions are used for indiscernibility of missing values. One assumption is that a missing value and an exact value are indiscernible with each other [14,15,16,17]. Another assumption is that indiscernibility is directional [5,6,38,40]. Each missing value is discernible with any exact values, whereas each exact value is indiscernible with any missing value, under indiscernibility or discernibility between missing values. In the method, it is not clarified why the assumptions are compromise to real data tables.

We focus on the third method. We adopt results from the method of possible worlds as a correctness criterion, when extended methods of rough sets are applied to a data table containing imprecise values [20,21,22]. This kind of criterion is commonly applied to query expressions in the field of databases handling imprecise information [1,2,3,4,12,13,41]. The correctness criterion under applying extended methods of rough sets to a data table is as follows:

**Correctness criterion**
*Results obtained from applying an extended method to a data table containing imprecise values are the same as ones obtained from applying the conventional method to every possible table derived from that data table and aggregating the results created in the possible tables.*

This is formulated as follows:

*Suppose that operator rep creates extended set rep(t) of possible tables derived from data table t containing imprecise values. Let $q'$ be the conventional method applied to rep(t), where $q'$ corresponds to extended method $q$ directly applied to data table t. The two results is the same; namely,*

$$q(t) = q'(rep(t)).$$

---

[1] Ras and Joshi propose a query answering system that handles imprecise values probabilistically [32].

This condition is schematized in Figure 1.

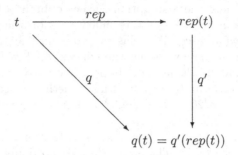

$$q(t) = q'(rep(t))$$

**Fig. 1.** Correctness criterion of extended method $q$

When this condition is valid, extended method $q$ gives correct results at the level of possible values. This correctness criterion is checked as follows:

- Derive the extended set of possible tables from a data table containing imprecise values.
- Apply the conventional method to each possible table.
- Aggregate the results obtained from possible tables.
- Apply the extended method to the original data table.
- Compare the aggregated results with ones obtained from the extended method.

The third method should be checked from the viewpoint of the correctness criterion. The first approach of the third method was attempted by Stefanowski and Tsoukiàs [38,39,40]. They calculate an inclusion degree between indiscernible classes for objects by using implication operators under handling missing values probabilistically. Their results do not satisfy the correctness criterion [21]. Recently, Nakata and Sakai has proposed the method that satisfies the correctness criterion when missing values are interpreted probabilistically [21,22]. However, the method has difficulties for definability, because approximations are defined by constructing sets from singletons. To overcome the difficulties, Nakata and Sakai have proposed a method of applying weighted equivalence classes to a data table containing probabilistic information or missing values interpreted probabilistically [23,24]. In this paper, we show that the method of weighted equivalence classes satisfies the correctness criterion, when a data table contains imprecise values expressed in a possibility distribution.

In section 2, we briefly address the conventional method of applying rough sets to a data table consisting of precise values. In section 3, methods of possible worlds are mentioned. This is the preparation for checking whether the method of weighted equivalence classes satisfies the correctness criterion. In section 4, the method of weighted equivalence classes is applied to a data table containing imprecise values expressed in a possibility distribution. The last section presents conclusions.

## 2    Applying Rough Sets to Precise Information

Data table $t$ having set $\mathcal{A}(=\{A_1,\ldots,A_n\})$ of attributes consists of set $U$ of objects. Binary relation $IND(\Psi_X)$ for indiscernibility of objects in subset $\Psi \subseteq U$ on subset $X \subseteq \mathcal{A}$ of attributes is:

$$IND(\Psi_X) = \{(o,o') \in \Psi \times \Psi \mid \forall A_i \in X \ \ o[A_i] = o'[A_i]\}, \tag{1}$$

where $o[A_i]$ and $o'[A_i]$ denote values of attribute $A_i$ for objects $o$ and $o'$, respectively. This relation is called an indiscernibility relation. Obviously, $IND(\Psi_X)$ is an equivalence relation. Family $\mathcal{E}(\Psi_X)$ $(=\{E(\Psi_X)_o \mid o \in \Psi\})$ of equivalence classes is obtained from the binary relation, where $E(\Psi_X)_o$ is the equivalence class containing object $o$ and is expressed in $E(\Psi_X)_o = \{o' \mid (o,o') \in IND(\Psi_X)\}$. All equivalence classes obtained from the indiscernibility relation do not cover with each other. This means that the objects are uniquely partitioned.

**Example 1**
We suppose that the following table is obtained:

| O | A | B |
|---|---|---|
| 1 | $x$ | $a$ |
| 2 | $y$ | $b$ |
| 3 | $x$ | $b$ |
| 4 | $x$ | $a$ |
| 5 | $z$ | $b$ |
| 6 | $y$ | $b$ |

The mark $O$ denotes the object identity and $U = \{o_1, o_2, o_3, o_4, o_5, o_6\}$. For set $\Psi(= \{o_2, o_3, o_4, o_5, o_6\})$ of objects, we obtain binary relation $IND(\Psi_A)$ for indiscernibility on attribute $A$:

$$IND(\Psi_A) = \{(o_2, o_2), (o_2, o_6), (o_3, o_3), (o_3, o_4), (o_4, o_4), (o_5, o_5), (o_6, o_6)\}.$$

From $IND(\Psi_A)$ we have family $\mathcal{E}(\Psi_A)$ of equivalence classes:

$$\mathcal{E}(\Psi_A) = \{\{o_3, o_4\}, \{o_2, o_6\}, \{o_5\}\}.$$

Similarly, for $\Phi = \{o_1, o_2, o_3, o_4, o_6\}$,

$$\mathcal{E}(\Phi_B) = \{\{o_1, o_4\}, \{o_2, o_3, o_6\}\}.$$

Using equivalence classes, lower approximation $\underline{Apr}(\Phi_Y, \Psi_X)$ and upper approximation $\overline{Apr}(\Phi_Y, \Psi_X)$ of $\mathcal{E}(\Phi_Y)$ by $\mathcal{E}(\Psi_X)$ are:

$$\underline{Apr}(\Phi_Y, \Psi_X) = \{E(\Psi_X) \mid \exists E(\Phi_Y) \ E(\Psi_X) \subseteq E(\Phi_Y)\}, \tag{2}$$

$$\overline{Apr}(\Phi_Y, \Psi_X) = \{E(\Psi_X) \mid \exists E(\Phi_Y) \ E(\Psi_X) \cap E(\Phi_Y) \neq \emptyset\}, \tag{3}$$

where $E(\Psi_X) \in \mathcal{E}(\Psi_X)$ and $E(\Phi_Y) \in \mathcal{E}(\Phi_Y)$ are equivalence classes on sets X and Y of attributes, respectively. These formulas are expressed in terms of a family of equivalence classes. When we express the approximations in terms of a set of objects, the following expressions are used:

$$\underline{apr}(\Phi_Y, \Psi_X) = \{o \mid o \in E(\Psi_X) \wedge \exists E(\Phi_Y)\ E(\Psi_X) \subseteq E(\Phi_Y)\}, \tag{4}$$

$$\overline{apr}(\Phi_Y, \Psi_X) = \{o \mid o \in E(\Psi_X) \wedge \exists E(\Phi_Y)\ E(\Psi_X) \cap E(\Phi_Y) \neq \emptyset\}. \tag{5}$$

**Example 2**
We check equivalence classes comprising families $\mathcal{E}(\Psi_A)$ and $\mathcal{E}(\Phi_B)$ in Example 1. For inclusion and intersection between equivalence classes, $\{o_2, o_6\} \subset \{o_2, o_3, o_6\}$, $\{o_3, o_4\} \not\subseteq \{o_1, o_4\}$, $\{o_3, o_4\} \not\subseteq \{o_2, o_3, o_6\}$, $\{o_3, o_4\} \cap \{o_1, o_4\} \neq \emptyset$, $\{o_5\} \cap \{o_1, o_4\} = \emptyset$, and $\{o_5\} \cap \{o_2, o_3, o_6\} = \emptyset$. Thus, for the lower approximation and the upper approximation,

$$\underline{Apr}(\Phi_B, \Psi_A) = \{\{o_2, o_6\}\},$$

$$\overline{Apr}(\Phi_B, \Psi_A) = \{\{o_2, o_6\}, \{o_3, o_4\}\}.$$

For the expressions in terms of a set of objects,

$$\underline{apr}(\Phi_B, \Psi_A) = \{o_2, o_6\},$$

$$\overline{apr}(\Phi_B, \Psi_A) = \{o_2, o_3, o_4, o_6\}.$$

## 3    Methods of Possible Worlds

In methods of possible worlds, data table $t$ is divided into extended set $rep(t)$ of possible tables, and then the conventional method $q'$ addressed in the previous section is applied to each possible table, and finally the results from possible tables are aggregated. When $m$ imprecise values expressed in a normal possibility distribution are contained in data table $t$, extended set $rep(t)$ of possible tables that is derived from $t$ is:

$$rep(t) = \{(t_1, \mu(t_1)), \ldots, (t_n, \mu(t_n))\}, \tag{6}$$

where $\mu(t_i)$ denotes the possibilistic degree to which possible table $t_i$ is the actual one, $n$ is equal to $\Pi_{j=1,m} l_j$, and each imprecise value is expressed in a possibility distribution having $l_j (j = 1, m))$ elements. We suppose that values from imprecise attribute values in $(t_i, \mu(t_i))$ are expressed in terms of $v_{i1}$, $v_{i2}$, $\ldots$, $v_{im}$. Possibilistic degree $\pi(v_{ik})$ of element $v_{ik}$ comes from the possibility distribution of the imprecise attribute value whose support has $v_{ik}$ as an element. Then,

$$\mu(t_i) = \min_{k=1,m} \pi(v_{ik}). \tag{7}$$

## Example 3

We suppose that data table $t$ is obtained:

$$t$$

| $O$ | $A$ | $B$ |
|---|---|---|
| 1 | $x$ | $a$ |
| 2 | $y$ | $\{(a, 0.2), (b, 1)\}_p$ |
| 3 | $\{(x, 1), (y, 0.7)\}_p$ | $b$ |
| 4 | $\{(x, 1), (y, 0.4)\}_p$ | $a$ |

$U = \{o_1, o_2, o_3, o_4\}$ and subscript $p$ of $\{(x, 1), (y, 1)\}_p$ denotes a possibility distribution. Extended set $rep(t)$ of possible tables is:

$$rep(t) = \{(t_1, \mu(t_1)), \cdots, (t_8, \mu(t_8))\},$$

where $t_i$ and $\mu(t_i)$ are:

| | $t_1$ | | | $t_2$ | | | $t_3$ | | | $t_4$ | | | $t_5$ | | | $t_6$ | | | $t_7$ | | | $t_8$ | |
|---|---|---|---|---|---|---|---|---|---|---|---|---|---|---|---|---|---|---|---|---|---|---|---|
| $O$ | $A$ | $B$ | $O$ | $A$ | $B$ | $O$ | $A$ | $B$ | $O$ | $A$ | $B$ | $O$ | $A$ | $B$ | $O$ | $A$ | $B$ | $O$ | $A$ | $B$ | $O$ | $A$ | $B$ |
| 1 | $x$ | $a$ | 1 | $x$ | $a$ | 1 | $x$ | $a$ | 1 | $x$ | $a$ | 1 | $x$ | $a$ | 1 | $x$ | $a$ | 1 | $x$ | $a$ | 1 | $x$ | $a$ |
| 2 | $y$ | $a$ | 2 | $y$ | $b$ | 2 | $y$ | $a$ | 2 | $y$ | $b$ | 2 | $y$ | $a$ | 2 | $y$ | $b$ | 2 | $y$ | $a$ | 2 | $y$ | $b$ |
| 3 | $x$ | $b$ | 3 | $x$ | $b$ | 3 | $x$ | $b$ | 3 | $x$ | $b$ | 3 | $y$ | $b$ | 3 | $y$ | $b$ | 3 | $y$ | $b$ | 3 | $y$ | $b$ |
| 4 | $x$ | $a$ | 4 | $x$ | $a$ | 4 | $y$ | $a$ | 4 | $y$ | $a$ | 4 | $x$ | $a$ | 4 | $x$ | $a$ | 4 | $y$ | $a$ | 4 | $y$ | $a$ |

$$\mu(t_1) = \min(1, 1, 0.2) = 0.2,$$
$$\mu(t_2) = \min(1, 1, 1) = 1,$$
$$\mu(t_3) = \min(1, 0.4, 0.2) = 0.2,$$
$$\mu(t_4) = \min(1, 0.4, 1) = 0.4,$$
$$\mu(t_5) = \min(0.7, 1, 0.2) = 0.2,$$
$$\mu(t_6) = \min(0.7, 1, 1) = 0.7,$$
$$\mu(t_7) = \min(0.7, 0.4, 0.2) = 0.2,$$
$$\mu(t_8) = \min(0.7, 0.4, 1) = 0.4.$$

The conventional method is applied to each possible table, which consists of crisp values. The family of equivalence classes on set $X$ of attributes is obtained from possible table $t_i$. These equivalence classes are a possible equivalence class on set $X$ of attributes and have possibilistic degree $\mu(t_i)$ to which they are an actual equivalence class. Thus, the family of possible equivalence classes accompanied by a possibilistic degree is obtained for each possible table.

## Example 4

Binary relations $IND(A)_{t_1}$ and $IND(B)_{t_1}$[2] for indiscernibility on attributes $A$ and $B$ in possible table $t_1$ of Example 3 are:

$$IND(A)_{t_1} = \{(o_1, o_3), (o_1, o_4), (o_3, o_4), (o_1, o_1), (o_2, o_2), (o_3, o_3), (o_4, o_4)\},$$
$$IND(B)_{t_1} = \{(o_1, o_2), (o_1, o_4), (o_2, o_4), (o_1, o_1), (o_2, o_2), (o_3, o_3), (o_4, o_4)\}.$$

---

[2] Attributes $A$ and $B$ are used in place of $U_A$ and $U_B$ for $U$.

Families of equivalence classes for attributes $A$ and $B$ are:

$$\mathcal{E}(A)_{t_1} = \{\{o_1, o_3, o_4\}, \{o_2\}\},$$
$$\mathcal{E}(B)_{t_1} = \{\{o_1, o_2, o_4\}, \{o_3\}\}.$$

These equivalence classes have possibilistic degree $0.2(= \mu(t_1))$ to which they are actual ones. Thus, families of possible equivalence classes accompanied by a possibilistic degree for attributes $A$ and $B$ are:

$$\{((\{o_1, o_3, o_4\}, 0.2), (\{o_2\}, 0.2)\},$$
$$\{((\{o_1, o_2, o_4\}, 0.2), (\{o_3\}, 0.2)\}.$$

The conventional method addressed in the previous section is applied to each possible table. Let $\underline{Apr(Y, X)}_{t_i}$ and $\overline{Apr(Y, X)}_{t_i}$ denote the lower approximation and the upper approximation of $\mathcal{E}(Y)_{t_i}$ by $\mathcal{E}(X)_{t_i}$ in possible table $t_i$ having possibilistic degree $\mu(t_i)$. Possibilistic degree $\kappa(E(X) \in \underline{Apr(Y, X)})_{t_i}$ to which equivalence class $E(X)$ is contained in $\underline{Apr(Y, X)}$ for possible table $t_i$ is obtained:

$$\kappa(E(X) \in \underline{Apr(Y, X)})_{t_i} = \begin{cases} \mu(t_i) & \text{if } E(X) \in \underline{Apr(Y, X)}_{t_i}, \\ 0 & \text{otherwise.} \end{cases} \tag{8}$$

This shows that the possibilistic degree to which equivalence class $E(X)$ is contained in $\underline{Apr(Y, X)}$ is equal to $\mu(t_i)$ for possible table $t_i$, if the equivalence class is an element in $\underline{Apr(Y, X)}_{t_i}$. Similarly, possibilistic degree $\kappa(E(X) \in \overline{Apr(Y, X)})_{t_i}$ to which equivalence class $E(X)$ is contained in $\overline{Apr(Y, X)}$ for possible table $t_i$ is obtained:

$$\kappa(E(X) \in \overline{Apr(Y, X)})_{t_i} = \begin{cases} \mu(t_i) & \text{if } E(X) \in \overline{Apr(Y, X)}_{t_i}, \\ 0 & \text{otherwise.} \end{cases} \tag{9}$$

The results from possible tables are aggregated. This is done by the union of the approximations obtained from possible tables. Note that the maximum possibilistic degree is adopted in the union if there exists the same equivalence class accompanied by a different possibilistic degree. Therefore, possibilistic degrees $\kappa(E(X) \in \underline{Apr(Y, X)})$ and $\kappa(E(X) \in \overline{Apr(Y, X)})$ to which equivalence class $E(X)$ is contained in $\underline{Apr(Y, X)}$ and $\overline{Apr(Y, X)}$ are:

$$\kappa(E(X) \in \underline{Apr(Y, X)}) = \max_{i=1,n} \kappa(E(X) \in \underline{Apr(Y, X)})_{t_i}, \tag{10}$$

$$\kappa(E(X) \in \overline{Apr(Y, X)}) = \max_{i=1,n} \kappa(E(X) \in \overline{Apr(Y, X)})_{t_i}. \tag{11}$$

Finally,

$$\underline{Apr(Y, X)} = \{(E(X), \kappa(E(X) \in \underline{Apr(Y, X)})) \mid \kappa(E(X) \in \underline{Apr(Y, X)}) > 0\}, \tag{12}$$

$$\overline{Apr(Y, X)} = \{(E(X), \kappa(E(X) \in \overline{Apr(Y, X)})) \mid \kappa(E(X) \in \overline{Apr(Y, X)}) > 0\}. \tag{13}$$

These approximations are $q'(rep(t))$ in the correctness criterion.

**Proposition 1**
We suppose that $(E(X), \kappa(E(X) \in \underline{Apr(Y, X)}))$ is an element of $\underline{Apr(Y, X)}$ and $(E(X), \kappa(E(X) \in \overline{Apr(Y, X)}))$ is an element of $\overline{Apr(Y, X)}$ in table $t$. There exist possible tables $t_i$ and $t_j$ where $\underline{Apr(Y, X)}_{t_i}$ contains $E(X)$ and $\mu(t_i)$ is equal to $\kappa(E(X) \in \underline{Apr(Y, X)})$ and where $\overline{Apr(Y, X)}_{t_j}$ contains $E(X)$ and $\mu(t_j)$ is equal to $\kappa(E(X) \in \overline{Apr(Y, X)})$.

*Proof*
We focus on the lower approximation. From formula (10), if $E(X) \in \underline{Apr(Y, X)}$, there are possible tables $t_i$ where $\kappa(E(X) \in \underline{Apr(Y, X)})_{t_i} > 0$. This means $\underline{Apr(Y, X)}_{t_i}$ contains $E(X)$. Then, $\kappa(E(X) \in \overline{Apr(Y, X)})_{t_i}$ is equal to $\mu(t_i)$ from formula (8). The maximum of these $\mu(t_i)$ is $\kappa(E(X) \in \underline{Apr(Y, X)})$. Thus, there exists possible table $t_i$ where $\underline{Apr(Y, X)}_{t_i}$ contains $E(X)$ and $\mu(t_i)$ is equal to $\kappa(E(X) \in \underline{Apr(Y, X)})$. The proof for the upper approximation is similar to that of the lower approximation.

When the lower approximation and the upper approximation are expressed in terms of a set of objects,

$$\underline{apr(Y, X)} = \{(o, \kappa(o \in \underline{apr(Y, X)})) \mid \kappa(o \in \underline{apr(Y, X)}) > 0\}, \quad (14)$$
$$\kappa(o \in \underline{apr(Y, X)}) = \max_{E(X) \ni o} \kappa(E(X) \in \underline{Apr(Y, X)}), \quad (15)$$

$$\overline{apr(Y, X)} = \{(o, \kappa(o \in \overline{apr(Y, X)})) \mid \kappa(o \in \overline{apr(Y, X)}) > 0\}, \quad (16)$$
$$\kappa(o \in \overline{apr(Y, X)}) = \max_{E(X) \ni o} \kappa(E(X) \in \overline{Apr(Y, X)}). \quad (17)$$

The most crucial factor in the computational complexity of the method of possible worlds is the number of possible tables. When a data table has $m$ imprecise values expressed in a possibility distribution whose support consists of $l$ elements averagely, the number of possible tables is $O(l^m)$.

**Example 5**
For inclusion and intersection of possible equivalence classes on attributes $A$ and $B$ of possible table $t_1$ in Example 4.

$$\{o_1, o_3, o_4\} \not\subseteq \{o_1, o_2, o_4\},$$
$$\{o_1, o_3, o_4\} \not\subseteq \{o_3\},$$
$$\{o_2\} \subseteq \{o_1, o_2, o_4\},$$
$$\{o_1, o_3, o_4\} \cap \{o_1, o_2, o_4\} \neq \emptyset,$$
$$\{o_2\} \cap \{o_1, o_2, o_4\} \neq \emptyset.$$

Thus,

$$\underline{Apr(B, A)}_{t_1} = \{o_2\},$$
$$\overline{Apr(B, A)}_{t_1} = \{\{o_2\}, \{o_1, o_3, o_4\}\},$$

and $\kappa(\{o_2\} \in \underline{Apr}(B,A))_{t_1}$, $\kappa(\{o_1, o_3, o_4\} \in \overline{Apr}(Y,X))_{t_1}$, and $\kappa(\{o_2\} \in \overline{Apr}(B,A))_{t_1}$ are equal to possibilistic degree 0.2 that $t_1$ has. Similarly for the other possible tables,

$$\underline{Apr}(B,A)_{t_2} = \{o_2\},$$

$$\overline{Apr}(B,A)_{t_2} = \{\{o_2\}, \{o_1, o_3, o_4\}\},$$

and $\kappa(\{o_2\} \in \underline{Apr}(B,A))_{t_2}$, $\kappa(\{o_1, o_3, o_4\} \in \overline{Apr}(B,A))_{t_2}$, and $\kappa(\{o_2\} \in \overline{Apr}(B,A))_{t_2}$ are equal to 1;

$$\underline{Apr}(B,A)_{t_3} = \{o_2, o_4\},$$

$$\overline{Apr}(B,A)_{t_3} = \{\{o_1, o_3\}, \{o_2, o_4\}\},$$

and $\kappa(\{o_2, o_4\} \in \underline{Apr}(B,A))_{t_3}$, $\kappa(\{o_1, o_3\} \in \overline{Apr}(B,A))_{t_3}$, and $\kappa(\{o_2, o_4\} \in \overline{Apr}(B,A))_{t_3}$ are equal to 0.2;

$$\underline{Apr}(B,A)_{t_4} = \emptyset,$$

$$\overline{Apr}(B,A)_{t_4} = \{\{o_1, o_3\}, \{o_2, o_4\}\},$$

and $\kappa(\{o_1, o_3\} \in \overline{Apr}(B,A))_{t_4}$ and $\kappa(\{o_2, o_4\} \in \overline{Apr}(B,A))_{t_4}$ are equal to 0.4;

$$\underline{Apr}(B,A)_{t_5} = \{o_1, o_4\},$$

$$\overline{Apr}(B,A)_{t_5} = \{\{o_1, o_4\}, \{o_2, o_3\}\},$$

and $\kappa(\{o_1, o_4\} \in \underline{Apr}(B,A))_{t_5}$, $\kappa(\{o_1, o_4\} \in \overline{Apr}(B,A))_{t_5}$, and $\kappa(\{o_2, o_3\} \in \overline{Apr}(B,A))_{t_5}$ are equal to 0.2;

$$\underline{Apr}(B,A)_{t_6} = \{\{o_1, o_4\}, \{o_2, o_3\}\},$$

$$\overline{Apr}(B,A)_{t_6} = \{\{o_1, o_4\}, \{o_2, o_3\}\},$$

and $\kappa(\{o_1, o_4\} \in \underline{Apr}(B,A))_{t_6}$, $\kappa(\{o_2, o_3\} \in \underline{Apr}(B,A))_{t_6}$, $\kappa(\{o_1, o_4\} \in \overline{Apr}(B,A))_{t_6}$, and $\kappa(\{o_2, o_3\} \in \overline{Apr}(B,A))_{t_6}$ are equal to 0.7;

$$\underline{Apr}(B,A)_{t_7} = \{o_1\},$$

$$\overline{Apr}(B,A)_{t_7} = \{\{o_1\}, \{o_2, o_3, o_4\}\},$$

and $\kappa(\{o_1\} \in \underline{Apr}(B,A))_{t_7}$, $\kappa(\{o_1\} \in \overline{Apr}(B,A))_{t_7}$, and $\kappa(\{o_2, o_3, o_4\} \in \overline{Apr}(B,A))_{t_7}$ are equal to 0.2;

$$\underline{Apr}(B,A)_{t_8} = \{o_1\},$$

$$\overline{Apr}(B,A)_{t_8} = \{\{o_1\}, \{o_2, o_3, o_4\}\},$$

and $\kappa(\{o_1\} \in \underline{Apr(B, A)})_{t_8}$, $\kappa(\{o_1\} \in \overline{Apr(B, A)})_{t_8}$, and $\kappa(\{o_2, o_3, o_4\} \in \overline{Apr(B, A)})_{t_8}$ are equal to $0.4$.

We aggregate the results obtained from possible tables. The union of the results from possible tables is made. Thus, the possible equivalence classes that satisfy $\kappa(E(A) \in \underline{Apr(B, A)}) > 0$ are $\{o_1\}$, $\{o_2\}$, $\{o_1, o_4\}$, $\{o_2, o_3\}$, and $\{o_2, o_4\}$. The possible equivalence classes that satisfy $\kappa(E(A) \in \overline{Apr(B, A)}) > 0$ are $\{o_1\}$, $\{o_2\}$, $\{o_1, o_3\}$, $\{o_1, o_4\}$, $\{o_2, o_3\}$, $\{o_2, o_4\}$, $\{o_1, o_3, o_4\}$, $\{o_2, o_3, o_4\}$. For lower approximation $\underline{Apr(B, A)}$,

$$\kappa(\{o_1\} \in \underline{Apr(B, A)}) = \max(0, 0, 0, 0, 0, 0, 0.2, 0.4) = 0.4,$$
$$\kappa(\{o_2\} \in \underline{Apr(B, A)}) = \max(0.2, 1, 0, 0, 0, 0, 0, 0) = 1,$$
$$\kappa(\{o_1, o_4\} \in \underline{Apr(B, A)}) = \max(0, 0, 0, 0, 0.2, 0.7, 0, 0) = 0.7,$$
$$\kappa(\{o_2, o_3\} \in \underline{Apr(B, A)}) = \max(0, 0, 0, 0, 0, 0.7, 0, 0) = 0.7,$$
$$\kappa(\{o_2, o_4\} \in \underline{Apr(B, A)}) = \max(0, 0, 0.2, 0, 0, 0, 0, 0) = 0.2.$$

Finally,

$$\underline{Apr(B, A)} = \{(\{o_1\}, 0.4), (\{o_2\}, 1), (\{o_1, o_4\}, 0.7), (\{o_2, o_3\}, 0.7), (\{o_2, o_4\}, 0.2)\}.$$

Similarly, for upper approximation $\overline{Apr(B, A)}$,

$$\overline{Apr(B, A)} = \{(\{o_1\}, 0.4), (\{o_2\}, 1), (\{o_1, o_3\}, 0.4), (\{o_1, o_4\}, 0.7), (\{o_2, o_3\}, 0.7),$$
$$(\{o_2, o_4\}, 0.4), (\{o_1, o_3, o_4\}, 1), (\{o_2, o_3, o_4\}, 0.4)\}.$$

These approximations are $q'(rep(t))$ in the correctness criterion.

When the lower approximation is expressed in terms of a set of objects,

$$\kappa(o_1 \in \underline{apr(B, A)}) = \max(0.4, 0.7) = 0.7,$$
$$\kappa(o_2 \in \underline{apr(B, A)}) = \max(1, 0.7, 0.2) = 1,$$
$$\kappa(o_3 \in \underline{apr(B, A)}) = 0.7,$$
$$\kappa(o_4 \in \underline{apr(B, A)}) = \max(0.7, 0.2) = 0.7.$$

Similarly for the upper approximation,

$$\kappa(o_1 \in \overline{apr(B, A)}) = 1,$$
$$\kappa(o_2 \in \overline{apr(B, A)}) = 1,$$
$$\kappa(o_3 \in \overline{apr(B, A)}) = 1,$$
$$\kappa(o_4 \in \overline{apr(B, A)}) = 1.$$

Thus,

$$\underline{apr(B, A)} = \{(o_1, 0.7), (o_2, 1), (o_3, 0.7), (o_4, 0.7)\},$$
$$\overline{apr(B, A)} = \{(o_1, 1), (o_2, 1), (o_3, 1), (o_4, 1)\}.$$

# 4  Applying Methods of Weighted Equivalence Classes to Possibilistic Information

When a data table contains imprecise values for attributes, we calculate the degree to which two objects are equal in the attribute values. The degree is the indiscernibility degree of the two objects on the attributes. In this case, a binary relation for indiscernibility is,

$$IND(X) = \{((o, o'), \kappa(o[X] = o'[X])) \mid (\kappa(o[X] = o'[X]) \neq 0) \wedge (o \neq o')\}$$
$$\cup \{((o, o), 1)\}, \quad (18)$$

where $\kappa(o[X] = o'[X])$ denotes the indiscernibility degree of objects $o$ and $o'$ on set $X$ of attributes and is equal to $\kappa((o, o') \in IND(X))$,

$$\kappa(o[X] = o'[X]) = \bigotimes_{A_i \in X} \kappa(o[A_i] = o'[A_i]), \quad (19)$$

where operator $\bigotimes$ depends on properties of imprecise attribute values. When the imprecise attribute values are expressed in a possibility distribution, the operator is min.

$$\kappa(o[A_i] = o'[A_i]) = \max_{u \in dom(A_i)} \min(\pi_{o[A_i]}(u), \pi_{o'[A_i]}(u)), \quad (20)$$

where $dom(A_i)$ denotes the domain of attribute $A_i$ and $\pi_{o[A_i]}(u)$ and $\pi_{o'[A_i]}(u)$ denote possibility distributions expressing attribute values $o[A_i]$ and $o'[A_i]$, respectively.

**Example 6**
In table $t$ of Example 3, binary relation $IND(A)$ for indiscernibility on attribute $A$ is obtained:

$$IND(A) = \{((o_1, o_3), 1), ((o_1, o_4), 1), ((o_2, o_3), 0.7), ((o_2, o_4), 0.4),$$
$$((o_3, o_4), 1), ((o_1, o_1), 1), ((o_2, o_2), 1), ((o_3, o_3), 1), ((o_4, o_4), 1)\}.$$

From binary relation $IND(X)$ for indiscernibility, family $\mathcal{E}(X)$ of weighted equivalence classes is obtained via indiscernible classes. Indiscernible class $S(X)_o$ consisting of objects that are paired with object $o$ among the elements of $IND(X)$ is:

$$S(X)_o = \{o' \mid \kappa((o, o') \in IND(X)) > 0\}. \quad (21)$$

$S(X)_o$ is the greatest possible equivalence class among the possible equivalence classes containing objects $o$. Let $PS(X)_o$ denote the power set of $S(X)_o$. From $PS(X)_o$, family $Poss\mathcal{E}(X)_o$ of possible equivalence classes containing object $o$ is obtained:

$$Poss\mathcal{E}(X)_o = \{E(X) \mid E(X) \in PS(X)_o \wedge o \in E(X)\}. \quad (22)$$

Whole family $Poss\mathcal{E}(X)$ of possible equivalence classes is:

$$Poss\mathcal{E}(X) = \cup_o Poss\mathcal{E}(X)_o. \tag{23}$$

Possibilistic degree $\kappa(E(X) \in \mathcal{E}(X))$ to which possible equivalence class $E(X) \in Poss\mathcal{E}(X)$ is an actual one is:

$$\kappa(E(X) \in \mathcal{E}(X)) = \kappa(\wedge_{o \in E(X) \; and \; o' \in E(X)}(o[X] = o'[X])$$
$$\wedge_{o \in E(X) \; and \; o' \notin E(X)}(o[X] \neq o'[X])), \tag{24}$$

where $o \neq o'$, $\kappa(f)$ is the possibilistic degree to which formula $f$ is satisfied, and $\kappa(f) = 1$ when there exists no $f$. Now, imprecise values are expressed in a normal possibility distribution. Suppose that a data table and $E(X)$ consist of $k$ objects and $l$ objects, respectively. The above possibilistic degree is calculated as follows:

$$\kappa(E(X) \in \mathcal{E}(X)) = \max_{u, v_1, \cdots, v_{k-l}} \min(\min_{o \in E(X)} \pi_{o[X]}(u), \min_{i=1,k-l} \pi_{o'_i[X]}(v_i)), \tag{25}$$

where $o'_i \notin E(X)$, $X = \{A_1, \cdots, A_m\}$, $u \in dom(X)(= dom(A_1) \times \cdots \times dom(A_m))$, $v_i \in dom(X)$ and $v_i \neq u$ for all $i$. Finally,

$$\mathcal{E}(X) = \{(E(X), \kappa(E(X) \in \mathcal{E}(X))) \mid \kappa(E(X) \in \mathcal{E}(X)) > 0\}. \tag{26}$$

**Proposition 2**
When $(E(X), \kappa(E(X) \in \mathcal{E}(X)))$ is an element of $\mathcal{E}(X)$ in data table $t$, there exists possible table $t_i$ where $\mathcal{E}(X)_{t_i}$ contains $E(X)$ and $\mu(t_i)$ is equal to $\kappa(E(X) \in \mathcal{E}(X))$.

*Proof*
Suppose that $u$ and $v_i$ are expressed in $(u_1, \cdots, u_m)$ and $(v_{i1}, \cdots, v_{im})$, respectively. $\min(\min_{o \in E(X)} \pi_{o[X]}(u), \min_{i=1,k-l} \pi_{o'_i[X]}(v_i))$ for $u, v_1, \cdots, v_{k-l}$ in formula (25) is equal to the possibilistic degree to which each object $o \in E(X)$ takes the same possible value $u_j$ and each object $o'_i \notin E(X)$ takes possible value $v_{ij}$ as the value of attribute $A_j$ for $j = 1, m$. The possibilistic degree is equal to that of the possible table where each object $o \in E(X)$ takes the same possible value $u_j$ and each object $o'_i \notin E(X)$ takes possible value $v_{ij}$ as the value of attribute $A_j$ for $j = 1, m$ and all objects take a possible value with the maximum degree 1 for the attributes except $X$. Thus, there exists possible table $t_i$ where $\mathcal{E}(X)_{t_i}$ contains $E(X)$ and $\mu(t_i)$ is equal to $\kappa(E(X) \in \mathcal{E}(X))$.

**Proposition 3**
$\mathcal{E}(X)$ in a data table is equal to the union of the families of possible equivalence classes accompanied by a possibilistic degree, where each family of possible equivalence classes is obtained from a possible table created from the data table.

*Proof*
From Proposition 2, if there exists $(E(X), \kappa(E(X) \in \mathcal{E}(X)))$, there exist possible tables having the family of possible equivalence classes containing $E(X)$. From formula (25), $\kappa(E(X) \in \mathcal{E}(X))$ is equal to the maximum degree that $E(X)$ has among the possible tables. The maximum degree is taken as the possibilistic degree when the same equivalence class accompanied by a different possibilistic degree is obtained in the union. So, $(E(X), \kappa(E(X) \in \mathcal{E}(X)))$ is equal to one obtained from the union of the families of possible equivalence classes.

**Proposition 4**
For any object $o$,

$$\max_{E(X) \ni o} \kappa(E(X) \in \mathcal{E}(X)) = 1. \tag{27}$$

*Proof*
Every imprecise value is expressed in a normal possibility distribution. So, there is a possible table where all imprecise values are replaced by an element having the maximum possibilistic degree 1 for any data table. Each object belongs to either of the equivalence classes obtained from the possible table. Thus, the above formula holds.

**Example 7**
We use the binary relation for indiscernibility in Example 6. Each indiscernible class for object $o_i$, the greatest possible equivalence class containing object $o_i$, is, respectively:

$$S(A)_{o_1} = \{o_1, o_3, o_4\},$$
$$S(A)_{o_2} = \{o_2, o_3, o_4\},$$
$$S(A)_{o_3} = \{o_1, o_2, o_3, o_4\},$$
$$S(A)_{o_4} = \{o_1, o_3, o_3, o_4\}.$$

Each power set of these sets is, respectively:

$$PS(A)_{o_1} = \{\emptyset, \{o_1\}, \{o_3\}, \{o_4\}, \{o_1, o_3\}, \{o_1, o_4\}, \{o_3, o_4\}, \{o_1, o_3, o_4\}\},$$
$$PS(A)_{o_2} = \{\emptyset, \{o_2\}, \{o_3\}, \{o_4\}, \{o_2, o_3\}, \{o_2, o_4\}, \{o_3, o_4\}, \{o_2, o_3, o_4\}\},$$
$$PS(A)_{o_3} = PS(A)_{o_4}$$
$$= \{\emptyset, \{o_1\}, \{o_2\}, \{o_3\}, \{o_4\}, \{o_1, o_2\}, \{o_1, o_3\}, \{o_1, o_4\}, \{o_2, o_3\},$$
$$\{o_2, o_4\}, \{o_3, o_4\}, \{o_1, o_2, o_3\}, \{o_1, o_2, o_4\}, \{o_1, o_3, o_4\}, \{o_2, o_3, o_4\},$$
$$\{o_1, o_2, o_3, o_4\}\}.$$

Each family of possible equivalence classes containing object $o_i$ is respectively:

$$Poss\mathcal{E}(A)_{o_1} = \{\{o_1\}, \{o_1, o_3\}, \{o_1, o_4\}, \{o_1, o_3, o_4\}\},$$
$$Poss\mathcal{E}(A)_{o_2} = \{\{o_2\}, \{o_2, o_3\}, \{o_2, o_4\}, \{o_2, o_3, o_4\}\},$$

$$Poss\mathcal{E}(A)_{o_3} = \{\{o_3\}, \{o_1, o_3\}, \{o_2, o_3\}, \{o_3, o_4\}, \{o_1, o_2, o_3\}, \{o_1, o_3, o_4\},$$
$$\{o_2, o_3, o_4\}, \{o_1, o_2, o_3, o_4\}\},$$
$$Poss\mathcal{E}(A)_{o_4} = \{\{o_4\}, \{o_1, o_4\}, \{o_2, o_4\}, \{o_3, o_4\}, \{o_1, o_2, o_4\}, \{o_1, o_3, o_4\},$$
$$\{o_2, o_3, o_4\}, \{o_1, o_2, o_3, o_4\}\}.$$

The whole family of possible equivalence classes is:

$$Poss\mathcal{E}(A) = \{\{o_1\}, \{o_2\}, \{o_3\}, \{o_4\}, \{o_1, o_2\}, \{o_1, o_3\}, \{o_1, o_4\}, \{o_2, o_3\}, \{o_2, o_4\},$$
$$\{o_3, o_4\}, \{o_1, o_2, o_3\}, \{o_1, o_2, o_4\}, \{o_1, o_3, o_4\}, \{o_2, o_3, o_4\},$$
$$\{o_1, o_2, o_3, o_4\}\}.$$

The possibilistic degree $\kappa(\{o_1\} \in \mathcal{E}(A))$ to which possible equivalence class $\{o_1\}$ is an actual one is:

$$\kappa(\{o_1\} \in \mathcal{E}(A)) = \kappa((o_1[A] \neq o_2[A]) \wedge (o_1[A] \neq o_3[A]) \wedge (o_1[A] \neq o_4[A]))$$
$$= \min(1, 0.7, 0.4)$$
$$= 0.4.$$

Similarly,

$$\kappa(\{o_2\} \in \mathcal{E}(A)) = 1,$$
$$\kappa(\{o_3\} \in \mathcal{E}(A)) = 0,$$
$$\kappa(\{o_4\} \in \mathcal{E}(A)) = 0,$$
$$\kappa(\{o_1, o_2\} \in \mathcal{E}(A)) = 0,$$
$$\kappa(\{o_1, o_3\} \in \mathcal{E}(A)) = 0.4,$$
$$\kappa(\{o_1, o_4\} \in \mathcal{E}(A)) = 0.7,$$
$$\kappa(\{o_2, o_3\} \in \mathcal{E}(A)) = 0.7,$$
$$\kappa(\{o_2, o_4\} \in \mathcal{E}(A)) = 0.4,$$
$$\kappa(\{o_3, o_4\} \in \mathcal{E}(A)) = 0,$$
$$\kappa(\{o_1, o_2, o_3\} \in \mathcal{E}(A)) = 0,$$
$$\kappa(\{o_1, o_2, o_4\} \in \mathcal{E}(A)) = 0,$$
$$\kappa(\{o_1, o_3, o_4\} \in \mathcal{E}(A)) = 1,$$
$$\kappa(\{o_2, o_3, o_4\} \in \mathcal{E}(A)) = 0.4,$$
$$\kappa(\{o_1, o_2, o_3, o_4\} \in \mathcal{E}(A)) = 0.$$

Thus, the family of weighted equivalence classes on attribute $A$ is:

$$\mathcal{E}(A) = \{(\{o_1\}, 0.4), (\{o_2\}, 1), (\{o_1, o_3\}, 0.4), (\{o_1, o_4\}, 0.7), (\{o_2, o_3\}, 0.7),$$
$$(\{o_2, o_4\}, 0.4), (\{o_1, o_3, o_4\}, 1), (\{o_2, o_3, o_4\}, 0.4)\}.$$

Similarly, the family of weighted equivalence classes on attribute $B$ is:

$$\mathcal{E}(B) = \{(\{o_3\}, 0.2), (\{o_1, o_4\}, 1), (\{o_2, o_3\}, 1), (\{o_1, o_2, o_4\}, 0.2)\}.$$

Using families of weighted equivalence classes, we can obtain lower approximation $\underline{Apr(Y,X)}$ and upper approximation $\overline{Apr(Y,X)}$ of $\mathcal{E}(Y)$ by $\mathcal{E}(X)$. For the lower approximation,

$$\underline{Apr(Y,X)} = \{(E(X), \kappa(E(X) \in \underline{Apr(Y,X)})) \mid \kappa(E(X) \in \underline{Apr(Y,X)}) > 0\}, \quad (28)$$
$$\kappa(E(X) \in \underline{Apr(Y,X)}) = \max_{E(Y)} \min(\kappa(E(X) \subseteq E(Y)),$$

$$\kappa(E(X) \in \mathcal{E}(X)), \kappa(E(Y) \in \mathcal{E}(Y))), \quad (29)$$

where

$$\kappa(E(X) \subseteq E(Y)) = \begin{cases} 1 \text{ if } E(X) \subseteq E(Y), \\ 0 \text{ otherwise.} \end{cases} \quad (30)$$

For the upper approximation,

$$\overline{Apr(Y,X)} = \{(E(X), \kappa(o \in \overline{Apr(Y,X)})) \mid \kappa(E(X) \in \overline{Apr(Y,X)}) > 0\}, \quad (31)$$
$$\kappa(E(X) \in \overline{Apr(Y,X)}) = \max_{E(Y)} \min(\kappa(E(X) \cap E(Y) \neq \emptyset),$$

$$\kappa(E(X) \in \mathcal{E}(X)), \kappa(E(Y) \in \mathcal{E}(Y))), \quad (32)$$

where

$$\kappa(E(X) \cap E(Y) \neq \emptyset) = \begin{cases} 1 \text{ if } E(X) \cap E(Y) \neq \emptyset, \\ 0 \text{ otherwise.} \end{cases} \quad (33)$$

**Proposition 5**
We suppose that $(E(X), \kappa(E(X) \in \underline{Apr(Y,X)}))$ is an element of $\underline{Apr(Y,X)}$ and $(E(X), \kappa(E(X) \in \overline{Apr(Y,X)}))$ is an element of $\overline{Apr(Y,X)}$ in data table $t$. There exist possible tables $t_i$ and $t_j$ where $\underline{Apr(Y,X)}_{t_i}$ contains $E(X)$ and $\mu(t_i)$ is equal to $\kappa(E(X) \in \underline{Apr(Y,X)})$ and where $\overline{Apr(Y,X)}_{t_j}$ contains $E(X)$ and $\mu(t_j)$ is equal to $\kappa(E(X) \in \overline{Apr(Y,X)})$.

*Proof*
We focus on the lower approximation. We suppose to obtain $E(X) \subseteq E(Y)$, $\kappa(E(X) \in \mathcal{E}(X)) > 0$, and $\kappa(E(Y) \in \mathcal{E}(Y)) > 0$ for $E(Y)$ that gives $\kappa(E(X) \in \underline{Apr(Y,X)}) > 0$ in formula (29). From Proposition 2, there is possible tables $t_j$ and $t_k$ accompanied by $\mu(t_j)$ and $\mu(t_k)$ equal to $\kappa(E(X) \in \mathcal{E}(X))$ and $\kappa(E(Y) \in \mathcal{E}(Y))$, respectively. In possible table $t_j$ we suppose that every object $o \in E(X)$ takes the same value $x_j$ and every object $o' \notin E(X)$ takes different value $x'_j$ from $x_j$ for $X$, and similarly in possible table $t_k$ every object $o \in E(Y)$ takes the same value $y_k$ and every object $o' \notin E(Y)$ takes different value $y'_k$ from $y_k$ for $Y$. Clearly, there is possible table $t_i$ where every object $o \in E(X)$ takes the same value $x_j$ and every object $o' \notin E(X)$ takes different value $x'_j$ from $x_j$ for $X$ and every object $o \in E(Y)$ takes the same value $y_k$ and every object $o' \notin E(Y)$ takes different value $y'_k$ from $y_k$ for $Y$. And in possible table $t_i$ $\underline{Apr(Y,X)}_{t_i}$ contains $E(X)$ and possible table $t_i$ is accompanied by possibilistic

degree $\mu(t_i)$ equal to $\min(\mu(t_j), \mu(t_k))(= \kappa(E(X) \in \underline{Apr(Y, X)}))$. The proof for the upper approximation is similar to that of the lower approximation.

## Proposition 6
The lower approximation and the upper approximation that are obtained by the method of weighted equivalence classes coincide with ones obtained by the method of possible worlds.

*Proof*
This is obvious because the lower approximation and the upper approximation that are obtained by the method of weighted equivalence classes are equal to the union of the ones obtained from possible tables, which is easily proved similarly to that of Proposition 3.

This proposition means that the method of weighted equivalence classes satisfies the correctness criterion.

For expressions in terms of a set of objects, the same expressions as in section 3 are used.

The most crucial factor in the computational complexity of the method of weighted equivalence classes is the number of weighted equivalence classes that are accompanied by a non-zero possibilistic degree. When $m$ is the maximum number of imprecise values whose support has non-empty intersection on an attribute, the number of weighted equivalence classes is $O(2^m)$ for the attribute. $m$ is usually much smaller than the total number of imprecise values in a data table. Furthermore, $m$ decreases as possibility distributions are restricted by data cleansing, etc.

## Example 8
Using the families of weighted equivalence classes in Example 7, we derive the lower approximation and the upper approximation of $\mathcal{E}(B)$ by $\mathcal{E}(A)$. For the lower approximation, the possibilistic degree to which equivalence class $\{o_1\}$ is contained in $\underline{Apr(B, A)}$ is:

$$\kappa(\{o_1\} \in \underline{Apr(B, A)}) = \max(\min(1, 0.4, 1), \min(1, 0.4, 0.2)) = 0.4.$$

Similarly,

$$\kappa(\{o_2\} \in \underline{Apr(B, A)}) = 1,$$
$$\kappa(\{o_1, o_3\} \in \underline{Apr(B, A)}) = 0,$$
$$\kappa(\{o_1, o_4\} \in \underline{Apr(B, A)}) = 0.7,$$
$$\kappa(\{o_2, o_3\} \in \underline{Apr(B, A)}) = 0.7,$$
$$\kappa(\{o_2, o_4\} \in \underline{Apr(B, A)}) = 0.2,$$
$$\kappa(\{o_1, o_3, o_4\} \in \underline{Apr(B, A)}) = 0,$$
$$\kappa(\{o_2, o_3, o_4\} \in \underline{Apr(B, A)}) = 0.$$

Thus,

$$\underline{Apr}(B, A) = \{(\{o_1\}, 0.4), (\{o_2\}, 1), (\{o_1, o_4\}, 0.7), (\{o_2, o_3\}, 0.7), (\{o_2, o_4\}, 0.2)\}.$$

From this expression, possibilistic degrees to which each object is contained in the lower approximation are:

$$
\begin{aligned}
\kappa(o_1 \in \underline{apr}(B, A)) &= \max(0.4, 0.7) = 0.7, \\
\kappa(o_2 \in \underline{apr}(B, A)) &= \max(1, 0.7, 0.2) = 1, \\
\kappa(o_3 \in \underline{apr}(B, A)) &= 0.7, \\
\kappa(o_4 \in \underline{apr}(B, A)) &= \max(0.7, 0.2) = 0.7.
\end{aligned}
$$

Thus,

$$\underline{apr}(B, A) = \{(o_1, 0.7), (o_2, 1), (o_3, 0.7), (o_4, 0.7)\}.$$

Similarly, for the upper approximation,

$$
\begin{aligned}
\overline{Apr}(B, A) = \{&(\{o_1\}, 0.4), (\{o_2\}, 1), (\{o_1, o_3\}, 0.4), (\{o_1, o_4\}, 0.7), (\{o_2, o_3\}, 0.7), \\
&(\{o_2, o_4\}, 0.4), (\{o_1, o_3, o_4\}, 1), (\{o_2, o_3, o_4\}, 0.4)\}, \\
\overline{apr}(B, A) = \{&(o_1, 1), (o_2, 1), (o_3, 1), (o_4, 1)\}.
\end{aligned}
$$

Indeed, the lower approximation and the upper approximation coincide with ones obtained from the method of possible worlds in Example 5.

Using families of weighted equivalence classes, we can obtain the lower approximation and the upper approximation for two sets $\Phi$ and $\Psi$. We suppose that families $\mathcal{E}(\Psi)$ and $\mathcal{E}(\Phi)$ of weighted equivalence classes are obtained for sets $\Psi$ and $\Phi$, respectively. Let $(E(\Psi), \kappa(E(\Psi) \in \mathcal{E}(\Psi)))$ denote an element of $\mathcal{E}(\Psi)$ and $(E(\Phi), \kappa(E(\Phi) \in \mathcal{E}(\Phi)))$ denote an element of $\mathcal{E}(\Phi)$. Lower approximation $\underline{Apr}(\Phi, \Psi)$ and upper approximation $\overline{Apr}(\Phi, \Psi)$ of $\mathcal{E}(\Phi)$ by $\mathcal{E}(\Psi)$ are:

$$\underline{Apr}(\Phi, \Psi) = \{(E(\Psi), \kappa(E(\Psi) \in \underline{Apr}(\Phi, \Psi))) \mid \kappa(E(\Psi) \in \underline{Apr}(\Phi, \Psi)) > 0\}, \quad (34)$$
$$\kappa(E(\Psi) \in \underline{Apr}(\Phi, \Psi)) = \max_{E(\Phi)} \min(\kappa(E(\Psi) \subseteq E(\Phi)), \kappa(E(\Psi) \in \mathcal{E}(\Psi)),$$

$$\kappa(E(\Phi) \in \mathcal{E}(\Phi))), \quad (35)$$

where

$$\kappa(E(\Psi) \subseteq E(\Phi)) = \begin{cases} 1 \text{ if } E(\Psi) \subseteq E(\Phi), \\ 0 \text{ otherwise.} \end{cases} \quad (36)$$

$$\overline{Apr}(\Phi, \Psi) = \{(E(\Psi), \kappa(E(\Psi) \in \overline{Apr}(\Phi, \Psi))) \mid \kappa(E(\Psi) \in \overline{Apr}(\Phi, \Psi)) > 0\}, \quad (37)$$
$$\kappa(E(\Psi) \in \overline{Apr}(\Phi, \Psi)) = \max_{E(\Phi)} \min(\kappa(E(\Psi) \cap E(\Phi) \neq \emptyset), \kappa(E(\Psi) \in \mathcal{E}(\Psi)),$$

$$\kappa(E(\Phi) \in \mathcal{E}(\Phi))), \quad (38)$$

where

$$\kappa(E(\Psi) \cap E(\Phi) \neq \emptyset) = \begin{cases} 1 \text{ if } E(\Psi) \cap E(\Phi) \neq \emptyset, \\ 0 \text{ otherwise.} \end{cases} \tag{39}$$

For expressions in terms of a set of objects,

$$\underline{apr(\Phi, \Psi)} = \{(o, \kappa(o \in \underline{apr(\Phi, \Psi)})) \mid \kappa(o \in \underline{apr(\Phi, \Psi)}) > 0\}, \tag{40}$$

$$\kappa(o \in \underline{apr(\Phi, \Psi)}) = \max_{E(\Psi) \ni o} \kappa(E(\Psi) \in \underline{Apr(\Phi, \Psi)}). \tag{41}$$

$$\overline{apr(\Phi, \Psi)} = \{(o, \kappa(o \in \overline{apr(\Phi, \Psi)})) \mid \kappa(o \in \overline{apr(\Phi, \Psi)}) > 0\}, \tag{42}$$

$$\kappa(o \in \overline{apr(\Phi, \Psi)}) = \max_{E(\Psi) \ni o} \kappa(E(\Psi) \in \overline{Apr(\Phi, \Psi)}). \tag{43}$$

## 5  Conclusions

We have described the extended method where weighted equivalence classes are used in order to deal with imprecise values expressed in a possibility distribution. The lower approximation and the upper approximation by the method of weighted equivalence classes coincide with ones by the method of possible worlds. In other words, this method satisfies the correctness criterion that is used in the field of incomplete databases. This is justification of the method of weighted equivalence classes.

**Acknowledgment.** We would express our regret for the death of Professor Zdzislaw Pawlak,@who would show us the research area on rough sets. This research has been partially supported by the Grant-in-Aid for Scientific Research (C), Japan Society for the Promotion of Science, No. 18500214.

## References

1. Abiteboul, S., Hull, R., and Vianu, V. [1995] Foundations of Databases, Addison-Wesley Publishing Company, 1995.
2. Bosc, P., Duval, L., and Pivert, O. [2003] An Initial Approach to the Evaluation of Possibilistic Queries Addressed to Possibilistic Databases, Fuzzy Sets and systems, **140**, 151-166.
3. Bosc, P., Liétard, N., and Pivert, O. [2006] About the Processing of Possibilistic Queries Involving a Difference Operation, Fuzzy Sets and systems, **157**, 1622-1640.
4. Grahne, G. [1991] The Problem of Incomplete Information in Relational Databases, Lecture Notes in Computer Science, **554**, Springer.
5. Greco, S., Matarazzo, B., and Slowinski, R. [1999] Handling Missing Values in Rough Set Analysis of Multi-attribute and Multi-criteria Decision Problem, Lecture Notes in Artificial Intelligence, **1711**, 146-157.
6. Greco, S., Matarazzo, B., and Slowinski, R. [2001] Rough Sets Theory for Multi-criteria Decision Analysis, European Journal of Operational Research, **129**, 1-47.

7. Grzymala-Busse, J. W. [1991] On the Unknown Attribute Values in Learning from Examples, Lecture Notes in Artificial Intelligence **542**, Springer-Verlag, 368-377.
8. Grzymala-Busse, J. W. [2004] Data with Missing Attribute Values: Generalization of Indiscernibility Relation and Rule Induction, Transactions on Rough Sets **1**, 78-95.
9. Grzymala-Busse, J. W. [2005] Characteristic Relations for Incomplete Data: A Generalization of the Indiscernibility Relation, Transactions on Rough Sets **4**, 58-68.
10. Grzymala-Busse, J. W. and Hu, M. [2000] A Comparison of Several Approaches to Missing Attribute Values in Data Mining, Lecture Notes in Artificial Intelligence **2005**, 378-385.
11. Grzymala-Busse, J. W. and Rzasa, W. [2006] Local and Global Approximations for Incomplete Data, Lecture Notes in Artificial Intelligence, **4259**, 244-253.
12. Imielinski, T. [1989] Incomplete Information in Logical Databases, Data Engineering, **12**, 93-104.
13. Imielinski, T. and Lipski, W. [1984] Incomplete Information in Relational Databases, Journal of the ACM, **31**:4, 761-791.
14. Kryszkiewicz, M. [1998] Rough Set Approach to Incomplete Information Systems, Information Sciences, **112**, 39-49.
15. Kryszkiewicz, M. [1998] Properties of Incomplete Information Systems in the framework of Rough Sets, in L. Polkowski and A. Skowron, (ed.), Rough Set in Knowledge Discovery 1: Methodology and Applications, Studies in Fuzziness and Soft Computing 18, Physica Verlag, 422-450.
16. Kryszkiewicz, M. [1999] Rules in Incomplete Information Systems, Information Sciences, **113**, 271-292.
17. Kryszkiewicz, M. and Rybiński, H. [2000] Data Mining in Incomplete Information Systems from Rough Set Perspective, in L. Polkowski, S. Tsumoto, and T. Y. Lin, (eds.), Rough Set Methods and Applications, Studies in Fuzziness and Soft Computing 56, Physica Verlag, 568-580.
18. Latkowski, R. [2003] On Decomposition for Incomplete Data, Fundamenta Informaticae, **54**, 1-16.
19. Latkowski, R. [2005] Flexible Indiscernibility Relations for Missing Values, Fundamenta Informaticae, **67**, 131-147.
20. Nakata, N. and Sakai, H. [2005] Rough-set-based approaches to data containing incomplete information: possibility-based cases, in K. Nakamatsu and J. M. Abe, (eds.), Advances in Logic Based Intelligent Systems, Frontiers in Artificial Intelligence and Applications 132, IOS Press, 234-241.
21. Nakata, N. and Sakai, H. [2005] Checking Whether or Not Rough-Set-Based Methods to Incomplete Data Satisfy a Correctness Criterion, Lecture Notes in Artificial Intelligence, **3558**, 227-239.
22. Nakata, N. and Sakai, H. [2005] Rough Sets Handling Missing Values Probabilistically Interpreted, Lecture Notes in Artificial Intelligence, **3641**, 325-334.
23. Nakata, N. and Sakai, H. [2005] Applying Rough Sets to Data Tables Containing Probabilistic Information, in Proceedings of 4th Workshop on Rough Sets and Kansei Engineering, Tokyo, Japan, pp. 50-53.
24. Nakata, N. and Sakai, H. [2006] Applying Rough Sets to Data Tables Containing Imprecise Information Under Probabilistic Interpretation, Lecture Notes in Artificial Intelligence, **4259**, 213-223.
25. Orłowska, E. and Pawlak, Z. [1984] Representation of Nondeterministic Information, Theoretical Computer Science, **29**, 313-324.

26. Parsons, S. [1996] Current Approaches to Handling Imperfect Information in Data and Knowledge Bases, IEEE Transactions on Knowledge and Data Engineering, **83**, 353-372.
27. Pawlak, Z. [1991] Rough Sets: Theoretical Aspects of Reasoning about Data, Kluwer Academic Publishers 1991.
28. Pawlak, Z. [2004] Some Issues on Rough Sets, Transactions on Rough Sets, **1**, 1-58.
29. Pawlak, Z. and Skowron, A. [2007] Rudiments of Rough Sets, Information Sciences, **177**, 3-27.
30. Pawlak, Z. and Skowron, A. [2007] Rough Sets: Some Extensions, Information Sciences, **177**, 28-40.
31. Pawlak, Z. and Skowron, A. [2007] Rough Sets and Boolean Reasoning, Information Sciences, **177**, 41-73.
32. Ras, Z. W. and Joshi, S. [1997] Query Approximate Answering System for an Incomplete DKBS, Fundamenta Informaticae, **30**, 313-324.
33. Sakai, H. [1998] Some Issues on Nondeterministic Knowledge Bases with Incomplete Information, Lecture Notes in Artificial Intelligence **1424**, 424-431.
34. Sakai, H. [2001] Effective Procedures for Handling Possible Equivalence Relation in Non-deterministic Information Systems, Fundamenta Informaticae, **48**, 343-362.
35. Sakai, H. and Nakata, M. [2006] An Application of Discernibility Functions to Generating Minimal Rules in Non-deterministic Information Systems, Journal of Advanced Computational Intelligence and Intelligent Informatics, **10**, 695-702.
36. Sakai, H. and Okuma, A. [2004] Basic Algorithms and Tools for Rough Non-deterministic Information Systems, Transactions on Rough Sets, **1**, 209-231.
37. Słowiński, R. and Stefanowski, J. [1989] Rough Classification in Incomplete Information Systems, Mathematical and Computer Modelling, **12**:10/11, 1347-1357.
38. Stefanowski, J. and Tsoukiàs, A. [1999] On the Extension of Rough Sets under Incomplete Information, Lecture Notes in Artificial Intelligence, **1711**, 73-81.
39. Stefanowski, J. and Tsoukiàs, A. [2000] Valued Tolerance and Decision Rules, Lecture Notes in Artificial Intelligence **2005**, 212-219.
40. Stefanowski, J. and Tsoukiàs, A. [2001] Incomplete Information Tables and Rough Classification, Computational Intelligence, **17**:3, 545-566.
41. Zimányi, E. and Pirotte, A. [1997] Imperfect Information in Relational Databases, in A. Motro and P. Smets, (eds.), Uncertainty Management in Information Systems: From Needs to Solutions, Kluwer Academic Publishers, pp. 35-87.

# Hybrid Rough Sets-Population Based System

Puntip Pattaraintakorn[1] and Nick Cercone[2]

[1] Department of Mathematics and Computer Science, Faculty of Science, King
Mongkut's Institute of Technology Ladkrabang (KMITL), Bangkok, Thailand
kppuntip@kmitl.ac.th
[2] Faculty of Science and Engineering, York University, Toronto, Ontario, Canada
ncercone@yorku.ca

**Abstract.** The integration of mathematical and statistical data analysis
research can engender a novel and better approach, especially for survival analysis. This paper is devoted to Professor Pawlak and his ideas
about rough sets and its applications. We propose MULTIHYRIS, an
alternative hybrid intelligent system with a rough sets and population
based approach for survival analysis. MULTIHYRIS is designed to increase the versatility and efficiency of survival analysis techniques. The
MULTIHYRIS architecture incorporates mathematics - rough sets (with
discernibility relations and individual patient consideration) - with statistics - Kaplan-Meier and Cox methods (with population estimates). The
central idea behind MULTIHYRIS is to perform univariate analysis by
using rough sets, database management and the Kaplan-Meier method
with soft computing.

All results from the univariate analysis are subsequently used in further mulitvariate analysis. In this stage, we provide two optional approaches to serve different requirements; rough sets integrated with
database management and the Cox method. The former approach is
able to produce decision rules while the latter generates a Cox model.
Furthermore, set operations are used to unite these two outcomes and
generate new reducts - *hybrid reducts* based on our rough sets-population
based system. The informativeness of the rules and models can be verified within this analysis by validation processes and statistical tests.
To demonstrate MULTIHYRIS, we have implemented it on a real-world
geriatric data set, collected from the Dalhousie Medical School.

## 1   Introduction

> *"... rough set theory it is not an alternative*
> *to classical set theory but it is embedded in it."*
> *Zdzislaw Pawlak (2005), A Treatise on Rough Sets*

Survival analysis describes time-to-event analysis. Statistics yield useful survival
analysis data and theoretical tests to provide solutions. Typically, researchers
accomplish this analysis by computing the probability that the event will occur
within a specific time and include a comparison of several risk factors. Frequently,
however, the prediction of whether the event will eventually occur or not is of

J.F. Peters et al. (Eds.): Transactions on Rough Sets VII, LNCS 4400, pp. 190–205, 2007.
© Springer-Verlag Berlin Heidelberg 2007

primary importance. It often happens that the study does not span enough time in order to observe the event for all patients.

Two extra factors we consider for survival analysis include: (i) survival time, commonly misleading, we mean the time patients are admitted to the study until the time to death as well as the time to particular events (e.g., recurrence of disease or time until metastatis to another organ); (ii) if any patient leaves the study for any reason, use of a censor variable is required, indicating the period of observation was cut off before the event of interest occurred. To properly address censoring, modeling techniques must take into account that for these patients the event does not occur during the follow-up period. For this reason, statistics which can be considered as a population based approach is inevitable.

In the early 1980's, Pawlak originally introduced rough sets theory [1,2]. During the years 1980–2006, he has continued his remarkable study. Rough set theory and its research has a long and rich history. Recently, Pawlak introduced the relations between rough sets and flow graphs [3,4]. He illustrated its applicability on conflict analysis and voting problems [5]. Pawlak and Skowron discussed some issues on rough sets and philosophical observations of rough set theory in [6,7,8,9,10]. Within the past two decades research into theoretical aspects of rough sets explored the complementary nature of its properties with other mathematical theories. At the same time, rough sets applications became larger, more complex and systematic. Rough sets, a modern mathematics which was originally devised by Pawlak, makes for balanced approaches between theory and practice possible.

## 1.1  Related Works

Applications of rough sets are widening and emerging and are continuously marked with advancements, for example, survival analysis. Some studies have been conducted using rough sets [11,12]. These studies utilized inconsistent data that occurred in 246 records out of 557 records in throat cancer patients. The authors set the threshold different for attribute values among such inconsistent records. They used the Kaplan-Meier survival function [13] to cluster groups of patients into approximately similar Kaplan-Meier characteristics. The notion of using Kaplan-Meier with rough set theory is used successfully. However, the Kaplan-Meier curve is used under the condition that the resulting clusters are representative of the data by visualization. Indeed, the visual determination of such curves requires special treatment. Statistical hypothesis testing techniques are required to prove a significant difference between the curves. Our work provides an analysis schema to accomplish this elsewhere.

In [14,15] the geriatric data set has been analyzed and rules for the *notification status* (the target function) were generated. To be able to analyze survival data and fit a model that can describe the notification status of each patient based on condition attributes (or explanatory variables) is useful for medical diagnosis. Nonetheless, the analysis process for fitting the models to describe the *survival time* (the target function) when considering notification status as censoring is of greater importance and inevitable.

Rough sets and decision trees have been used to study kidney dialysis patients' survival [16]. Their rough sets algorithm and decision tree can produce correct predictions on survival time 56% and 67% for the test data sets, respectively. The most significant result is to demonstrate that rough sets, decision tree and data mining approaches are useful for survival prediction of dialysis patients.

A number of rough sets applications to medical survival data have been proposed. Primarily, these studies simply apply rough sets methods to data that happens to have a medical origin, without much regard for the underlying medical problem at hand. These studies contrast our approach that contains domain knowledge consideration with probe attributes and probe reducts [15]. Furthermore, our previous study [17] analyzed a number of survival data sets and captured the necessary semantic information embedded in the data.

In this paper, we extended on previous analysis of multivariate data analysis by offering the optional multivariate analysis. We introduce a framework to obtain such models for predicting survival time from our proposed hybrid intelligent system. We aim at the important aspects of data analysis, to offer a system that is the most versatile and efficient for survival analysis.

This article is organized as follows. We discuss in Sect. 2 the role of soft computing to devise new systems for survival analysis. We introduce preliminaries of the components we amalgamated in our system architecture, rough set theory and some statistical approaches. New reducts are provided in the end of this section. In Sect. 3 we propose a multivariate hybrid rough sets intelligent system for survival analysis architecture (MULTIHYRIS). We demonstrate the applicability of MULTIHYRIS on the geriatric data set described in Sect. 4. Section 5 contains experimental results. In Sect. 6 we provide a summary and some general remarks of what next steps will be taken.

## 2    The Role of Soft Computing and System Components

*The rough set approach seems to be of fundamental importance to AI and cognitive sciences, especially in the areas of machine learning, knowledge acquisition, decision analysis, knowledge discovery from databases, expert systems, inductive reasoning and pattern recognition.*
*Zdzislaw Pawlak (2004), Some Issues on Rough Sets.*

It is difficult to gain insight into which method is best suited for survival data. One technique may generate very accurate results for one data set and poor results for another data set. Moreover, each technique has underlying advantages and disadvantages. The disadvantages represent difficult data analysis problems to solve. A stand-alone technique inevitably reveals characteristics of the problem. This reasoning leads us to consider hybridization of methods which complement each other to overcome disadvantages.

Thus, much current work tends to hybridize diverse methods. Studies indicate advantages of hybridization over conventional techniques [18]. The amount of survival data requires that data analysis approaches have tractable time

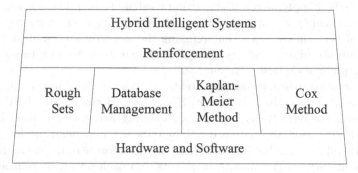

**Fig. 1.** The system components in soft computing for hybrid intelligent system (adapted from [18])

complexity, and simultaneously provide satisfactory outcome. Research in 'soft computing' has demonstrated data analysis successes. Soft computing methodology works synergistically with other methods to provide flexible analytical tools in real situations. Medsker [18] stated that soft computing differs from traditional computing in that it is tolerant of imprecision, uncertainty and partial truth. We extend his idea to include a broader view of hybrid intelligent systems (Fig. 1).

In this paper, we specialize our system to include four components (layer 2): rough sets, database management, the Kaplan-Meier method and the Cox method. We provide the preliminaries of such components below.

### 2.1 Rough Set Theory and Database Management

Rough set theory is the first and most important technique to turn our proposed system into a hybrid system. In order to accommodate errorful, imprecise, and uncertain data, use of rough sets is expedient.

According to Pawlak [2], the power of rough sets *"... is that it does not need any preliminary or additional information about the data, such as probability distributions in statistics, basic probability assignment in the Dempster-Shafer theory, or grade of membership or the value of possibility in fuzzy set theory"*. Most of the time, rigid use of thresholds may seem unrealistic; in contrast rough sets provides more flexible use of lower and upper approximations with discernibility relation.

The results of using rough sets systems are comparable with those obtained by using other systems under wide domain varieties [19]. Research and systematic developments for integrating rough sets to other intelligent systems are at an initial stage. Especially, hybrid intelligent systems based on rough sets for survival analysis are scarce.

The purely statistical measurement gives reasonable evidence to support the hypothesis and can be considered as population-based method. When considering real-world data, however, purely statistical measures can be less meaningful.

The initial study applying rough sets to survival analysis is [11]. They used rough sets for a medical expert system. The rough sets principle can perform attribute selection of decision concepts that remains the same over all information. The primary purpose of our study is to explore the survival data by using a hybrid rough sets-population based system.

Since computer performance has greatly increased, database management is another choice to obtain information from data. However, database management approaches tend to use traditional filtering and query approaches. This indicates that major alternative developments are indispensable for performing meaningful data analysis. Rough sets can incorporate database management to increase its utility. Traditional rough sets approaches in real applications are time-consuming, thus rendering rough sets less efficient. One reason for this phenomenon is data resides in flat files for most of the time. Furthermore, selecting attributes and rules require computing all equivalence classes according to values of the condition attributes and decision attribute. This computation greatly increases complexity with real data. Therefore, studies to reduce complexity remain necessary.

In [17,20], rough sets are redefined using database operations and systems linked to the database directly. The computational time is improved remarkably. Hence, benefits of database set operations Count ($Card$) and Projection ($\prod$) permit the computation to scale up in our study.

Let us assume that a *decision table* is denoted by $T(U, C, D)$, where $C$ is the set of *condition attributes* and $D$ is a singleton set of *target function* (decision). For simplicity, we write $C$ and $D$ instead of $\{C\}$ and $\{D\}$. The terms core attribute, dispensable attribute and reduct are provided as the following.

**Definition 1.** *An attribute $C_i$ is a core attribute if*

$$Card(\prod(C - C_i + D)) \neq Card(\prod(C - C_i)).$$

**Definition 2.** *An attribute $C_i \in C$ is a dispensable attribute with respect to $D$ if*

$$Card(\prod(C - C_i + D)) = Card(\prod(C - C_i)).$$

**Definition 3.** *The subset of attributes $RED \subseteq C$ is a reduct of $C$ with respect to $D$ if*

$$K(RED, D) = K(C, D) \quad and$$

$$K(RED, D) \neq K(RED', D) \text{ for all } RED' \subset RED.$$

The modeling results from rough sets are the *decision rules* that can be read, interpreted and further used without requiring any medical professional knowledge. The decision rules can be described as the following.

$$\text{"If } C \text{ is } c_1 \text{ then } D \text{ is } d_1\text{"}$$

where $c_1$, $c_2$ and $d_1$ are values corresponding to attributes $C$ and $D$, respectively.

## 2.2   Kaplan-Meier Survival Analysis

In survival analysis, the standard univariate analysis is the Kaplan-Meier survival analysis method [13]. This method creates the Kaplan-Meier survival curves. These curves provide insight into the survival function for each group. The proportion of the population of patients who would survive a given length of time under the same circumstances is given by the Kaplan-Meier method as shown in (1). $S$ is based on the probability that each patient survives at the end of a time interval, on the condition that the patient was present at the start of the time interval.

$$\hat{S}(t) = \prod_{t_i \leq t} \left(1 - \frac{d_i}{n_i}\right) \tag{1}$$

where $t_i$ is the period of study at point $i$, $d_i$ is number of events up to point $i$ and $n_i$ is number of patients at risk just prior to $t_i$.

The method produces a table and a graph, referred to as the life time table and survival curve. There are initial assumptions required to make use of the Kaplan-Meier method, they appear in [13].

## 2.3   Cox Model

Cox (1972) proposed a semi-parametric model for the hazard function that allows the addition of explanatory variables but keeps the baseline hazard as an arbitrary, unspecified, nonnegative function of time. While the Kaplan-Meier method focused on a single risk factor or attribute, the Cox proportional hazard model is used for multiple attributes. This model assumes a relationship between the dependent and explanatory variables and uses fine-tuned tests (cf. [21]). We analyze multiple attributes with system hybridization using rough set theory (Sect. 2.1).

Using the method of maximum partial likelihood, we estimate the parameters in Cox's model. Partial likelihood is remarkable in that you can estimate the coefficients without having to specify the baseline hazard function $h_0$. The Cox hazard function for fixed-time covariates, $X$, is

$$h(t) = h_0(t)e^{b_1 X_1 + b_2 X_2 + ... + b_k X k} \tag{2}$$

where $h(t)$ refers to the hazard function at time $t$, $h_0(t)$ refers to the baseline hazard or hazard for an individual when the value of all the independent variables equal zero. The $X_1, X_2, ..., X_k$ refer to explanatory variables, $b_1, b_2, ..., b_k$ refer to Cox regression coefficients determined by partial likelihood estimation while $k$ refers to the number of explanatory variables.

## 2.4   Hybrid Reducts

The outcomes from rough sets and Cox methods, reducts and significant condition attributes, will be integrated to yield the new set. This new set contain the informative attributes from rough sets and significant attributes from statistics and is defined as follows:

**Definition 4.** *(Hybrid Reducts) Let $REDU = \{redu_1, redu_2, ...redu_m\} \neq \emptyset$ be a reducts set, where $m$ is a number of attributes contained in the $REDU$ set. Let $SIG = \{sig_1, sig_2, ...sig_n\} \neq \emptyset$ be a significant condition attribute set, where $n$ is a number of attributes contained in the $SIG$ set. We define hybrid reducts as follows:*

$$hybrid\ reducts= REDU \cup SIG$$

*where $\cup$ denotes set union operation.*

## 3    Hybrid Intelligent System

> *"... data analysis can be perceived as a part of inductive reasoning, and therefore it can be understood as a kind of reasoning about data methods ..."*
> *Zdzislaw Pawlak (1998), Reasoning about Data - A Rough Set Perspective.*

Survival data can be analyzed using several methods and the results are affected by both the analysis algorithm and the problem studied. From our architecture

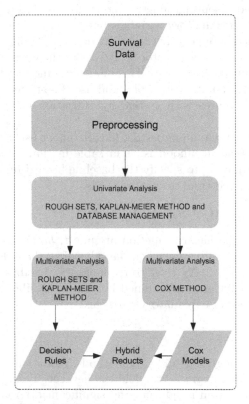

**Fig. 2.** MULTIHYRIS system architecture

and previous studies [15,17], we will expand the hybrid rough sets intelligent system architecture for survival analysis (HYRIS). We develop and add components to HYRIS to perform comprehensive survival analysis.

The objective is to expand the utility of HYRIS to *MULTIvariate HYbrid Rough sets Intelligent System* architecture for survival analysis (MULTIHYRIS). The system architecture of MULTIHYRIS is depicted in Fig. 2.

MULTIHYRIS can analyze survival data and generate decision rules prior to the implementation of HYRIS (see more [17]). The additional process is the multivariate analysis with the Cox method to generate Cox models. Furthermore, hybrid reducts that integrated the outcome from both rough sets and Cox method will be generated. The details of this process will be described in the Sect. 5.

# 4   Description of the Geriatric Data Set

*Rough set theory provides a variety of set functions that can be studied relative to various measure spaces.*
*Z. Pawlak, J. Peters, A. Skowron, Z. Suraj, S. Ramanna, and M. Borkowski (2001), Rough Measures and Integrals: A Brief Introduction.*

**Table 1.** The geriatric data description

| Attribute | Description | Attribute | Description |
|---|---|---|---|
| *edulevel* | Education level | *hbp* | High blood pressure |
| *eyesight* | Eyesight | *heart* | Heart |
| *hearing* | Hearing | *stroke* | Stroke |
| *eat* | Eat | *arthriti* | Arthritis or rheumatism |
| *dress* | Dress and undress yourself | *parkinso* | Parkinson's disease |
| *takecare* | Take care of your appearance | *eyetroub* | Eye trouble |
| *walk* | Walk | *eartroub* | Ear trouble |
| *getbed* | Get in and out of bed | *dental* | Dental |
| *shower* | Take a bath or shower | *chest* | Chest |
| *bathroom* | Go to the bathroom | *stomach* | Stomach or digestive |
| *phoneuse* | Use the telephone | *kidney* | Kidney |
| *walkout* | Get places out of walking distance | *bladder* | Bladder |
| *shopping* | Go shopping for groceries etc. | *bowels* | Bowels |
| *meal* | Prepare your own meals | *diabetes* | Diabetes |
| *housewk* | Do your housework | *feet* | Feet |
| *takemed* | Take your own medicine | *nerves* | Nerves |
| *money* | Handle your own money | *skin* | Skin |
| *health* | Health | *fracture* | Fractures |
| *trouble* | Trouble | *age* | Age group |
| *livealon* | Live alone | *studyage* | Age at investigation |
| *cough* | Cough | *sex* | Gender |
| *tired* | Tired | *livedead* | Notification status |
| *sneeze* | Sneeze | *livmonth* | Time lived (in month) after interview |

In a medical data table, each patient record represents an example or object. Each field is treated as an attribute or feature. Each column contains patients' information (e.g., patients' symptoms, clinical information) which can be treated as the attribute values provided in an attribute set. Such a data set is treated as the training set. Predicting the known outcome (target function or decision) is a goal of analysis. The task can be to predict a patient's survival time.

Our case study is conducted on the geriatric data set from Dalhousie Medical School. It is discretized and preprocessed as in [17]. The preprocessed geriatric data set contains 8546 patient records with 44 clinical information, notification statuses and survival times. This data set is analyzed to determine the *notification status* and *survival time* (the target functions) of a patient given all the clinical information. Our previous study for the target function - notification status can be found in [15]. In this study, the survival time of patients is treated as the target function, and the 44 clinical information of each patient are used as condition attributes, which include for example, *diabetes, Parkinson's disease, the age of the patient at investigation* as described in Table 1.

## 5   Experiments

*"... knowledge is deep-seated in the classificatory abilities of human being and other species."*
*Zdzislaw Pawlak (1991), Rough Sets. In Theoretical Aspects of Reasoning about Data.*

To illustrate that the system designs provided in Sect. 3 are indeed feasible and applicable in real situations, we implement system prototypes in this section. Let us start from the results of univariate analysis (Fig. 2). In our previous study the attribute mining process from [17] produced core and dispensable attributes with rough sets and statistical calculations as shown in Tables 2. The terms probe, core and dispensable attribute, reducts and probe reducts we use in this paper are introduced in [15].

**Table 2.** Core and dispensable attributes experimental results

| Data sets | Core attributes | Dispensable attributes |
|---|---|---|
| geriatric | *edulevel eyesi hear shower phoneuse shopping meal housew money health trouble livealo cough tired sneeze hbp heart stroke arthriti eyetroub dental chest stomac kidney bladder bowels diabetes feet nerves skin fracture age sex* | *eartroub walk* |

Hypotheses are chosen in order to answer the risk factor significant questions e.g. *diabetes* risk factor,

*"Is diabetes a significant risk factor for survival time in geriatric patients?"*.

We explore whether or not to include each attribute (risk factor) in the prediction survival model constructions. For example, the hypotheses for *diabetes* risk factor in geriatric data set are:

$H_0$: There will be no significant difference in survival times between diabetes groups.

$H_1$: There will be significant differences in survival times between diabetes groups.

The statistical analysis process described in [17] generated life time table (Table 3).

**Table 3.** Life time table (or the calculation of the Kaplan-Meier estimate) of the survival function of the geriatric data set

| Time (months) | Status | Cumulative survival | Standard error | Cumulative event | Number at risk |
|---|---|---|---|---|---|
| 1 | 1 | | | 1 | 8,545 |
| 1 | 1 | 0.9998 | .0002 | 2 | 8,544 |
| | | | . | | |
| | | | . | | |
| | | | . | | |
| 1 | 0 | | | 2 | 8,526 |
| 2 | 1 | | | 3 | 8,525 |
| 2 | 1 | | | 4 | 8,524 |
| 2 | 1 | | | 5 | 8,523 |
| 2 | 1 | 0.9993 | .0003 | 6 | 8,522 |
| | | | . | | |
| | | | . | | |
| | | | . | | |
| 71 | 1 | | | 6,685 | 2 |
| 71 | 1 | 0.0010 | .0010 | 6,686 | 1 |
| 73 | 0 | | | 6,686 | 0 |

Subsequently, we generated the Kaplan-Meier survival functions and survival functions with risk factors as depicted in Fig(s). 3-4.

We consider the log-rank, Brewslow and Tarone-Ware tests which explore whether or not to include the attribute in the prediction survival model constructions. All of these are interpreted and the significance is tested with three statistical tests; log-rank, Breslow and Tarone-Ware with $df=1$ and $p-value$ (less than 0.2) to answer the hypotheses.

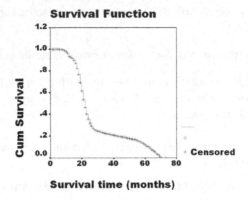

**Fig. 3.** Survival function of geriatric data

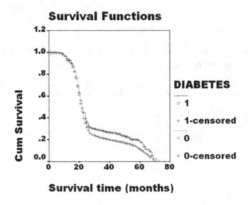

**Fig. 4.** Survival functions of geriatric data with diabetes factor

The example of risk factor diabetes test values:

|            | Statistic | Significance |
|------------|-----------|--------------|
| Log Rank   | 12.77     | .0004        |
| Breslow    | 5.78      | .0162        |
| Tarone-Ware| 8.46      | .0036        |

By considering characteristics and shapes of survival curves we can choose the probe attribute reasonably. The conclusion is therefore that diabetes has the most impact on survival time and $H_1$ is accepted.

The next step is multivariate analysis (Fig. 2). Let us begin with multivariate analysis by using rough set theory. The probe attribute selection process of [17] generated reducts or probe reducts as illustrated in Table 4.

The models generated from this module are in the form of decision rules (Sect. 2.1). We can describe particular tendencies of survival time with the decision rules generated from ELEM2 (version 3) [22]. ELEM2 uses a new heuristic

**Table 4.** Reducts from HYRIS

| Data sets | Reducts |
|-----------|---------|
| geriatric | *edulevel    eyesi    hear shower   phoneuse   shopping meal housew money health    trouble    livealo cough   tired   sneeze   hbp heart    stroke    arthriti eyetroub   dental   chest stomac   kidney   bladder bowels    diabetes    feet nerves skin fracture age sex* |

for evaluating the relevance of an attribute value pair to a target function during induction. A new measure based on information entropy and consideration of rule distribution provides excellent results. ELEM2 provides more accurate rules than C4.5 and CN2 for both artificial and real data sets [22]. Rule induction with ELEM2 generated decision rules for survival tendencies prediction from the above results, e.g., two exemplary prognostic rules:

***Decision Rule 1:*** *If (edlevel = 2) and (eyesi ≤ 0.25) and (hear > 0) and (meal = 0) and (housew = 0) and (0 < health ≤ 0.25) and (trouble = 0) and (livealo = 0) and (hbp>0) and (heart=0) and (stomac = 0) and (bladder = 0) and (diabetes = 0) and (skin = 0) and (age ≤ 3) and (sex = 2) then (survival time = 19-22 months)*

and

***Decision Rule 2:*** *If (sex = 0) and (edlevel = 2) and (eyesi > 0) and (0 < health ≤ 1) and (0 < hear ≤ 0.25) and (diabetes = 1) and (tired = 0) and (feet = 0) then (survival time = 56-73 months).*

The second medical diagnosis rule can be interpreted below:

- if female patients have a low education level and
- eyesight problem from low to serious type and
- health problem from low to serious type and
- can hear quite well and
- do not have diabetes experience and
- are easily tired and
- have foot problems
- then patients are likely to have survival time between 56-73 months.

The 3-fold cross validation is used to guarantee the correctness of the decision rules. The results are the desired classification accuracy from 91% to 100%.

**Table 5.** Case processing summary

| Cases available in analysis | Number | Percent |
|---|---|---|
| Event(LIVMONTH) | 6686 | 78.2% |
| Censored | 1860 | 21.8% |
| Cases dropped | 0 | 0.0% |
| Total | 8546 | 100.0% |

**Table 6.** Test of the coefficients; -2 Log likelihood of the null model $(-2LL_0)$

| $-2LL_0$ |
|---|
| 110591.558 |

**Table 7.** Test of the coefficients; -2 Log likelihood of the model with *diabetes* $(-2LL_1)$

| $-2LL_1$ | Change from the previous step | | |
|---|---|---|---|
| | Chi-square | df | Sig. |
| 110579.417 | 12.141 | 1 | .000 |

**Table 8.** Explanatory variable *diabetes* in the equation with $df = 1$

| Var. | b | SE | Wald | Sig | $Exp(b)$ | 95% CI L/U |
|---|---|---|---|---|---|---|
| DIABETES | 148 | .043 | 1.654 | .001 | .862 | .792/.939 |

The optional mulitvariate analysis of MULTIHYRIS is Cox mothod. We took the significant variables from Sect. 2.2 and treated them as the explanatory variables when constructing the Cox proportional hazard model. Our analysis results in the following.

As Table 5 shows, the geriatric data set we used contains 8,546 records, with 1,860 censoring with no missing values, no cases with negative time and no cases before the earliest event in the stratum.

In Table 6, the $-2LL_0$ value of the model without any explanatory variable included is 110591.558.

The enter method is used and results in Table 7. *Diabetes* risk factor is entered as the first variable. The likelihood of the model with *diabetes* $(-2LL_1)$ is 110579.417. The $-2LL$ decreases by 12.141. This decline 12.141 is significant when considering the last column, $p - value = 0.0004932$.

The interpretation of this risk factor in the Cox model is the hazard of a patient death is decreased 0.862 times if a patient has experienced diabetes diagnosis (Table 8). We concluded that *diabetes* is a significant risk factor.

The stepwise enter method is used until the last step when all explanatory variables are included as demonstrated in Tables 9-11.

In the last step, the $-2LL$ of the last step differ from the one in Fig. 6, 547.298 which is significant. In Table 9, the overall $-2LL$ is 110032.119.

The results in Table 11 demonstrate the utility and versatility of MULTI-HYRIS for analyzing survival data efficiently. MULTIHYRIS is also versatile

**Table 9.** Test of the coefficients; -2 Log likelihood of the model with all explanatory variables

| $-2LL$ | Overall (score) | | |
|---|---|---|---|
| | Chi-square | df | Sig. |
| 110032.119 | 513.312 | 32 | 3.85e-088 |

**Table 10.** Explanatory variables *heart, sex* and *shower* in the equation with $df = 1$

| Var. | b | SE | Wald | Sig | $Exp(b)$ | 95% CI L/U |
|---|---|---|---|---|---|---|
| HEART | -.072 | .030 | 5.765 | .016 | .931 | .878/.987 |
| SEX | .165 | .023 | 7.551 | .000 | 1.179 | 1.119/1.243 |
| SHOWER | .017 | .094 | .033 | .856 | 1.017 | .846/1.224 |

**Table 11.** Significant condition attributes produced from Cox model in the last step

| Data sets | Significant attributes |
|---|---|
| geriatric | *diabetes heart trouble* |
| | *getbed walk age sex* |

with the comprehensive univariate and multivariate analysis processes. Nonetheless, the results from Tables 4 and 11 illustrate two different outcomes from two different aspects; discernibility relation from mathematics and population estimates from statistics. They are clearly not the disjoint sets or subset of each other, but they both are of inexpedient from both point of views.

One most important concept of this research is to integrate mathematics and statistics to generate the better outcomes that best suit for multidiscipline study. Thus, our new reducts (Definition 4) is constructed from the set operation as the following.

**hybrid reducts** = {*edulevel eyesi hear shower phoneuse shopping meal housew money health trouble livealo cough tired sneeze hbp heart stroke arthriti eyetroub dental chest stomac kidney bladder bowels diabetes feet nerves skin fracture age sex getbed walk*}

These hybrid reducts is constituted both mathematics and statistics importance. It should be considered to be the most informative and significant attributes for the survival data and for further analysis.

# 6   Concluding Remarks and Future Works

The most significant result obtained from this research was to demonstrate that mathematics - rough sets and statistics - Kaplan-Meier and Cox appraoches are useful for survival analysis. Our system, MULTIHYRIS, can be used as the practical tool in survival analysis, as it fulfils the typical needs of survival analysis.

Our preliminary experiment illustrated that MULTIHYRIS can perform versatile and efficient survival analysis. The generated survival prediction rules and Cox models can be used by anyone concerned about the progression of patients.

Several issues presented in this paper require further research. These features include: (i) From theoretical viewpoint, we should pay more attention to many advances in rough sets e.g., rough mereology, rough inclusion or decision logic. It is interesting to explore the feasibility of augmenting advanced rough sets to our hybrid intelligent system to increase its utility. (ii) The study should also continue the development of background knowledge learning to reach a dynamic rough sets based framework. Interesting and practical applications of this should be addressed with further research. (iii) The consecutive use of hybrid reducts for further analysis should be conducted.

## Acknowledgment

Thank you Zdzislaw Pawlak who developed rough set theory, the inspiration for our research. This research was supported by King Mongkut's Institute of Technology Ladkrabang grant (KMITL), Thailand and Natural Sciences and Engineering Research Council of Canada (NSERC), Canada. Thanks are also due to Arnold Mitnitski and to Gregory Zaverucha.

## References

1. Z. Pawlak: Rough Sets. Int. J. Inform. Comput. Sc., 11(5), 1982, 341–356.
2. Z. Pawlak: Rough Sets. In Theoretical Aspects of Reasoning about Data, Kluwer Academic Publishers, Dordrecht, 1991.
3. Z. Pawlak: Decision Networks. Rough Sets and Current Trends in Computing, 2004, 1–7.
4. Z. Pawlak: Rough Sets and Flow Graphs. in Lect. Notes. Artif. Int., vol. 3642, D. Slezak et al. Eds. Springer-Verlag, Berlin, Heidelberg, 2005, 1–11.
5. Z. Pawlak: Some Remarks on Conflict Analysis. European Journal of Operational Research 166, 2005, 649–654.
6. Z. Pawlak: Some Issues on Rough Sets. T. Rough Sets, 2004, 1–58.
7. Z. Pawlak: A Treatise on Rough Sets. T. Rough Sets, 2005, 1–17.
8. Z. Pawlak, A. Skowron: Rudiments of Rough Sets, Inform. Sciences 177(1), 2007, 3–27.
9. Z. Pawlak, A. Skowron: Rough Sets: Some Extensions, Inform. Sciences 177(1), 2007, 28–40.
10. Z. Pawlak, A. Skowron: Rough Sets and Boolean Reasoning , Inform. Sciences 177(1), 2007, 41–73.
11. J. Bazan, A. Osmolski, A. Skowron, D. Slezak, M. S. Szczuka, J. Wroblewski: Rough Set Approach to the Survival Analysis, in Lect. Notes. Artif. Int., vol. 2475, Springer-Verlag, Berlin Heidelberg, 2002, 522–529.
12. J. Bazan, A. Skowron, D. Slezak, J. Wroblewski: Searching for the Complex Decision Reducts: The Case Study of the Survival Analysis, in Proc. of the IS-MIS, Maebashi, Japan, Lect. Notes. Artif. Int., vol. 2871, Springer-Verlag Berlin, Heidelberg, 2003, 160–168.

13. E. L. Kaplan, P. Meier: Nonparametric Estimation from Incomplete Observations, J. of the Amer. Stat. Asso., vol. 53, 457–481, 1958.
14. J. Li, N. Cercone: Discovering and Ranking Important Rules, in Proc. of the IEEE GrC, Beijing, China, 2005.
15. P. Pattaraintakorn, N. Cercone, K. Naruedomkul: Hybrid Intelligent Systems: Selecting Attributes for Soft-Computing Analysis, in Proc. of COMPSAC, 2005, 319–325.
16. A. Kusiak, B. Dixon, S. Shah: Predicting Survival Time for kidney Dialysis Patients: A Data Mining Approach, Computers in Biology and Medicine 35, 2005, 311–327.
17. P. Pattaraintakorn, N. Cercone, K. Naruedomkul: Selecting Attributes for Soft-Computing Analysis in Hybrid Intelligent Systems, in Lect. Notes. Artif. Int., vol. 3642, D. Slezak et al. Eds. Springer-Verlag, Berlin, Heidelberg, 2005, 698–708.
18. M. R. Larry: Hybrid Intelligent System, Kluwer Academic Publishers, Boston, 1995.
19. J. Komorowski, L. Polkowski, A. Skowron: Rough Sets: A Tutorial, in Rough Fuzzy Hybridization: A New Trend in Decision-Making, S. K. Pal, A. Showorn, Eds. Springer, Berlin, 1999, 3–98.
20. X. Hu, J. Han, T. Y. Lin: A New Rough Sets Models Based on Database Systems. Fund. Inform. 59(2-3), 2004, 1–18.
21. D. R. Cox: The Analysis of Exponentially Distributed Life-times with Two Types of Failure, J. of the Royal Statistical Society, vol. 21, 1959, 411–422.
22. A. An, N. Cercone: ELEM2: A Learning System for More Accurate Classifications, in Lect. Notes Comput. Sc., vol. 1418, 1998, 426–441.

# Hybrid Rough Sets Intelligent System Architecture for Survival Analysis

Puntip Pattaraintakorn[1], Nick Cercone[2], and Kanlaya Naruedomkul[3]

[1] Department of Mathematics and Computer Science, Faculty of Science, King Mongkut's Institute of Technology Ladkrabang (KMITL), Bangkok, Thailand
kppuntip@kmitl.ac.th
[2] Faculty of Science and Engineering, York University, Toronto, Ontario, Canada
ncercone@yorku.ca
[3] Department of Mathematics, Faculty of Science, Mahidol University, Thailand
scknr@mahidol.ac.th

**Abstract.** Survival analysis challenges researchers because of two issues. First, in practice, the studies do not span wide enough to collect all survival times of each individual patient. All of these patients require censor variables and cannot be analyzed without special treatment. Second, analyzing risk factors to indicate the significance of the effect on survival time is necessary. Hence, we propose "Enhanced Hybrid Rough Sets Intelligent System Architecture for Survival Analysis" (Enhanced HYRIS) that can circumvent these two extra issues.

Given the survival data set, Enhanced HYRIS can analyze and construct a life time table and Kaplan-Meier survival curves that account for censor variables. We employ three statistical hypothesis tests and use the $p-value$ to identify the significance of a particular risk factor. Subsequently, rough set theory generates the probe reducts and reducts. Probe reducts and reducts include only a risk factor subset that is large enough to include all of the essential information and small enough for our survival prediction model to be created. Furthermore, in the rule induction stage we offer survival prediction models in the form of decision rules and association rules. In the validation stage, we provide cross validation with ELEM2 as well as decision tree. To demonstrate the utility of our methods, we apply Enhanced HYRIS to various data sets: geriatric, melanoma and primary biliary cirrhosis (PBC) data sets. Our experiments cover analyzing risk factors, performing hypothesis tests and we induce survival prediction models that can predict survival time efficiently and accurately.

**Keywords:** Rough sets, Survival analysis, Kaplan-Meier method, Hybrid intelligent systems, Reducts, Soft computing.

## 1 Introduction

Among prognostic modeling techniques that induce models from medical data, survival analysis warrants special treatment in the type of data required and its

J.F. Peters et al. (Eds.): Transactions on Rough Sets VII, LNCS 4400, pp. 206–224, 2007.

modeling. The data required for medical analysis include demographic, symptoms, laboratory tests, treatment information etc. The special features for survival data are the events of interest, censoring, follow-up time and survival time specific for each type of disease that we will discuss. For modeling, there exist techniques for processing synthetic survival data. Nonetheless, innovative approaches that consider survival data challenge researchers. We propose a novel approach to address these complexities.

Recent hybrid system research tends to hybridize diverse methods which complement each other to overcome underlying individual hindrances. Soft computing methodology can work synergistically with other data analysis methods to provide flexible information processing in real situations. Medsker [1] stated that soft computing differs from traditional computing in that it is tolerant of imprecision, uncertainty and partial truth. Studies have shown that soft computing in medical applications is sometimes more appropriate than conventional techniques. We are encouraged by our experiences with CDispro [2] and HYRIS [3] to utilize the soft computing methods for survival analysis.

We introduce in Sect. 2, challenges and preliminary notation of some statistics and rough set theory for survival analysis. In Sect. 3, we propose the Enhanced Hybrid Rough Sets Intelligent System for Survival Analysis Architecture (Enhanced HYRIS). We demonstrate the applicability of Enhanced HYRIS on several data sets and their experimental results in Sect. 4. In Sect. 5, survival prediction models are constructed with the informative attributes from Sect. 4 and their validation results. In Sect. 6, we add some general remarks of what steps will be taken next.

## 2 Preliminaries and Notation

### 2.1 Survival Analysis

Survival analysis describes time-to-event analysis. Survival analysis is called *reliability analysis* in engineering, and *duration analysis* in economics. Survival analysis is the study of the time between entry to a study and a subsequent event (e.g., death, recurrence of cancer).

We accomplish this analysis by computing the probability that the event will occur within a specific time and include a comparison of several risk factors. Frequently however, the prediction of whether the event will eventually occur or not is of primary importance. It often happens that the study does not span enough time in order to observe the event for all patients.

Two extra factors we consider for survival analysis include: (i) survival time (which is commonly misleading), the time patients are admitted to the study until the time to death as well as the time to particular events (e.g., recurrence of disease or time until metastatis to another organ); (ii) if any patient leaves the study for any reason, use of a censor variable is required, indicating the period of observation was cut off before the event of interest occurred. To properly address censoring, modeling techniques must take into account that for these patients the event does not occur during the follow-up period. The following are some example questions of our study that will be answered:

"Is diabetes a significant risk factor for geriatric patients?"
"What are the rules for survival time predictions of geriatric patients?"
"What is the survival tendency of a geriatric patient?"

## 2.2   Kaplan-Meier Survival Analysis

In survival analysis it is highly recommended to use the Kaplan-Meier survival analysis method [4]. Kaplan-Meier survival analysis offers Kaplan-Meier survival curves, which provide insight into the survival function for each group. The proportion of the population of such patients who would survive a given length of time under the same circumstances is given by the Kaplan-Meier method or the product limit (PL) as shown in equation (1). $S$ is based on the probability that each patient survives at the end of a time interval, on the condition that the patient was present at the start of the time interval.

$$\hat{S}(t) = \prod_{t_i \leq t} \left(1 - \frac{d_i}{n_i}\right) \tag{1}$$

where $t_i$ is the period of study at point $i$, $d_i$ is number of events up to point $i$ and $n_i$ is number of patients at risk prior to $t_i$.

The method produces a table and a graph, referred to as the life time table and survival curve. There are initial assumptions to make use of Kaplan-Meier method that appear in [4]. While the Kaplan-Meier method focused on a single risk factor or attribute, the Cox proportional hazard model is used for multiple attributes. This model assumes a relationship between the dependent and explanatory variables and uses fine-tuned tests (cf. [5]). In this paper, we propose multiple attributes analysis with system hybridization using rough set theory that does not require any initial assumption (Sect. 2.4).

## 2.3   Log-Rank Test

An important part of survival analysis is to analyze the risk factor on a plot of the survival curves (Sect. 2.2) for each group of interest. The comparison of the survival curves for two groups cannot based on visual impressions. Statistics yield useful survival analysis data and theoretical tests to provide solutions. Thus, we consider the log-rank [6], Brewslow [7] and Tarone-Ware tests [8] which explore whether or not to include the attribute (or risk factor) in the prediction survival model constructions. These tests calculate their *p-values* that test the *null hypothesis* ($H_0$–the survival curves has no significant difference in survival times in two groups of interest) against the *alternative hypothesis* ($H_1$–the survival curves has significant difference in survival times in two groups of interest).

For example, we can consider *diabetes* risk factor in geriatric data set. These three hypothesis testing approaches can answer the question: "Is diabetes a significant risk factor for geriatric patients?", under the example hypotheses:

$H_0$: No significant difference in survival times between diabetes groups.
$H_1$: Significant difference in survival times between diabetes groups.

The three statistical tests differ in how they weight the examples. The log-rank test weights all examples equally, the Breslow test weights earlier periods more heavily and the Tarone-Ware test weights earlier examples less heavily than the Breslow test.

The early studies applying rough set theory to survival analysis are [9,10]. They used rough sets to discover relevant patterns for complex decisions successfully. A case study considered is the postsurgery survival analysis for the head and neck cancer cases. Nonetheless, hypothesis testing for each risk factor was not included the analysis and is still an open problem.

## 2.4 Rough Sets

In the early 1980's, Pawlak [11] introduced rough sets theory. Rough set theory is the last and most important technique to turn our proposed system into a hybrid system. The purely statistical measurement gives reasonable evidence to support the hypothesis. When considering noisy real-world data, however, purely statistical measures can be less meaningful. Furthermore, the Cox proportional hazard model (for multiple attributes analysis) requires a relationship between the dependent and explanatory variables and uses fine-tuned tests (cf. [5]). Rough sets can perform this task efficiently [2,3]. For these reasons, we propose the hybrid rough sets approach with the integration of rough sets and statistics.

The primary purpose of our study is to explore individual attributes by statistics (univariate) while simultaneously exploring the effects of several attributes (multivariate) on survival by using a hybrid rough sets approach. The rough sets principle can perform attribute selection of decision concepts that remains the same over all information. Finding a heuristic method for attribute selection that is feasible for large data sets is an open problem. Skowron et al. [12] showed that the lower and upper approximations, positive regions, short reducts, etc. can be computed in a straightforward manner from the discernibility matrix with $O(kn^2)$ time complexity where $n$ is the number of examples and $k$ is the number of attributes of the data set, which is not feasible for large data sets.

Nguyen et al. [13,14] proposed several algorithms that do not require storing the discernibility matrix in the calculation step. Their algorithm for generating *short reducts* by using Johnson strategy has $O(k^2 n \log n)$ time complexity. This algorithm is an efficient way to compute reducts without using a discernibility matrix.

Wroblewski [15] proposed a hybrid algorithm for generating reducts. His proposed approach is more efficient compared to a classical GA. Bazan et al. [16] reported a method to search for reducts that generates a minimal number of rules. The authors also introduced several measures for reduct quality. Fewer rules were generated from these reducts, occupied less memory and classified new examples faster.

In several studies, the effects of a certain attribute are the main goal of analysis. This attribute is not necessarily included in the reduct sets of the subtables. For example, the risk factor that will impact the progression of disease is the important candidate component in the reducts. Such risk factors should be further analyzed for some problems in the medical domain.

In [13,14,17], rough sets were redefined using database operations. The computing times were improved remarkably by using database operations and the database system directly. The straightforward approach using databases in the implementation is a promising approach for rough sets. Hence, benefits of database set operations `Count` (*Card*) and `Projection` ($\prod$) permit the computation to scale up in our study. The terms probe attribute and probe reducts were introduced in [2] as the following.

Let us assume that a *decision table* is denoted by $T(U, C, D)$, where $C$ is the set of *condition attributes* and $D$ is a singleton set of *target function*. For simplicity, we write $C$ and $D$ instead of $\{C\}$ and $\{D\}$.

**Definition 1.** *An attribute $C_i$ is a core attribute if*

$$Card(\prod(C - C_i + D)) \neq Card(\prod(C - C_i)).$$

**Definition 2.** *An attribute $C_i \in C$ is a dispensable attribute with respect to $D$ if*

$$Card(\prod(C - C_i + D)) = Card(\prod(C - C_i)).$$

**Definition 3.** *The degree of dependency, $K(R, D)$, between the attribute subset $R \subseteq C$ and attribute $D$ in decision table $T(U, C, D)$ is*

$$K(R, D) = \frac{Card(\prod(R + D))}{Card(\prod(C + D))}.$$

**Definition 4.** *The subset of attributes $RED \subseteq C$ is a reduct of $C$ with respect to $D$ if*

$$K(RED, D) = K(C, D) \quad and$$

$$K(RED, D) \neq K(RED', D) \text{ for all } RED' \subset RED.$$

**Definition 5.** *A probe attribute $P \in C$ corresponding to $T(U, C, D)$ is defined as an attribute of concern in $T(U, C, D)$ for each domain by an expert.*

**Definition 6.** *A probe reduct corresponding to decision table $T(U, C, D)$ is defined as a reduct consisting of a selected before attribute of concern.*

*Example 1.* The notion of probe attribute and probe reducts can be described with the following example; in survival analysis, *survival time* is the decision attribute while *patient's symptoms, surgery type* and so on describe the condition attributes. If we want to know about the survival time for each patient, the risk of radical surgery or mild surgery becomes the significant part of this determination. Hence, we consider the *surgery type* attribute as a probe attribute. The probe reducts are the reducts constructed from the probe attribute.

## 2.5   Rule Quality - Measure of Discrimination

In order to gauge the quality of the generated decision rules, *measure of discrimination* [18] will be used in our study. Let $Q_{MD}$ denote the measure of discrimination, $R$ denotes rule or a query term in an information retrieval, $D$ denotes the target function or class of relevant documents and $D'$ denotes the class of non-relevant documents. $Q_{MD}$ can be expressed as follows:

$$Q_{MD} = log \frac{P(R|D)(1 - P(R|D'))}{P(R|D')(1 - P(R|D))},$$ (2)

where $P$ denotes probability.

# 3   Methodology

Survival data can be analyzed using several methods and the results are affected by both the analysis algorithm and the problem studied. From our architecture and previous studies [2,3], we will expand hybrid rough sets intelligent system architecture for survival analysis (HYRIS). We develop and add components to HYRIS to perform comprehensive survival analysis. The system architecture of Enhanced HYRIS is depicted in Fig. 1.

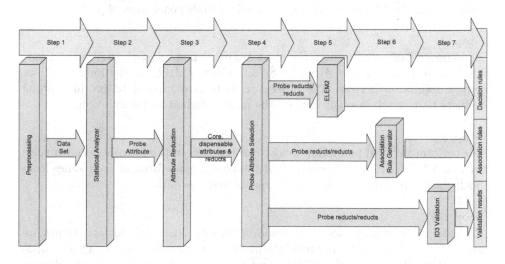

**Fig. 1.** Enhanced HYRIS system overview

The Enhanced HYRIS implementation is an extension of HYRIS [3] in order to offer an optional rule generator and validation process. Enhanced HYRIS is a seven step system, providing comprehensive survival analysis approach according to the statistics and rough set theory. Given our survival data sets, Enhanced HYRIS can analyze survival data as follows:

**Step 1**

Preprocess survival data sets into a usable format for entry to the *Statistical Analyzer* module e.g., discretization, inconsistency removal.

**Step 2**

In the *Statistical Analyzer* module, the Kaplan-Meier method analyzes the entire survival data and generates statistical summaries, e.g., life time table (Table 1) and survival curve of overall data (Figs. 2(a)-2(b)) Subsequently, a particular risk factor is included in the Kaplan-Meier method (Sect. 2.2) to determine the survival curves (Figs. 3-4) with respect to the survival time attribute. Significance levels are tested by three statistical techniques (Sect. 2.3). Overall outcomes are then considered and the probe attribute is identified for all survival data set and sent to next step.

**Step 3**

Rough set theory extracts core attributes, dispensable attributes and reducts. Due to uncertainty, survival data can have no core attributes or use all attributes as core. *Attribute Reduction* will always complete despite this issue. The probe attribute from the *Statistical Analyzer* is used to guide generation of probe reducts.

**Step 4**

If *Attribute Reduction* returns the most distinguished selected attributes to predict survival time, *Probe Attribute Selection* sets reducts as the final attribute subset results. Otherwise, the probe attribute is employed to produce probe reducts that simultaneously extract the most informative information.

**Step 5**

We perform survival prediction model construction in the form of decision rules by ELEM2 [18] in *Model Construction* and compare performance outcomes from both the entire data and the reducts/probe reducts data.

**Step 6**

We induce association rules in *Association Rule Generator*. Association rules are generated to explore the relative information for each risk factor. This component allows the degree of flexibility and generality that was designed, whereas most existing systems tend to be highly specialized toward a particular kind of rule generation technique.

**Step 7**

We employ ID3 [21] to run 10-fold cross validation process in the last component, *ID3 Validation*. The validation process for real-world data is simulated and will be illustrated later on.

## 4    Experimental Results

Enhanced HYRIS was applied to several data sets - both benchmark and actual data sets; geriatric, melanoma and PBC. The description of data sets and data preparation can be found in [3].

We will explain in detail the geriatric data from Dalhousie Medical School. As a real-world case study, the geriatric data set contains 8546 patient records with *status* (dead or alive) as the censor attribute and *survival time* as the target function. All data is discretized based on percentile groups, with each group containing approximately the same number of patients (equal density).

**Table 1.** Life time table (or the calculation of the Kaplan-Meier estimate) of the survival function of the geriatric data set from Enhanced HYRIS

| Time (months) | Status | Cumulative survival | Standard error | Cumulative event | Number at risk |
|---|---|---|---|---|---|
| 1 | 1 | | | 1 | 8,545 |
| 1 | 1 | 0.9998 | .0002 | 2 | 8,544 |
| | | | . | | |
| | | | . | | |
| | | | . | | |
| 2 | 1 | | | 5 | 8,523 |
| 2 | 1 | 0.9993 | .0003 | 6 | 8,522 |
| | | | . | | |
| | | | . | | |
| | | | . | | |
| 71 | 1 | | | 6,685 | 2 |
| 71 | 1 | 0.0010 | .0010 | 6,686 | 1 |
| 73 | 0 | | | 6,686 | 0 |

For the geriatric data, we assigned a specification of 4 groups for survival time. Group 1 describes 1,908 patients with survival time 7-17 months, 2,411 patients with 18-22 months, 2,051 patients with 23-48 months and 2,176 patients with 49-73 months, respectively.

To determine the Kaplan-Meier estimate of the survival function, we took geriatric data and formed a series of time intervals. Each of these intervals is constructed in such a way that one observed death is contained in the interval. *Status* = 0 indicates that the example has been censored and *status* = 1 indicates death. Table 1 presents the life time table or the calculation of Kaplan-Meier estimate of the geriatric survival function.

We generate the Kaplan-Meier curves by calculating the Kaplan-Meier estimate of the survival function. A plot of this curve is a step function, in which the estimated survival probabilities are constant between adjacent death times and only decrease at each death. In Fig. 2(a), no risk factor is included, it displays one Kaplan-Meier survival curve in which all geriatric data are considered to belong to one group. The important aspect of the survival function is to understand how

SURVIVAL FUNCTION

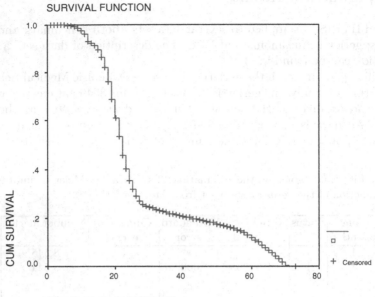

SURVIVAL TIME (MONTHS)

(a) Geriatric survival function: no risk factor is included, the geriatric data are considered to belong to one group.

SURVIVAL FUNCTION

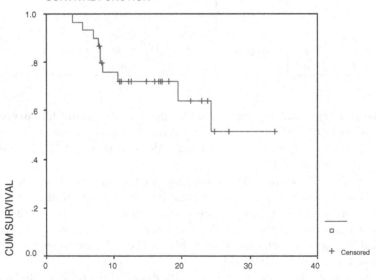

SURVIVAL TIME

(b) Melanoma survival function: no risk factor is included, melanoma data are considered to belong to one group.

**Fig. 2.** Survival functions

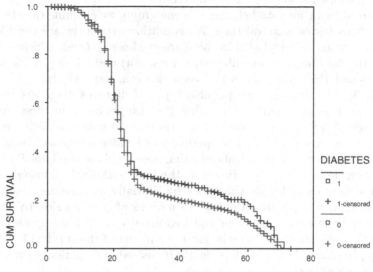

(a) Survival function of the *diabetes* factor from geriatric:
two types of diabetes diagnosis {0,1} are compared.

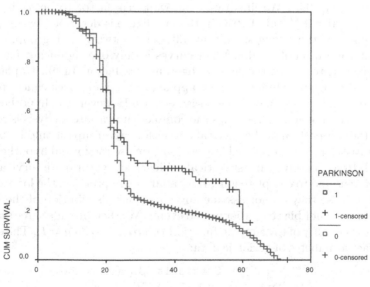

(b) Survival function of the *Parkinson's* factor from
geriatric: two types of Parkinson's are compared.

**Fig. 3.** Survival functions

it influences survival time. Fig. 2(b) depicts the Kaplan-Meier survival curve in which all melanoma data belong to one group.

After considering all data belonging to one group, we examine the effect of all suspect attributes on survival time. All condition attributes are considered to be risk factors and are included in the Kaplan-Meier method, which is used to separate the data into several subgroups. Figs. 3(a) and (b) depict the analysis of *diabetes* and *Parkinson's* factors of geriatric data, respectively.

In Fig. 3(a) we consider two possible types of diabetes diagnosis {0,1} and the same for Fig. 3(b). One can see that the two curves from these groups of patients reveal different survival characteristics. We notice a slight difference between the two groups of diabetes patients and wider differences between two groups of Parkinson's. The patients who diagnosed diabetes and not Parkinson's seem to provide better results. However, the type without Parkinson's seem to have few censor cases left as time goes by. Initially at time zero, cumulative survival is 1. In both figures, the first 15 months after admission to our study reveals very little chance of dying and two groups are visually close together. During the 15-25 month period (the most steep part of the survival functions in Fig. 3(a)), the hazard of death has clearly increased. The patients start to have less risk of dying in the following 3 years.

Figures 4(a) and 4(b) display the analysis of risk factor *walk* (whether patient can walk?) of the geriatric data and risk factor *alkal* (alkaline phosphatase in unit/liter) of the PBC data respectively. Fig. 4(a) illustrates the walk risk factor is ambiguous between patients who cannot walk (group 1) and can walk but need help (group 0.5). We illustrated the walk risk factor to be a dispensable attribute in Table 2. Alkal of PBC in the last figure is described by six possible groups and show the strong significant difference between each group.

The interpretation of Kaplan-Meier curves is only one significant part and it is not sufficient to design the most dangerous risk factor. In [9], Kaplan-Meier method and Prognostic Index (PI) are applied to head and neck cancer patients. Afterward, rough sets generate the decision rules. However, the hypothesis tests and *p-value* are not considered, thus the complete univariate analysis is required.

Since the comparison of the survival curves for two groups should be based on formal statistical tests, not visual impressions, we use the formal hypothesis tests (Sect. 2.3) to see if there is any statistical evidence for two or more survival curves being different and to complete univariate analysis. In practice, the log-rank tests are used because they do not assume any particular distribution of the survival function and are not bias to earlier period events. We then provide three statistical tests for the equality of the survival function *(degree of freedom = 1)*. The example for risk factor diabetes of geriatric data:

|             | Statistic | Significance |
|-------------|-----------|--------------|
| Log Rank    | 12.77     | .0004        |
| Breslow     | 5.78      | .0162        |
| Tarone-Ware | 8.46      | .0036        |

The effects of all risk factors on the survival curves are compared. This allows us to confirm which risk factor impacts survival time of patients significantly

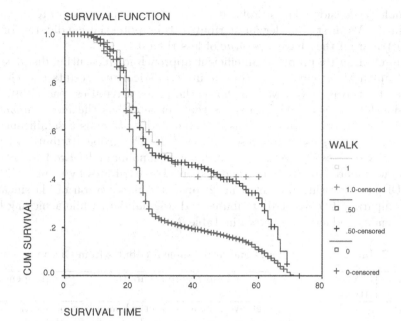

(a) Survival function of the *walk* factor from geriatric:
3 levels of whether patients can walk are compared.

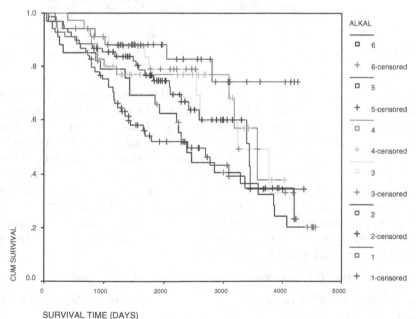

(b) Survival function of the *akal* factor from PBC: 6 levels
of alkaline phospatase in unit/liter are compared.

**Fig. 4.** Survival functions

and should be considered as a probe attribute. An example series of tests can be found in [3]. We first consider all attributes to be potential candidates for the probe attribute if they have a *p-value* of less than 0.2.

Bazan et al. [9,10] proposed an efficient approach for measuring distance between Kaplan-Meier curves. By considering statistical test results and characteristics of survival curves, we can choose the probe attributes reasonably. The conclusion for the geriatric data set is that diabetes has the most impact on survival time. In other words, $H_1$ is accepted and *diabetes* is a significant risk for geriatric patients (see hypotheses in Sect. 2.3). The probe attributes we generated are: {*diabetes*} for geriatric, {*ini2*} for melanoma and {*alkal*} for PBC.

Next, we found core attributes and dispensable attributes by using CDispro. The CDispro algorithm presented in [2] provides preservation of classification when comparing its extracted attributes and original data while achieving high dimensionality reduction as shown in Table 2.

**Table 2.** Core attributes and dispensable attributes from CDispro

| Data sets | # of original attributes | CDispro core attributes | Dispensable |
|---|---|---|---|
| geriatric | 44 | edulevel eyesi hear shower phoneuse shopping meal housew money health trouble livealo cough tired sneeze hbp heart stroke arthriti eyetroub dental chest stomac kidney bladder bowels diabetes feet nerves skin fracture age6 sex | eartroub, walk |
| melanoma | 7 | age, sex, trt | none |
| PBC | 17 | none | none |

Table 2 shows the high number of condition attributes are reduced: 44 original condition attributes are reduced to 33 attributes for the geriatric data set and 7 original condition attributes to 3 for the melanoma data set. The CDispro algorithm produces dispensable attributes as depicted in the last column. The absence of these attributes loses no dependency relationship information nor do we lose any predictive ability from the original data set. Normally, in the medical domain, the exclusion of dispensable attributes can minimize an expensive series of laboratory tests, drop high risk treatments and animal/human clinical trial.

We perform association rule generation to explore the strong relationship in the geriatric data set. The following are the strong relationship association rule examples we obtained from the geriatric data set.

***Association rule 1:*** *If (getbed=0) and (takemed=0) then (eat=0).*

***Association rule 2:*** *If (dress=0) and (takecare=0) and (getbed=0) and (parkinso=0) then (eat=0).*

Both association rules have support numbers 7921 and 7941, respectively. Both association rules have confidence equal to 1. The interpretation of these association rules are as the following.

**Table 3.** Reducts and probe reducts from Enhanced HYRIS

| Data sets | Reducts | Probe reducts |
|---|---|---|
| geriatric | edulevel eyesi hear shower phoneuse shopping meal housew money health trouble livealo cough tired sneeze hbp heart stroke arthriti eyetroub dental chest stomac kidney bladder bowels diabetes feet nerves skin fracture age6 sex | N/A |
| melanoma | age, sex, trt | age, sex, trt, ini2, ini3a |
| PBC | none | alkali, drug |

First rule: if patients can get in and out of bed and take medicine by themselves very well then these patients can eat very well.

Second rule: if patients can dress and undress, take care of their appearances and can get in and out of bed by themselves very well, (even patients that experience Parkinson's) then they are likely to eat very well.

The Probe Attribute Selection module is applied to distil traditional reducts or probe reducts (Table 3). Enhanced HYRIS produces probe reducts using probe attributes if it returns reducts that do not clarify pattern groups of survival. For example, the PBC data set use {*alkal*} as probe attribute and {*alkal, drug*} as probe reducts for handling situation with no reducts. For the method used to generate probe reducts (in Table 3) from the probe attribute, please refer to [2].

## 5    Model Construction and Validation

ELEM2 [18] is used for survival prediction model construction. We can describe particular tendencies in the survival outcome of patients, e.g., three exemplary survival prediction rules from geriatric data:

**Decision Rule 1:** If (health>0.25) and (hear=0) and (nerves=0) and (feet=0) and (heart=0) and (dental=0) and (stomac=0) and (hbp=0) and (diabet=0) and (age≤2) then (survival time = 7-18 months)

**Decision Rule 2:** If (sex=0) and (edlevel=2) and (eyesi>0) and (0<health≤ 1) and (0<hear≤0.25) and (diabet=1) and (tired=0) and (feet=0) then (survival time = 56-73 months)

**Decision Rule 3:** If (sex=0) and (age>2) and (phoneuse=0) and (bladder=1) and (trouble=1) and (health≤0.25) and (livealo=0) and (tired=0) and (hbp=0) and (diabet=0) and (kidney=0) and (nerves=0) and (skin≤0) then (survival time =56-73 months).

The first survival prediction rule for severe patients can be interpreted as the following.

- If patients are unhealthy and
- have severe hearing damage and
- nerve problem and
- foot problem and
- heart disease and
- dental disease and
- stomach disease and
- high blood pressure and
- especially those who experience diabetes
- then patients have a tendency of survival time around 7-18 months after being admitted to our study.

The rule quality (Sect. 2.5)of this rule is 1.8598. The second rule interpretation:

- if female patients have a low education level and
- eyesight problem from low to serious type and
- health problem from low to serious type and
- can hear quite well and
- do not have diabetes experience and
- are easily tired and
- have foot problems
- then patients are likely to have survival time between 56-73 months.

The rule quality is 1.5761. The interpretation of last example rule for geriatric:

- if female elderly patients and
- can use the telephone and
- have the ability to control their bladder and
- don't have trouble in their liver and
- are unhealthy and
- live alone now and
- are easily to feel tired and
- have high blood pressure and
- kidney problem and
- experience diabetes and
- have skin problem
- then they are likely to have survival time between 56-73 months.

Final rule quality is 1.8613. For more details on all generated rules and further work for rule analysis please refer to [19,20].

The sample survival prediction rules from PBC data with the rule quality 2.0810 is:

**Decision Rule:** If (age>2) and (biliru≤3) and (albumi>3) and (alkal>2) and (sgot>1) and (proth>3) then (survival time = 1,361-1,781 days).

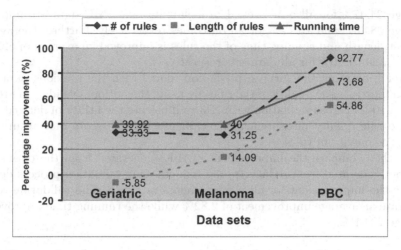

**Fig. 5.** Improved performance of the generated rules from the geriatric, melanoma and PBC data sets

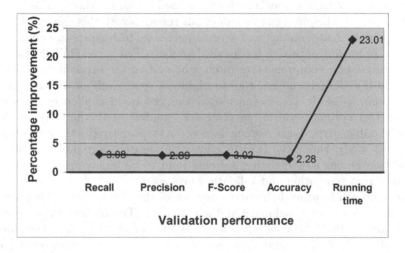

**Fig. 6.** Improved performance from 10-fold cross validation by ID3

The interpretation of this rule is: if patients are middle aged or elderly and serum bilirubin is low and albumin, alkaline phosphatase (in unit/liter), SGOT (in unit/ml) are high and prothrombin time (in seconds) are long then the survival time is approximately 4 years.

The results illustrate a compact and easy interpretation of survival prediction rules with no medical expertise required. The improvements of all rule performance compared to rule constructions from entire data and from reducts/probe reducts are depicted in Fig. 5.

Almost all rule performance outcomes are improved (except the average number of geriatric survival prediction rules). The geriatric data is improved on average 24.47% for all outcomes. The melanoma and PBC data are improved average 28.45% and 73.77% for all outcomes respectively. Further, the average number, length and running time of the rules is improved an average of 52.45%, 21.03% and 51.20% for all data sets respectively.

After generating the survival prediction model with the previous paradigm, we will illustrate the quality of the rules by using the optional validation process. We run 10-fold cross validation with ID3 to illustrate the utility of derived rules. We use rule quality measurements: recall, precision, F-score and accuracy to gauge the quality of rules.

We then compare the improvement of rules generated from the entire data to those generated from reducts/probe reducts. Our validation process demonstrates the improvement for all measurements in Fig. 6. The validation results illustrate an average improvement of 2.82% while the running time improved on average 23.01%.

# 6   Concluding Remarks and Future Work

Bless Professor Zdzislaw Pawlak who developed rough set theory, the ingenious mathematics that inspired us to conduct our research with this theory. Rough set theory, modern mathematics and several pertinent techniques are combined to provide a promising approach for survival analysis, Enhanced HYRIS. Enhanced HYRIS presents a comprehensive intelligent system for survival data analysis. Enhanced HYRIS is an ad-hoc tool to highlight specifically on survival analysis domain knowledge. It also circumvented the two main challenges for survival analysis. Our system performs survival data analysis with the consideration of censor variable, hypothesis testing and reducts/probe reducts that are original to this study. Enhanced HYRIS allows the desired degree of flexibility and generality, whereas most existing systems tend to be highly specialized toward particular rule generation and validation processes.

Our approach improved the performances of the survival analysis procedure significantly, both for benchmark and actual data. The highest improvement is 73.77% for the PBC data set while average number and length of the rules are improved over 50%. The validation results also demonstrate better results for all rule quality measures while reducing the running time by 23.01%. We plan to improve this initial step and perform further tests using Enhanced HYRIS. We will continue the development of the remaining module in our proposed hybrid intelligent system [2]. Additional research is required to explore postprocessing of achieved rules. Future work will also extend the present analysis with the resulting survival prediction models and also complete multivariate analysis.

## Acknowledgements

This research was supported by Natural Sciences and Engineering Research Council of Canada (NSERC), Canada, the King Mongkut's Institute of Technology Ladkrabang (KMITL) and Thailand research fund. Thanks are also due to Arnold Mitnitski for the geriatric data set, and the anonymous reviewers for their helpful comments.

# References

1. Larry, M.R.: Hybrid Intelligent System. Kluwer Academic Publishers, Boston (1995).
2. Pattaraintakorn, P., Cercone, N., Naruedomkul, K.: Hybrid intelligent systems: selecting attributes for soft-computing analysis. In: Proceedings of the 29th Annual International Computer Software and Applications Conference (COMPSAC2005), Edinburgh, Scotland, UK, vol. 1, IEEE Computer Society (2005) 319–325.
3. Pattaraintakorn, P., Cercone, N., Naruedomkul, K.: Selecting attributes for soft-computing analysis in hybrid intelligent systems. In: Slezak, D., Yao, J.T., Peters, J.F., Ziarko, W., Hu, X. (Eds.), Rough Sets, Fuzzy Sets, Data Mining, and Granular Computing (RSFDGrC2005), Regina, Canada, Part II, Lecture Notes in Artificial Intelligence, vol. 3642. Springer-Verlag, Heidelberg (2005) 698–708.
4. Kaplan, E.L., Meier, P.: Nonparametric estimation from incomplete observations. Journal of the American Statistical Association 53 (1958) 457–481.
5. Cox, D.R.: The analysis of exponentially distributed life-times with two types of failure. Journal of the Royal Statistical Society 21 (1959) 411–412.
6. Peto, R., Peto, J.: Asymptotically efficient rank invariant procedures. Journal of the Royal Statistical Society 135 (1972) 185–207.
7. Gehan, E.A.: A Generalized Wilcoxon test for comparing arbitrarily singly-censored data. Biometrika 52 (1965) 203–223.
8. Tarone, R.E., Ware, J.: On distribution-free tests for equality of survival distributions. Biometrika 64 (1977) 156–160.
9. Bazan, J., Osmolski, A., Skowron, A., Slezak, D., Szczuka, M., Wroblewski, J.: Rough set approach to the survival analysis. In: Alpigini, J., Peters, J., Skowron, A., Zhong, N. (Eds.), Rough Sets and Current Trends in Computing (RSCTC 2002), Malvern, PA, USA, Lecture Notes in Artificial Intelligence, vol. 2475, Springer-Verlag, Berlin, Heidelberg (2002) 522–529.
10. Bazan, J., Skowron, A., Slezak, D., Wroblewski, J.: Searching for the complex decision reducts: the case study of the survival analysis. In: Zhong, N., Ras, Z., Tsumoto, S., Suzuki, E. (Eds.), Foundations of Intelligent Systems, Lecture Notes in Artificial Intelligence, vol. 2871, Springer-Verlag, Berlin, Heidelberg (2003) 160–168.
11. Pawlak, Z.: Rough sets. Theoretical aspects of reasoning about data. Kluwer Academic Publishers, Dordrecht (1991).
12. Skowron, A., Rauszer, C.: The discernibility matrices and functions in information systems. In: Slowinski, R. (Ed.), Intelligent Decision Suppport - Handbook of Applications and Advances of the Rough Sets Theory. Dordrecht, Kluwer (1992) 331–362.
13. Nguyen, S.H., Nguyen, H.S.: Some efficient algorithms for rough set methods. In: Proceedings of the Sixth International Conference on Information Processing and Management of Uncertainty Knowledge Based Systems (IPMU1996), Granada, Spain, vol. 3, (1996) 1451–1456.
14. Nguyen, H.S.: Approximate Boolean reasoning approach to rough sets and data mining. In: Slezak, D., Yao, J.T., Peters, J.F., Ziarko, W., Hu, X. (Eds.), Rough Sets, Fuzzy Sets, Data Mining, and Granular Computing (RSFDGrC2005), Regina, Canada, Part II, Lecture Notes in Artificial Intelligence, vol. 3642, Springer-Verlag, Heidelberg (2005) 12–22.

15. Wroblewski, J.: Theoretical foundations of order-based genetic algorithms. Fundamenta Informaticae 28 (1996) 423–430.
16. Bazan, J., Nguyen, H.S., Nguyen, S.H., Synak, P., Wroblewski, J.: Rough set algorithms in classification problems. In: Polkowski, L., Lin, T.Y., Tsumoto, S. (Eds.), Rough Set Methods and Applications: New Developments in Knowledge Discovery in Information Systems, Studies in Fuzziness and Soft Computing, vol. 56, Springer-Verlag/Physica-Verlag, Heidelberg (2000) 49–88.
17. Hu, X., Han, J., Lin, T.Y.: A new rough sets models based on database systems. Fundamenta Informaticae 59(2-3) (2004) 1–18.
18. An, A., Cercone, N.: ELEM2: a learning system for more accurate classifications. In: Proceedings of the 12th Biennial Conference of the Canadian Society for Computational Studies of Intelligence on Advances in Artificial Intelligence, Lecture Notes in Computer Science, vol. 1418, Springer-Verlag, Heidelberg (1998) 426–441.
19. Pattaraintakorn, P., Cercone, N., Naruedomkul, K.: Rule analysis with rough sets theory. In: Proceedings of the IEEE International Conference on Granular Computing (IEEEGrC2006), Atlanta, USA, IEEE Computer Society (2006) 582–585.
20. Pattaraintakorn, P., Cercone, N., Naruedomkul, K.: Rule learning: ordinal prediction based on rough set and soft–computing. Applied Mathematics Letters 19 (2006) 1300–1307.
21. Quinlan, J.R.: Induction of decision trees. Machine Learning 1 (1986) 81-106.

# Rough Sets in Bioinformatics

Torgeir R. Hvidsten and Jan Komorowski

The Linnaeus Centre for Bioinformatics, Uppsala University, Uppsala, Sweden
hvidsten@lcb.uu.se, janko@lcb.uu.se

**Abstract.** Rough set-based rule induction allows easily interpretable descriptions of complex biological systems. Here, we review a number of applications of rough sets to problems in bioinformatics, including cancer classification, gene and protein function prediction, gene regulation, protein-drug interaction and drug resistance.

## 1 Introduction

Molecular biology represents a fascinating and important application area for machine learning techniques in general and rough set-based methods in particular. Although biology traditionally has been a reductionistic discipline focusing on breaking living systems into increasingly smaller parts, and on studying these parts separately, the discovery of the remarkable order and structure of these systems at the molecular level has suggested the possibility of studying their holistic molecular operation. However, it is only in the last 10 years that technological breakthroughs have made it possible to obtain large scale data that can facilitate such research. The first complete genome was sequenced in 1995 (the bacteria *Haemophilus influenzae Rd* [1]) and has been followed by many others (including the human genome, see http://www.genomesonline.org, [2]). Although important, sequence information only gives us the static code inherited from individual to individual. Other insights such as identifying the functional elements (i.e. genes) of the genomic sequence, understanding how and under which conditions genes are transcribed and translated into protein(s) (i.e. gene regulation) and determining the tasks/interactions carried out by each protein (i.e. protein function) require data on the dynamic operation of biological systems under different conditions. One example of a technology that provides this type of data is DNA microarrays that can measure the transcription levels of thousands of genes in parallel [3,4]. Moreover, developing technology will soon be able to directly perform similar large-scale measurements of proteins (i.e. proteomics, [5]).

High-throughput experimental technologies have created the need for computer programs and techniques to analyze the resulting data. The field of bioinformatics has thus developed from being a discipline mainly associated with sequence databases and sequence analysis to a computational science that uses different types of data to describe biology [6]. The ultimate goal of this research is to allow computational simulations of complex living systems. This will presumable require that we determine the function of all sequenced proteins (functional

J.F. Peters et al. (Eds.): Transactions on Rough Sets VII, LNCS 4400, pp. 225–243, 2007.
© Springer-Verlag Berlin Heidelberg 2007

genomics) and that we understand the general principles orchestrating protein regulation and interaction (systems biology).

Over the years, biologists have accumulated a large amount of knowledge about the specific functions of individual proteins. This has been accomplished through carefully chosen experimental strategies that often start out with a hypothetical function that is then confirmed or rejected in the laboratory. As the sequence databases grew larger, experimental biology was revolutionized by computational sequence similarity search methods such as BLAST [7]. These programs can align functionally uncharacterized protein sequences with protein sequences of known function, and identify statistically significant sequence similarities. Such similarities indicate that the two proteins have a common evolutionary ancestor and that, although their amino acid sequences have diverged over time, their functions have remained similar. The relationship between growth proteins and cancer was discovered in this way (by Doolittle in the early 1980s). This first success story of bioinformatics further suggested a general strategy for using machine learning in molecular biology, that is, to take advantage of the assumed relationship between high-throughput data such as sequence data and available knowledge of, for example, protein function to induce general models that represent this relationship. These models may then be used to predict function for uncharacterized proteins and thus provide experimentalists with novel hypotheses that can be tested in the laboratory. Furthermore, these models may give us valuable biological insight such as, in the case of function prediction, the location of the functional site of a protein.

One problem with the machine learning strategy to functional genomics is that most knowledge of protein function only exists as plain text in scientific publications. For this reason, text mining has been and will continue to be an important part of bioinformatics [8,9]. Furthermore, large efforts have been put into developing controlled vocabularies and data structures for representing knowledge in a computer readable form [10]. A prominent example is Gene Ontology (GO) [11]. GO consists of three sub-ontologies that describe three different aspects of protein function. Molecular functions are tasks performed by single proteins, biological processes are ordered assemblies of molecular functions that together carry out broad biological goals in the cell and cellular components are subcellular locations where proteins are active. Each ontology is a directed acyclic graph (DAG) where nodes describe, for example, molecular functions at different levels of specificity (called GO terms) and edges represent the relationships between different GO terms. For example, GO tells us that the *cell cycle* is a part-of *cell proliferation* and that the *mitotic cell cycle* is-a *cell cycle*. The advantage of GO is that the current knowledge about the function of a gene or a protein may be represented by associating it with one or more terms in GO, and that the structure of GO makes it easy to write computer programs that can compare and organize these annotations for many or all genes in a genome.

Given a set of training examples, e.g. sequences with GO annotations, the application of supervised machine learning is far from trivial. High-throughput experimental data is inevitably obscured by a relatively large amount of noise. In addition, the training examples are reflections of the currently available knowledge about the function of a protein and may thus be incomplete or, in the worst case, wrong. This problem is made worse by the fact that the biological knowledge used to build the training examples is often automatically retrieved from text or even computationally inferred from e.g. high sequence similarity. Finally, functional genomics presents us with particularly difficult challenges related to learning, including an often large number of functional classes to discriminate, examples belonging to several different functional classes and classes with few examples. These challenges make it especially important to choose good methods for validating the statistical and biological significance of the induced models. Furthermore, it demands a lot from the applied machine learning method.

Rough set theory [12,13,14] is founded on the concept of discernibility, i.e. that data may be described only in terms of what differentiates relevant classes of observations. From the concept of discernibility, decision rules are constructed by extracting minimal information needed to uphold the discernibility structure in the data set [15,16]. The fact that the framework does not attempt do discern objects that are equal or objects that are from the same class (e.g. have the same function), makes it possible to describe incomplete and conflicting data in terms of easily interpretable decision rules. In this article, we will review some of the successful studies in which rough set-based rule induction has been used to describe biological systems at the molecular level. These studies include

- cancer classification using gene expression data,
- prediction of the participation of genes in biological processes based on temporal gene expression profiles,
- modeling of the combinatorial regulation of gene expression,
- prediction of molecular function from protein structure,
- prediction of protein-ligand interactions in drug discovery, and
- modeling of drug resistance in HIV-1

and are modeled using rules such as

- **IF** Gene A is up-regulated AND Gene D is down-regulated
  **THEN** Tissue is healthy
- **IF** Transcription factor F binds AND Transcription factor V binds
  **THEN** Gene is co-regulated with Gene H
- **IF** Protein contains motif J
  **THEN** Function is magnesium ion binding OR copper ion binding
- **IF** Protein contain motif D AND Ligand water-octanol coeff. $> c_1$
  **THEN** Binding affinity is high
- **IF** Change in frequency of alpha-helix at position X $> c_3$
  **THEN** Resistant to drug W

In particular, we will focus on how these application areas have been coded in a discrete manner to facilitate rule induction, how biological knowledge can be incorporated into this representation process and what can be read out of the rule model in terms of biological insight. Technical details will not be discussed here and may be found in the respective publications.

## 2   Gene Expression Analysis

The complementary nature of the DNA double helix is of great importance to replication and transcription in living organisms, and may also be utilized for the large-scale measurement of mRNA levels in cells. Two complementary nucleic acid molecules (i.e. strands) will combine under the right conditions to form double stranded helices. In a reaction vessel this is referred to as hybridization. Hence, it is possible to use identified DNA strands (probes) to query complex populations of unidentified, complementary strands (targets) by checking for hybridization. Microarrays are glass slides or wafers populated with large numbers of strands derived from identified genes. By applying a target sample of unidentified mRNA to the array, the expression level of each gene probe may be quantified from the extent of hybridization between the probes and the targets. Since one slide may contain probes from thousands of genes, one microarray experiment may determine the genome-wide expression state of a cell sample. Furthermore, systematic series of microarray experiments may reveal the specific changes in cellular gene expression associated with different physiological or pathophysiological responses. A microarray study comprises a number of steps including experimental design [17], filtering and normalization of the data [18] and high-level computational data analysis. The last step was in the early phase of microarray analysis mostly restricted to clustering analysis, and in particular, hierarchical clustering [19]. However, the limitations of clustering both in terms of interpretation and evaluation soon saw a shift in focus from unsupervised learning (i.e. clustering) to supervised learning [20,21].

### 2.1   Cancer Classification

Standard medical classification systems for cancer tumors are based on clinical observations and the microscopical appearance of the tumor. These systems fail to recognize the molecular characteristics of the cancer that often corresponds to subtypes that need different treatment. Studying the expression levels of genes in tumor tissue may reveal such subtypes and may also diagnose the disease before it manifests itself on a clinical level. Thus, the goal of data analysis of cancer microarray data is to develop models for earlier detection and better understanding and treatment of cancer.

**Gastric Carcinoma.** Midelfart *et al.* [22,23] used rough set-based classifiers to identify molecular markers that allow classification of gastric carcinoma. Gastric carcinoma is often not detected until at an advanced stage, which is one of the reasons why this is the second most frequent cause of cancer death world-wide. The study developed classifiers for six different clinical parameters; intestinal or diffuse types (also known as the Lauren classification), site of primary tumor (cardia, corpus or antrum), penetration of the stomach wall or not, lymph node metastasis or not, remote metastasis or not, and high or normal serum gastrin. The expression levels of 2504 genes were measured in tumor samples taken from only 17 patients. Rule models were induced in a leave-one-out cross-validation procedure for each of the six clinical parameters. In each iteration, the 10 to 40 most differentially expressed genes were identified using a bootstrap t-test [24]. By differentially expressed genes, we here mean genes that showed a consistently higher expression in e.g. intestinal samples compared to the diffuse samples as measured by the bootstrap t-test. The expression of these genes was discretized into e. g. low, medium and high expression and rules where induced. Classification accuracy and area under the receiver operating characteristics curve (AUC) [25] were reported for all six clinical parameters ranging from 0.79 to 1.00 (average 91.5) and 0.66 to 1.00 (average 0.89), respectively.

A particularly difficult challenge in cancer classification from microarray data is the large number of measured genes compared to the number of cancer patients. This is a problem because one is faced with a huge search space (i.e. subsets of 2504 genes) and only a few data points to restrict the search. A possible consequence could be overfitting, that is, decision rules that explain the training set, but fail to generalize to the test set. In this study, reduct computation was limited to a low number of differentially expressed genes. However, the number of genes compared to the number of patients still makes it difficult to exclude the possibility that some of these genes are discriminatory by chance. Thus, to further add robustness to the identification of gastric carcinoma markers, the study reported as a measure of strength the number of cross validations in which a particular gene was part of at least one decision rule in the rule model. This resulted in the identification of several genes known to be highly expressed in gastric carcinomas as well as several interesting new genes.

The rule induction process offer a number of algorithms for discretization and reduct computation. Combined with a low number of training examples, these options constitute a real risk that even cross validation estimates may be optimistic in the sense that they do not reflect a true ability to correctly classify unseen samples. The authors of the study realized this, and consecutively repeated the cross validation procedure for each of the six clinical parameters on 2000 dataset where the clinical parameter values were randomly shuffled. By recording the fraction of randomized data sets that resulted in a higher AUC value than the real data set, they obtained a p-value for each clinical parameter reflecting the probability that the reported AUC value could be obtained by chance [26]. Even though the initial cross validation estimates looked impressive, this careful analysis showed that the AUC value of three of the six clinical parameters were not statistically significant at p-value 0.05. However, location of tumor ($p < 0.031$), lymph node metastasis ($p < 0.007$) and the Lauren classification ($p < 0.007$) were shown to be adequately described by the rule model.

**Adenocarcinoma.** Dennis et al. [27] used rough sets to build a classification system for identifying the primary site of cancer based on expression levels in a sample taken from a secondary tumor. While it is the primary tumor that causes symptoms in most patients, about 10-15% of cancers are discovered as metastases in solid organs, body cavities or lymph nodes. Most of these secondary tumors are adenocarcinomas, for which the seven commonest primary sites are

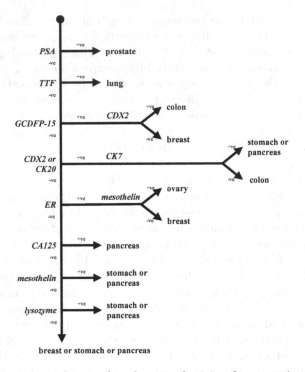

**Fig. 1.** Decision tree used to predict the site of origin of metastatic cancer from 10 molecular markers [27]

breast, colon, lung, ovary, pancreas, prostate and stomach. Because prognosis and therapy are linked to the site of origin, and because histologically such tumors appear similar, finding molecular markers for these sites could greatly improve treatment.

The study assessed the expression patterns of 27 markers in 452 adenocarcinoma patients. 12 markers were scored as either present or absent (+ or -), while the remaining markers were scored as absent, weak, intermediate or strong (0, 1, 2 or 3). Decision rules were induced from 352 adenocarcinomas and used to build a decision tree of 10 markers (see Figure 1). This tree was then used to predict the site of origin of 100 unseen adenocarcinomas with a success rate of 88%. This is a very high accuracy considering there were seven different sites to predict, and indicate a huge potential for molecular markers in identifying the primary site of these cancers.

## 2.2 Predicting Participation of Gene Products in Biological Processes

Hvidsten *et al.* [28,29,30] developed a method for modeling the participation of gene products in GO biological process from temporal expression profiles. Several publications had earlier used hierarchical clustering to illustrate the correspondence between expression similarity and gene function [31,32,19]. However, none of these studies actually quantified the relationship. Furthermore, it is known that functionally related genes often are anti-coregulated and that genes usually are associated with more than one function. These aspects are not well modeled by a set of broad, non-overlapping expression clusters. Brown et al. [21] was the first to approach the problem in a supervised manner by using support vector machines to predict a limited set of six functional categories from expression data.

In the 2001 paper [28], a template language was proposed to describe the discrete changes in expression over subsets of time points in an expression time profile. The idea behind this language was that the relative change in mRNA levels over limited periods of time is more important to distinguish one biological process from another than the absolute mRNA levels given by each time point. Furthermore, rough set-based rule induction was used to associate combinations of discrete changes in expression with one or a small number of GO biological processes. For example, the rule

**IF**      0 - 4(constant) AND 0 - 10(increasing)
**THEN** GO(protein metabolism and modification) OR
         GO(mesoderm development) OR GO(protein biosynthesis)

and the rule

**IF**      0h-4h (increasing) AND 6h-10h (decreasing)
         AND 14h-18h (constant)
**THEN** GO (cell proliferation) OR GO (cell-cell signaling) OR
         GO (intracellular signaling cascade) OR GO (oncogenesis)

describe the limited set of biological processes (THEN-part) associated with particular expression profile constraints (IF-part, e.g. 0h-4h (increasing) means increasing expression level from 0 to 4 hours). The first rule has a support of five genes, four of which are annotated to *protein metabolism and modification*. The second rule has support of four, three of which were annotated to *cell proliferation*. Thus the main reason for indeterministic rules is that genes are annotated to several different GO terms.

The predictive performance of the approach was tested using cross validation on all annotated genes in two expression time profile data sets with human genes [19,33]. Thus the correspondence between expression similarity and GO biological process was properly evaluated and quantified for 23 and 27 biological process, respectively. Each biological process was subjected to a permutation test that showed that most of the classes indeed could be predicted with a statistically significant AUC value not obtainable by chance.

The cross validation results may in general be considered estimates of the prediction quality one can expect when predicting functionally uncharacterized genes using a model induced from *all* training examples. However, predictions to uncharacterized genes were also evaluated directly by searching for homology information that could be used to make assumptions about the biological processes of these genes [30]. Of the 24 genes where such assumptions could be made, 11 genes had one or more classifications that matched this assumption.

In addition to predicting the biological process of uncharacterized genes, a model induced from all examples was also used to re-classify characterized genes [30]. The resulting false positives were then used to guide a second literature search for possible missing annotations (i.e. information on biological process annotations existing in the literature, but overlooked during the initial literature search). Of the 14 genes with a false positive re-classification to DNA metabolism, four were found to actually participate in this process. Furthermore, it was revealed that 12 of the 24 false positive re-classifications to oncogenesis also represented missing annotations. Thus, it was shown that computational models could be used directly both to guide new literature searches for partially characterized genes and to propose new functional hypotheses for unseen genes.

The studies described here all used a set of predefined biological processes as basis for learning. Midelfart *et al.* [71-73] later introduced rough set-based rule classifiers that actively learn in the Gene Ontology graph, dynamically selecting biological processes with the best predictive performance.

## 2.3   Gene Regulation

One of the major challenges faced by molecular biology is to dissect the regulatory circuitry of living cells. The ability of transcription factors to selectively bind specific DNA motifs (i.e. transcription factor binding sites) in the regulatory regions of genes is essential for the complex regulation systems observed in living organisms. The assumption that genes regulated by the same transcription factors (i.e. co-regulated) should contain common binding sites and exhibit similar expression (i.e. co-expressed) enables the study of gene regulation at a genome-wide scale using sequence and expression data.

Pilpel *et al.* [34] found that genes sharing pairs of binding sites are significantly more likely to be co-expressed than genes with only single binding sites in common. This result is in agreement with the hypothesis that a limited number of transcription factors combine in various ways in order to respond to a large number of various stress conditions.

Hvidsten *et al.* [35] used rough set-based rule induction to perform a comprehensive analysis of the combinatorial nature of gene regulation in yeast. The method extracted IF-THEN rules of minimal binding site combinations or modules (IF-part) shared by genes with a common expression profile (THEN-part). The rules hence described general, underlying relationships in an easily understandable format, providing hypotheses on combinatorial co-regulation that may later be experimentally validated.

The approach was tested on a database of known and putative regulatory sequence motifs in yeast [36] using six expression data sets including one cell cycle study and five studies including different stress conditions [34]. The rule learning framework was subsequently applied to each gene to obtain rules that associate the expression profile of that gene with a minimal binding site combination shared by similarly expressed genes. Rules were then discarded if they did not provide a clear and general pattern in terms of modules associated with several genes where a majority had similar expression. Only in these cases the evidence for actual co-regulation was considered sufficiently strong.

The discovered binding site modules were evaluated using transcription factor binding interactions provided by a genome-wide location analysis [37] and Gene Ontology annotations. The evaluation clearly showed that the retrieved binding site modules reflected actual co-regulation and furthermore showed that genes associated with these modules very often share biological roles in terms of biological process, molecular function and cellular component. The results were statistically significant compared to genes either associated with a randomly chosen set of binding sites, similar expression or neither of these constraints.

Two rules were discussed as a case study and had support in the literature. As an example, the rule

IF RAP1 AND MCM1 AND SWI5 **THEN** Similar expression

describes eight genes and suggests that the transcription factor RAP1 (that regulates genes that encode ribosomal proteins in growing yeast cells, but also other non-ribosomal genes) requires the cell cycle regulating transcription factors MCM1 and SWI5 to be present when specifically targeting ribosomal genes in growing yeast. That is, the ribosomal genes targeted by RAP1 are only regulated when the cell is in the cell cycle (i.e. growing) which is when MCM1 and SWI5 are present. RAP1 presumable combines with other transcription factors when regulating other non-ribosomal genes.

By applying the method to expression data obtained under several different conditions the authors were able to discover a number of binding site modules common to several of these responses in addition to modules that seem to be exclusive to a particular stress response. The overlap between modules clearly shows the large extent to which relatively few transcription factors combine to facilitate a much large number of expression outcomes.

A later follow-up study [38] used expression similarity restricted to subintervals of cell cycle time profiles (similar to the template language discussed in section 2.2), and showed that this improvement greatly increased the biological significance of the retrieved modules as well as making it possible to retrieve modules that were not detectable using expression similarity over the whole time profile. A second follow-up study [39] refrained from using expression similarity altogether. Instead, this study used prior knowledge of the cell cycle period time to detect different classes of periodically expressed genes in three different synchronization studies, and then used rough set-based rule learning to describe the regulatory mechanisms behind these classes. These mechanisms were then shown to be much more specific towards the cell cycle machinery than mechanisms discovered from expression clusters, and thus showed the advantage of incorporating biological knowledge into the data analysis process whenever it is possible.

## 3  Protein Analysis - Function and Interaction

It is believed that sequence similarity search methods can identify functionally characterized homologues for less than 50% of the proteins predicted from genome sequencing projects. However, even though global sequence similarity between distantly related proteins may be virtually undetectable, similarities may still be present in terms of conserved amino acids in the functional sites (functional sites are known to be more conserved than the overall sequence), conserved global structure (structure is known to be more conserved than sequence) or conserved local structure related to the functional site (again, functional sites are more conserved also in terms of structure). Thus more advanced sequence similarity methods and methods using structural similarities may represent a solution for proteins where functional hypotheses cannot be obtained from global sequence similarity [40]. Unfortunately, the Protein Data Bank (PDB) [2] only contains around 30 thousands protein structures while there are about 30 million protein sequences in UniProt (Universal Protein Resource) [41]. To remedy this situation, structural genomics projects systematically aim at solving protein structures for new protein families [42], using these structures as templates for in silico structure prediction methods (i.e. homology modeling) [43], and then applying the solved and predicted structures to infer function [44]. However, to be successful this strategy requires new and improved methods that utilize structure to predict function and interactions.

Here we will review research using rough set-based rule induction to model protein function and interaction. Two of these studies describe protein structure in terms of local descriptors of protein structure. A local descriptor is defined by A. Kryshtafovych and K. Fidelis as a set of short backbone fragments centered in three dimensional space around a particular amino acid [45]. By generating local descriptors for all amino acids and all proteins in PDB, and by clustering these descriptors into groups of structurally similar descriptors, it is possible to build a library of a few thousand local building blocks from which virtually all proteins in PDB may be assembled (see Figure 2). This library of recurring local substructures may then be used for representing and comparing protein structure.

## 3.1    Function Prediction from Structure

Although global structural similarity is often a sign of function similarities [46], many folds such as the TIM barrel and the Rossmann fold are found in proteins with many different functions. Thus local similarity methods are more powerful in these cases [47]. Recently, researchers have started building tools that use a large number of different features including both local and global structure [48,49]. These so-called meta-servers obtain functional predictions by allowing a large number of different evidence to vote, and then selecting the most likely function. However, such approaches do not construct explicit models that are often very useful in further analysis.

Hvidsten et al. [50] proposed a change in this paradigm by inducing IF-THEN rules that associate combinations of local substructures with specific protein functions (Figure 3). This approach differs from other studies in that the applied library of local substructures encompasses *all* recurring motifs and *all* annotated proteins using no prior knowledge of functional sites or any sequence information, and in that the structure-function relationship is explicitly represented in

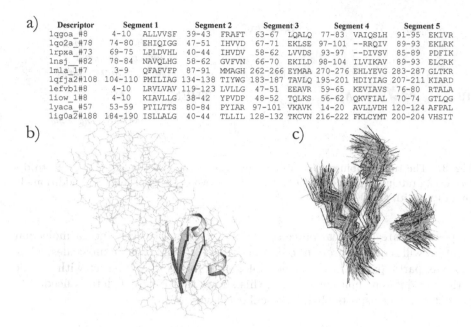

a)

| Descriptor | Segment 1 | | Segment 2 | | Segment 3 | | Segment 4 | | Segment 5 | |
|---|---|---|---|---|---|---|---|---|---|---|
| 1qgoa_#8 | 4-10 | ALLVVSF | 39-43 | FRAFT | 63-67 | LQALQ | 77-83 | VAIQSLH | 91-95 | EKIVR |
| 1qo2a_#78 | 74-80 | EHIQIGG | 47-51 | IHVVD | 67-71 | EKLSE | 97-101 | --RRQIV | 89-93 | EKLRK |
| 1rpxa_#73 | 69-75 | LPLDVHL | 40-44 | IHVDV | 58-62 | LVVDS | 93-97 | --DIVSV | 85-89 | PDFIK |
| 1nsj__#82 | 78-84 | NAVQLHG | 58-62 | GVFVN | 66-70 | EKILD | 98-104 | ILVIKAV | 89-93 | ELCRK |
| 1mla_1#7 | 3-9 | QFAFVFP | 87-91 | MMAGH | 262-266 | EYMAA | 270-276 | EHLYEVG | 283-287 | GLTKR |
| 1qfja2#108 | 104-110 | PMILIAG | 134-138 | TIYWG | 183-187 | TAVLQ | 195-201 | HDIYIAG | 207-211 | KIARD |
| 1efvb1#8 | 4-10 | LRVLVAV | 119-123 | LVLLG | 47-51 | EEAVR | 59-65 | KEVIAVS | 76-80 | RTALA |
| 1iow_1#8 | 4-10 | KIAVLLG | 38-42 | YPVDP | 48-52 | TQLKS | 56-62 | QKVFIAL | 70-74 | GTLQG |
| 1yaca_#57 | 53-59 | PTILTTS | 80-84 | PYIAR | 97-101 | VKAVK | 14-20 | AVLLVDH | 120-124 | AFPAL |
| 1ig0a2#188 | 184-190 | ISLLALG | 40-44 | TLLIL | 128-132 | TKCVN | 216-222 | FKLCYMT | 200-204 | VHSIT |

b)

c)

**Fig. 2.** An example of a local descriptor of protein structure (b), its structural neighbors (c) and the resulting sequence alignment (a)

a descriptive model. Moreover, the structure-function relationship in proteins was quantified by assessing the predictive performance of the model using cross validation and AUC analysis. The main conclusions that could be drawn from this study were as follows:

- A majority of the 113 molecular functions could be predicted with a statistically significant accuracy as assessed by a permutation study.
- GO molecular functions were better predicted than GO biological processes or GO cellular components.
- Combinations of local similarities allowed discerning proteins with different functions, but similar global structure (i.e. fold), e.g. the TIM barrel and the Rossmann fold.
- Catalytic activities were better predicted than most functions involving binding.
- Structure-based predictions complemented sequence-based predictions and also provided correct predictions when no significant sequence similarities to characterized proteins existed.

It has previously been observed that GO biological processes are better explained by expression data than are GO molecular functions [21,35]. This is intuitive, since genes participating in the same biological process need to be transcribed at the same point in time. However, it is interesting to observe in this study that molecular functions are better explained by specific structural

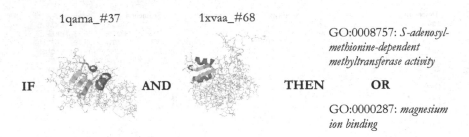

**Fig. 3.** The rule combines two substructure (i.e. 1qama_#37 and 1xvaa_#68) to describe 12 proteins annotated with GO:0008757. Two of these proteins are additionally annotated with GO:0000287.

shapes than are biological processes. Again, this is intuitive since a molecular function will require a protein to interact with a specific type of molecules, while proteins participating in the same biological process may interact with a wide variety of different molecules. Thus, this shows that different data is needed to predict different aspects of the molecular activity of proteins.

## 3.2   Protein – Ligand Interactions

An important goal of modern drug discovery is to develop computational models that can predict the interactions between drug targets (i.e. proteins) and ligands (i.e. drugs). A common approach in this research is QSAR (Quantitative Structure Activity Relationship), where the interaction between one protein and a series of ligands is modeled, and docking, where the three-dimensional structure of the protein is used to model the protein-ligand complex [51]. Proteochemometrics (PCM) [52] takes a different approach to molecular recognition in which the protein-ligand interaction space is modeled using series of *both* proteins and ligands. This approach greatly reduces the number of known interactions needed for modeling and may predict cross-interactions between drugs and other proteins in the proteome.

PCM uses machine learning methods to model the degree to which proteins and ligands interact (i.e. experimentally measured binding affinity) using chemical and structural descriptors to represent the proteins and the ligands. Strömbergsson *et al.* [53] used rough set-based rule learning to model interactions between G-Protein-coupled receptors (GPCR) and ligands. GPCRs are membrane-bound proteins that share a conserved structural topology of seven transmembrane helices. GPCRs are of particular interest, since about 50% of all recently launched drugs are targeted towards these receptors. The main novel result of this study was that rules allowed direct interpretation of the model, something that is not possible with the commonly used linear regression approach. For example, the rule model suggested that helix 2 was determinative for high and low binding affinity in three different data sets.

Previous approaches to PCM modeling have used protein descriptors that are calculated from a multiple alignment of the studied proteins. This limits modeling to closely related proteins in terms of sequence or structure. In Strömbergsson *et al.* [54], the authors showed that using local descriptors of protein structure one can model vastly different proteins both in terms of sequence and structure. It was shown that the induced rule model combined local substructures and ligand descriptors to generalize beyond the enzyme-ligand interactions present in the training set. An interesting interpretation from the rules was that strongly bound enzyme-ligand complexes were described in terms of the presence of specific local substructures, while weakly bound complexes were described by the absence of certain local substructures. This is intuitive, since there may be only one or a few ligands that geometrically fit the active site of a specific enzyme and form a strongly bound complex, while there may be many ligands that only form weakly bound complexes with the same enzyme. The preferred description of the latter is to point to the absence of the local substructure that, if present, would have resulted in a strongly bound complex.

## 3.3   HIV-1 Modeling

The HIV virus has a high rate of replication leading to mutations and the development of drug resistance. The HIV-1 protease plays an essential role in replication by cleaving the viral precursor Gag and Gag-Pol polyproteins into structural and functional elements. For this reason the HIV-1 protease has been an attractive targets for the design of drugs that inhibit the protease and thus stop the replication of the HIV virus.

The HIV-1 protease cleaves the viral polyprotein by recognizing a sequence represented by four amino acids on each side of the actual position of cleavage. Kontijevskis *et al.* [55] collected all the experimental data on cleavable and non-cleavable sites from 16 years of HIV research (374 cleavable and 1251 non-cleavable substrates). Decision rules were induced based on the physico-chemical properties of the amino acids in these substrates, and cross validation demonstrated high predictive performance (accuracy and AUC well above 0.90). While previous studies based on less comprehensive data sets have revealed some patterns of limited predictive ability, analysis of this model showed that the rules encompassed properties from at least three substrate positions indicating a more complex relation than previously assumed. Nonetheless, as the cross validation evaluation showed, the rough set-based approach did recover general patterns determining HIV-1 protease cleavage specificity and several novel patterns were reported in the paper.

The HIV-1 reverse transcriptase (RT) transforms the viral RNA into DNA that can be incorporated into the genetic material of the host cell. Kierczak *et al.* [56] used rough set-based rule induction to predict drug resistance to six different drugs for a large number of mutated RTs. Existing biochemical knowledge related to the sequence and structure of RT was used to build descriptors from 19 known resistance-related positions. Cross validation accuracy and AUC values were in the ranges of 0.82-0.94 and 0.70-0.97, respectively, for the different drugs. As for the study of Kontijevskis *et al.*, the rules were pruned and inspection revealed general and novel patterns important for drug resistance.

# 4   Discussion

In this paper, we have reviewed a number of publications where rough set-based rule learning has been used to predict and describe molecular properties of biological systems. And we have seen how discrete representations and legible rules allow interpretations that gain new insight into molecular biology. The ability to describe data in terms of legible rules is particularly important in biology where biologists are interested in understanding the mechanisms underlying the data just as much as they are interested in the predictions themselves. Moreover, discrete representations add to this readability and allow the models to combine different heterogenous data sources containing both continuous and categorical data. Furthermore, the elegant representation of indeterministic data in terms of disjunctions of decisions in rules makes otherwise difficult problems, such as proteins annotated to several GO terms, easy to handle. Finally, developments in rough set-based rule induction such as dynamic reducts [57] and approximate reducts [16,58] allow the description of noisy data. We believe that it is due to these properties that rough sets now have gained a wide acceptance as a powerful tool for data analysis in life sciences. Pawlak's ideas were simple, yet powerful and rich enough to be of outstanding practical use in biology, and also continue to stimulate theoretical research in computer science.

Several challenges are particularly interesting in the context of rough sets and molecular biology. The first challenge is that of developing methods for illustrating and pruning rules [59,60] in order to allow interpretation. Some methods were reviewed in this article. In Dennis *et al.* [27], rules were represented as a decision tree, something which is very familier to physicians. In Hvidsten *et al.* [35], predicted regulatory mechanisms with inconclusive evidence or low support were not considered. This is a simple yet powerful approach to rule filtering, but is also dangerous since potentially important discoveries may be lost [61,62]. Finally, the HIV-1 studies [55,56] used a group generalization method for rule pruning, where groups of rules with overlapping IF-parts and identical decisions are merged into generalized rules if the accuracy do not fall below a predefined threshold [60]. The second challenge is that of feature selection in order to avoid overfitting. In this review we saw a statistical method for selecting differentially expressed genes in a cancer classification study [22,23]. However, this procedure will exclude genes that individually are not significant, but that posses a significant discriminatory power in combination with other genes. Sampling methods such as random forests [63] might offer a solution to this problem in feature selection since they investigate the classification power of more than one gene at a time using subsets of all features. The third challenge is that of representing biological systems in a way that allows effective machine learning (i.e. feature synthesis). Examples discussed here were a template language for representing expression time profiles and local descriptors to represent protein structure. More than the development of computational methods themselves, we believe that the development of new ways to represent biological systems is the most important in order to successfully solve the puzzles of molecular biology. This also includes a final challenge, namely that of combining various sources of data in the representation process such as, for example, using both molecular markers and clinical data in cancer classification. The significance of the second and third challenges was already recognized my A. Skowron in 1995 [64].

PubMed (http://www.ncbi.nlm.nih.gov/PubMed) is the main database providing access to all published biomedical literature. Searching for "rough set(s)" in titles and abstracts of articles in this database gives 69 hits since 1988 and reveal a large number of application areas beyond those described here. The true number of articles using rough sets in life sciences, however, is probably much higher since this search was limited to title and abstract and since only four of the papers reviewed in this article were retrieved by the search. Google Scholar (http://scholar.google.com), which searches through the whole text of all available scientific publications online, returned 290 articles with "rough set(s)" and "bioinformatics", and 11900 articles with "rough set(s)".

The published studies reviewed here all used the ROSETTA system, which is a user friendly, freely available software package for rough set-based rule induction and model evaluations [65] (http://rosetta.lcb.uu.se/).

## Acknowledgements

We would like to thank co-authors of the reviewed articles for a stimulating collaboration. In particular Astrid Lægreid for her continuous help with all issues related to biology. The ROSETTA system has been an essential aid in this research. It was mainly developed by Alexander Øhrn under the supervision of Jan Komorowski, Trondheim, and in collaboration with Andrzej Skowron's group in Warsaw.

This research was supported by grants from the Knut and Alice Wallenberg Foundation (in part through the Wallenberg Consortium North), the Swedish Research Council, and the Swedish Foundation for Strategic Research.

## References

1. Fleischmann, R.D., Adams, M.D., White, O., Clayton, R.A., Kirkness, E.F., Kerlavage, A.R., Bult, C.J., Tomb, J.F., Dougherty, B.A., Merrick, J.M.: Whole-genome random sequencing and assembly of Haemophilus influenzae Rd. Science **269** (1995) 496–512
2. Berman, H.M., Westbrook, J., Feng, Z., Gilliland, G., Bhat, T.N., Weissig, H., Shindyalov, I.N., Bourne, P.: The protein data bank. Nucleic Acids Research **28** (2000) 235–242
3. Schena, M., Shalon, D., Davis, R., Brown, P.O.: Quantitative monitoring of gene expression patterns with a complementary dna microarray. Science **270** (1995) 467–470
4. Duggan, D.J., Bittner, M., Chen, Y., Meltzer, P., Trent, J.M.: Expression profiling using cDNA microarrays. Nat Genet **21** (1999) 10–14
5. Patterson, S.D., Aebersold, R.H.: Proteomics: the first decade and beyond. Nat Genet **33 Suppl** (2003) 311–323
6. Kanehisa, M., Bork, P.: Bioinformatics in the post-sequence era. Nat Genet **33 Suppl** (2003) 305–310

7. Altschul, S.F., Madden, T.L., Schaffer, A.A., Zhang, J., Zhang, Z., Miller, W., Lipman, D.J.: Gapped blast and psi-blast: a new generation of protein database search programs. Nucleic Acids Research **25** (1997) 3389–3402

8. Shatkay, H., Feldman, R.: Mining the biomedical literature in the genomic era: an overview. J Comput Biol **10** (2003) 821–855

9. Jenssen, T.K., Lægreid, A., Komorowski, J., Hovig, E.: A literature network of human genes for high-throughput analysis of gene expression. Nat Genet **28** (2001) 21–28

10. Brazma, A., Krestyaninova, M., Sarkans, U.: Standards for systems biology. Nat Rev Genet **7** (2006) 593–605

11. The Gene Ontology Consortium: Gene ontology: tool for the unification of biology. Nature Genetics **25** (2000) 25–29

12. Pawlak, Z.: Rough sets. International Journal of Information and Computer Science **11** (1982) 341–356

13. Pawlak, Z.: Rough Sets: Theoretical Aspects of Reasoning about Data. Volume 9 of Series D: System Theory, Knowledge Engineering and Problem Solving. Kluwer Academic Publishers, Dordrecht, The Netherlands (1991)

14. Komorowski, J., Pawlak, Z., Polkowski, L., Skowron, A.: Rough sets: A tutorial. In Pal, S.K., Skowron, A., eds.: Rough Fuzzy Hybridization: A New Trend in Decision-Making. Springer, Singapore (1999) 3–98

15. Skowron, A., Rauszer, C.: The discernibility matrices and functions in information systems. In Słowiński, R., ed.: Intelligent Decision Support: Handbook of Applications and Advances in Rough Sets Theory. Volume 11 of Series D: System Theory, Knowledge Engineering and Problem Solving. Kluwer Academic Publishers, Dordrecht, The Netherlands (1992) 331–362

16. Skowron, A., Nguyen, H.S.: Boolean reasoning scheme with some applications in data mining. In [66] 107–115

17. Churchill, G.A.: Fundamentals of experimental design for cDNA microarrays. Nat Genet **32 Suppl** (2002) 490–495

18. Quackenbush, J.: Microarray data normalization and transformation. Nat Genet **32 Suppl** (2002) 496–501

19. Iyer, V.R., Eisen, M.B., Ross, D.T., Schuler, G., Moore, T., Lee, J.C.F., Trent, J.M., Staudt, L.M., Jr., J.D., Boguski, M.S., Lashkari, D., Shalon, D., Botstein, D., Brown, P.O.: The transcriptional program in the response of human fibroblasts to serum. Science **283** (1999) 83–87

20. Golub, T., Slonim, D., Tamayo, P., Huard, C., Gaasenbeek, M., Mesirov, J., Coller, H., Loh, M., Downing, J., Caligiuri, M., Bloomfield, C., Lander, E.: Molecular classification of cancer: class discovery and class prediction by gene expression monitoring. Science **286** (1999) 531–537

21. Brown, M.P.S., Grundy, W.N., Cristianini, N., Sugnet, C.W., Furey, T.S., Ares, M., Haussler, D.: Knowledge-based analysis of microarray gene expression data by using support vector machines. Proc. Natl. Acad. Sci. USA **97** (2000) 262–267

22. Midelfart, H., Komorowski, J., Nørsett, K., Yadetie, F., Sandvik, A., Lægreid, A.: Learning rough set classifiers from gene expression and clinical data. Fundamenta Informaticae **53** (2002) 155–183

23. Nørsett, K.G., Lægreid, A., Midelfart, H., Yadetie, F., Erlandsen, S.E., Falkmer, S., Grønbech, J.E., Waldum, H.L., Komorowski, J., Sandvik, A.K.: Gene expression based classification of gastric carcinoma. Cancer Lett **210** (2004) 227–237

24. Efron, B., Tibshirani, R.J.: An Introduction to the Bootstrap. Chapman & Hall, London (1993)

25. Hanley, J.A., McNeil, B.J.: The meaning and use of the area under a receiver operating characteristic (ROC) curve. Radiology **143** (1982) 29–36
26. Manley, B.F.J.: Randomization, Bootstrap and Monte Carlo Methods in Biology. Chapman & Hall (2002)
27. Dennis, J.L., Hvidsten, T.R., Wit, E.C., Komorowski, J., Bell, A.K., Downie, I., Mooney, J., Verbeke, C., Bellamy, C., Keith, W.N., Oien, K.A.: Markers of adenocarcinoma characteristic of the site of origin: Development of a diagnostic algorithm. Clin Cancer Res **11** (2005) 3766–3772
28. Hvidsten, T.R., Komorowski, J., Sandvik, A.K., Lægreid, A.: Predicting gene function from gene expressions and ontologies. In Altman, R.B., Dunker, A.K., Hunter, L., Lauderdale, K., Klein, T.E., eds.: Pacific Symposium on Biocomputing, Mauna Lani, Hawai'i, World Scientific Publishing Co. (2001) 299–310
29. Hvidsten, T.R., Lægreid, A., Komorowski, J.: Learning rule-based models of biological process from gene expression time profiles using gene ontology. Bioinformatics **19** (2003) 1116–1123
30. Lægreid, A., Hvidsten, T.R., Midelfart, H., Komorowski, J., Sandvik, A.K.: Predicting gene ontology biological process from temporal gene expression patterns. Genome Res **13** (2003) 965–979
31. Eisen, M., Spellman, P., Brown, P., Botstein, D.: Cluster analysis and display of genome-wide expression pattern. Proc. Natl. Acad. Sci. USA **95** (1998) 14863–14868
32. Brown, P.O., Botstein, D.: Exploring the new world of the genome with DNA microarrays. Nat Genet **21** (1999) 33–37
33. Cho, R.J., Huang, M., Campbell, M.J., Dong, H., Steinmetz, L., Sapinoso, L., Hampton, G., Elledge, S.J., Davis, R.W., Lockhart, D.J.: Transcriptional regulation and function during the human cell cycle. Nature Genetics **27** (2001) 48–54
34. Pilpel, Y., Sudarsanam, P., Church, G.M.: Identifying regulatory networks by combinatorial analysis of promoter elements. Nature genetics **29** (2001) 153–159
35. Hvidsten, T.R., Wilczyński, B., Kryshtafovych, A., Tiuryn, J., Komorowski, J., Fidelis, K.: Discovering regulatory binding-site modules using rule-based learning. Genome Res **15** (2005) 856–866
36. Hughes, J.D., Estep, P.W., Tavazoie, S., Church, G.M.: Computational identification of cis-regulatory elements associated with groups of functionally related genes in Saccharomyces cerevisiae. J Mol Biol **296** (2000) 1205–1214
37. Lee, T.I., Rinaldi, N.J., Robert, F., Odom, D.T., Bar-Joseph, Z., Gerber, G.K., Hannett, N.M., Harbison, C.T., Thompson, C.M., Simon, I., Zeitlinger, J., Jennings, E.G., Murray, H.L., Gordon, D.B., Ren, B., Wyrick, J.J., Tagne, J.B., Volkert, T.L., Fraenkel, E., Gifford, D.K., Young, R.A.: Transcriptional regulatory networks in Saccharomyces cerevisiae. Science **298** (2002) 799–804
38. Wilczyński, B., Hvidsten, T.R., Kryshtafovych, A., Tiuryn, J., Komorowski, J., Fidelis, K.: Using local gene expression similarities to discover regulatory binding site modules. Accepted in BMC Bioinformatics (2006)
39. Andersson, C.R., Hvidsten, T.R., Isaksson, A., Gustafsson, M.G., Komorowski, J.: Revealing cell cycle control by combining model-based detection of periodic expression with novel *cis*-regulatory descriptors. Submitted (2006)
40. Skolnick, J., Fetrow, J.S.: From genes to protein structure and function: Novel applications of computational approaches in the genomic era. Trends Biotechnol **18** (2000) 34–39

41. Apweiler, R., Bairoch, A., Wu, C.H., Barker, W.C., Boeckmann, B., Ferro, S., Gasteiger, E., Huang, H., Lopez, R., Magrane, M., Martin, M.J., Natale, D.A., O'Donovan, C., Redaschi, N., Yeh, L.S.L.: UniProt: the Universal Protein knowledgebase. Nucleic Acids Res **32** (2004) D115–D119
42. Chandonia, J.M., Brenner, S.E.: The impact of structural genomics: Expectations and outcomes. Science **311** (2006) 347–351
43. Tress, M., Ezkurdia, I., Graña, O., López, G., Valencia, A.: Assessment of predictions submitted for the CASP6 comparative modeling category. Proteins **61 Suppl 7** (2005) 27–45
44. Zhang, C., Kim, S.H.: Overview of structural genomics: from structure to function. Curr Opin Chem Biol **7** (2003) 28–32
45. Hvidsten, T.R., Kryshtafovych, A., Komorowski, J., Fidelis, K.: A novel approach to fold recognition using sequence-derived properties from sets of structurally similar local fragments of proteins. Bioinformatics **19 Suppl 2** (2003) II81–II91
46. Pazos, F., Sternberg, M.J.E.: Automated prediction of protein function and detection of functional sites from structure. Proc Natl Acad Sci U S A **101** (2004) 14754–14759
47. Orengo, C.A., Todd, A.E., Thornton, J.M.: From protein structure to function. Curr Opin Struct Biol **9** (1999) 374–382
48. Laskowski, R.A., Watson, J.D., Thornton, J.M.: ProFunc: a server for predicting protein function from 3D structure. Nucleic Acids Res **33** (2005) W89–W93
49. Pal, D., Eisenberg, D.: Inference of protein function from protein structure. Structure **13** (2005) 121–130
50. Hvidsten, T.R., Lægreid, A., Kryshtafovych, A., Andersson, G., Fidelis, K., Komorowski., J.: High through-put protein function prediction using local substructures. Submitted (2006)
51. Terfloth, L.: Drug design. In Gasteiger, J., Engel, T., eds.: Chemoinformatics. Wiley-VCH, Weinheim (2003) 497–618
52. Wikberg, J.E.S., Maris, L., Peteris, P.: Proteochemometrics: A tool for modelling the molecular interaction space. In Kubinyi, H., Mller, G., eds.: Chemogenomics in Drug Discovery - A Medicinal Chemistry Perspective. Wiley-VCH, Weinheim (2004) 289–309
53. Strömbergsson, H., Prusis, P., Midelfart, H., Lapinsh, M., Wikberg, J.E.S., Komorowski, J.: Rough set-based proteochemometrics modeling of G-protein-coupled receptor-ligand interactions. Proteins **63** (2006) 24–34
54. Strömbergsson, H., Kryshtafovych, A., Prusis, P., Fidelis, K., Wikberg, J.E.S., Komorowski, J., Hvidsten, T.R.: Generalized modeling of enzyme-ligand interactions using proteochemometrics and local protein substructures. Accepted to Proteins (2006)
55. Kontijevskis, A., Wikberg, J.E.S., Komorowski, J.: Computational proteomics analysis of HIV-1 protease interactome. Submitted (2006)
56. Kierczak, M., Rudnicki, W.R., Komorowski, J.: Construction of rough set-based classifiers for predicting HIV resistance to non-nucleoside reverse transcriptase inhibitors. Manuscript (2006)
57. Bazan, J.G., Skowron, A., Synak, P.: Dynamic reducts as a tool for extracting laws from decision tables. In: Proc. International Symposium on Methodologies for Intelligent Systems. Volume 869 of Lecture Notes in Artificial Intelligence., Springer-Verlag (1994) 346–355
58. Vinterbo, S., Øhrn, A.: Minimal approximate hitting sets and rule templates. International Journal of Approximate Reasoning **25** (2000) 123–143

59. Ågotnes, T., Komorowski, J., Løken, T.: Taming large rule models in rough set approaches. In [66] 193–203
60. Makosa, E.: Rule tuning. Master thesis. The Linnaeus Centre for Bioinformatics, Uppsala University (2005)
61. Düntsch, I.: Statistical evaluation of rough set dependency analysis. Int. J. Human-Computer Studies 46 (1997) 589–604
62. Düntsch, I., Gediga, G.: Uncertainty measures of rough set prediction. Artificial Intelligence 106 (1998) 109–137
63. Breiman, L.: Random forests. Machine Learning 45 (2001) 5–32
64. Skowron, A.: Synthesis of adaptive decision systems from experimental data. In Aamodt, A., Komorowski, J., eds.: Fifth Scandinavian Conference on Artificial Intelligence, Trondheim, Norway, IOS Press (1995) 220–238
65. Komorowski, J., Øhrn, A., Skowron, A.: ROSETTA rough sets. In Klösgen, W., Żytkow, J., eds.: Handbook of Data Mining and Knowledge Discovery. Oxford University Press (2002) 554–559
66. Żytkow, J.M., Rauch, J., eds.: Proceedings of the Third European Symposium on Principles and Practice of Knowledge Discovery in Databases (PKDD'99). Volume 1704 of Lecture Notes in Artificial Intelligence. Springer-Verlag, Prague, Czech Republic (1999)

# Rough Feature Selection for Intelligent Classifiers

Qiang Shen

Department of Computer Science
The University of Wales
Aberystwyth SY23 3DB, UK
qqs@aber.ac.uk

**Abstract.** The last two decades have seen many powerful classification systems being built for large-scale real-world applications. However, for all their accuracy, one of the persistent obstacles facing these systems is that of data dimensionality. To enable such systems to be effective, a redundancy-removing step is usually required to pre-process the given data. Rough set theory offers a useful, and formal, methodology that can be employed to reduce the dimensionality of datasets. It helps select the most information rich features in a dataset, without transforming the data, all the while attempting to minimise information loss during the selection process. Based on this observation, this paper discusses an approach for semantics-preserving dimensionality reduction, or feature selection, that simplifies domains to aid in developing fuzzy or neural classifiers. Computationally, the approach is highly efficient, relying on simple set operations only. The success of this work is illustrated by applying it to addressing two real-world problems: industrial plant monitoring and medical image analysis.

## 1 Introduction

Knowledge-based classification systems have been successful in many application areas. However, complex application problems, such as reliable monitoring and diagnosis of industrial plants and trustworthy analysis and comparison of medical images, have emphasised the issue of large numbers of features present in the problem domain, not all of which will be essential for the task at hand. The applicability of most classification systems is often limited by the curse of dimensionality that imposes a ceiling on the complexity of the application domain. A method to allow generation of intelligent classifiers for such application domains is clearly desirable.

Dimensionality reduction is also required to improve the runtime performance of a classifier. For example, in industrial plant monitoring, by requiring less observations per variable, the dimensionality reduced system becomes more compact and its response time decreases. The cost of obtaining data drops accordingly, as fewer connections to instrumentation need be maintained. In the meantime, the overall robustness of the system can increase, since, with fewer instruments, the chances of instrumentation malfunctions leading to spurious readings may be reduced dramatically.

Inspired by such observations, numerous different dimensionality reduction methodologies have been proposed in the literature. Unfortunately, many of them remove redundancy by irretrievably destroying the original meaning of the data given for learning.

J.F. Peters et al. (Eds.): Transactions on Rough Sets VII, LNCS 4400, pp. 244–255, 2007.

This significantly reduces, if not completely loses, the potential expressive power of the classification systems for computing with clear semantics. This, in turn, leads to a lack of trust in such systems, while such trust is usually critical for the systems to be taken up by end users.

The work on rough set theory [7] offers an alternative, and formal, methodology (amongst many other possible applications, e.g. [6,8]) that can be employed to reduce the dimensionality of datasets, as a preprocessing step to assist the development of any type of classifiers via learning from data. It helps select the most information rich features in a dataset, without transforming the data, all the while attempting to minimise information loss during the selection process [14]. Computationally, the approach is highly efficient, relying on simple set operations, which makes it suitable as a preprocessor for techniques that are much more complex. Unlike statistical correlation-reducing approaches [1], it requires no human input or intervention and retains the semantics of the original data.

Combined with an intelligent classification system built by, say, a fuzzy system or a neural network, the feature selection approach based on rough set theory can not only retain the descriptive power of the overall classifier, but also allow simplified system structure. This helps enhance the interoperability and understandability of the resultant systems and their reasoning. Drawing on the initial results previously presented in [12,13,14], this paper demonstrates the applicability of this approach in supporting transparent fuzzy or neural classifiers, with respect to two distinct application domains.

The remainder of this paper is structured as follows. The rough set-assisted feature selection mechanism is summarised in section 2 for self-containedness. This is followed by an illustration of the two example applications, demonstrating how different classification tasks can benefit from rough set-assisted semantics-preserving dimensionality reduction. The paper is concluded in section 5, with interesting further work pointed out.

## 2 Feature Selection

This section shows the basic ideas of rough sets [7] that are relevant to the present work and describes an efficient computational algorithm, named Rough Set Attribute Reduction (RSAR), for feature selection.

### 2.1 Rough Sets

A rough set is an approximation of a vague concept by a pair of precise concepts, called lower and upper approximations. The lower approximation is a description of the domain objects which are known with absolute certainty to belong to the subset of interest, whereas the upper approximation is a description of the objects which possibly belong to the subset.

Rough sets have been employed to remove redundant conditional attributes from discrete-valued datasets, while retaining their information content. Central to this work is the concept of indiscernibility. Without losing generality, let $I = (U, A)$ be an information system, where $U$ is a non-empty set of finite objects (the universe of discourse), and $A$ is a non-empty finite set of variables such that $a : U \rightarrow V_a \; \forall a \in A$, $V_a$ being the value set of variable $a$. In building a classification system, for example, $A = \{C \cup D\}$

where $C$ is the set of input features and $D$ is the set of class indices. Here, a class index $d \in D$ is itself a variable $d : U \rightarrow \{0, 1\}$ such that for $a \in U, d(a) = 1$ if $a$ has class $d$ and $d(a) = 0$ otherwise.

With any $P \subseteq A$ there is an associated equivalence relation $IND(P)$:

$$IND(P) = \{(x, y) \in U \times U \mid \forall a \in P,\ a(x) = a(y)\} \tag{1}$$

Note that this corresponds to the equivalence relation for which two objects are equivalent if and only if they have the same vectors of attribute values for the attributes in $P$. The partition of $U$, determined by $IND(P)$ is denoted $U/P$, which is simply the set of equivalence classes generated by $IND(P)$.

If $(x, y) \in IND(P)$, then $x$ and $y$ are indiscernible by features in $P$. The equivalence classes of the $P$-indiscernibility relation are denoted $[x]_P$. Let $X \subseteq U$, the P-*lower* and P-*upper* approximations of a classical crisp set are respectively defined as:

$$\underline{P}X = \{x \mid [x]_P \subseteq X\} \tag{2}$$

$$\overline{P}X = \{x \mid [x]_P \cap X \neq \varnothing\} \tag{3}$$

Let $P$ and $Q$ be subsets of $A$, then the important concept of *positive region* is defined as:

$$POS_P(Q) = \bigcup_{X \in U/Q} \underline{P}X \tag{4}$$

For tasks like classification with feature patterns, the positive region contains all objects of $U$ that can be classified into classes of $U/Q$ using the knowledge conveyed by the features of $P$.

## 2.2 Feature Dependency and Significance

The important issue here is to discover dependencies of object classes upon given features. Intuitively, a set of classes $Q$ depends totally on a set of features $P$, denoted $P \Rightarrow Q$, if all class indices from $Q$ are uniquely determined by values of features from $P$. Dependency can be measured in the following way [14]:

For $P, Q \subseteq A$, $Q$ depends on $P$ in a degree $k$ ($0 \le k \le 1$), denoted $P \Rightarrow_k Q$, if

$$k = \gamma_P(Q) = \frac{|POS_P(Q)|}{|U|} \tag{5}$$

where $|S|$ stands for the cardinality of set $S$.

If $k = 1$, $Q$ depends totally on $P$; if $0 < k < 1$, $Q$ depends partially (in a degree $k$) on $P$; and if $k = 0$, $Q$ does not depend on $P$.

By calculating the change in dependency when a feature is removed from the set of considered possible features, an estimate of the significance of that feature can be obtained. The higher the change in dependency, the more significant the feature is. If the significance is 0, then the feature is dispensable. More formally, given $P, Q$ and a feature $x \in P$, the significance of feature $x$ upon $Q$ is defined by

$$\sigma_P(Q, x) = \gamma_P(Q) - \gamma_{P-\{x\}}(Q) \tag{6}$$

## 2.3   Feature Selection Algorithm

The selection of features is achieved by reducing the dimensionality of a given feature set, without destroying the meaning conveyed by the individual features selected. This is, in turn, achieved by comparing equivalence relations generated by sets of features with regard to the underlying object classes, in the context of classification.

Features are removed so that the reduced set will provide the same quality of classification as the original. For easy reference, the concept of *retainer* is introduced as a subset $R$ of the initial feature set $C$ such that $\gamma_R(D) = \gamma_C(D)$. A minimal retainer is termed a *reduct* in the literature [9]. That is, a further removal of any feature from a reduct will make it violate the constraint $\gamma_R(D) = \gamma_C(D)$.

Thus, a given dataset may have many feature retainers, and the collection of all retainers is denoted by

$$R = \{X \mid X \subseteq C, \gamma_X(D) = \gamma_C(D)\} \tag{7}$$

The intersection of all the sets in $R$ is called the *core*, the elements of which are those features that cannot be eliminated without introducing more contradictions to the representation of the dataset. Clearly, for feature selection, an attempt is to be made to locate a minimal retainer, or a single reduct, $R_{min} \subseteq R$:

$$R_{min} = \{X \mid X \in R, \forall Y \in R, |X| \leq |Y|\} \tag{8}$$

A basic way of achieving this is to calculate the dependencies of all possible subsets of $C$. Any subset $X$ with $\gamma_X(D) = 1$ is a retainer; the smallest subset with this property is a reduct. However, for large datasets with a large feature set this method is impractical and an alternative strategy is required.

The RSAR feature selection algorithm given in Figure 1 attempts to calculate a reduct without exhaustively generating all possible subsets. It starts off with an empty set and adds in turn, one at a time, those features that result in the greatest increase in $\gamma_P(Q)$, until the maximum possible value of $\gamma_P(Q)$, usually 1, results for the given dataset. Note that this method does not always generate a *minimal* retainer (or reduct), as $\gamma_P(Q)$ is not a perfect heuristic. However, it does result in a close-to-minimal retainer, which is still useful in greatly reducing feature set dimensionality. It is also

1. $R \leftarrow \{\}$
2. do
3.     $T \leftarrow R$
4.     $\forall x \in (C - R)$
5.        if $\gamma_{R \cup \{x\}}(D) > \gamma_T(D)$
6.           $T \leftarrow R \cup \{x\}$
7.     $R \leftarrow T$
8. until $\gamma_R(D) = \gamma_C(D)$
9. return $R$

**Fig. 1.** The RSAR feature selection algorithm

worth noting that one way to guarantee the generation of a reduct is to apply RSAR in conjunction with a selection strategy that works in reverse order (i.e., starting with a full set of features and then deleting one at a time). Nevertheless, such an approach has a significant practical limit when the original feature set is of a high dimensionality.

RSAR works in a greedy manner, not compromising with a set of features that contains a large part of the information of the initial set. It attempts to reduce the feature set without loss of information significant to solving the problem at hand. The way it works is clearly dependent upon features being represented in nominal values. However, this does not necessarily give rise to problems in the use of the overall classification system which includes such a feature selection preprocessor. This is because the real feature values are only required to be temporarily discretised for feature selection itself. The classifier will use the original real-valued features directly. In this regard, it is independent of the classification methods adopted. When used in conjunction with an explicit descriptive classifier, the resulting system will be defined in terms of only the significant features of the data, retaining the desirable transparency. The training process is accelerated, while the runtime operation of the system is sped up since fewer attributes are required.

## 3   Application I: Industrial Plant Monitoring

This application concerns the task of monitoring a water treatment plant [14]. To illustrate the generality of the presented approach and its independence from any specific classification system, this first application involves the use of a fuzzy system based classifier. This domain was chosen because of its realism. A large plant is likely to involve a number of similar features, not all of which will be essential in determining the operational status. Interrelations between features are unavoidable as the plant is a single system with interconnections, leading to a fair degree of redundancy.

### 3.1   Problem Case

The Water Treatment dataset comprises a set of historical data obtained over a period of 521 days, with one series of measurements per day. Thirty eight different feature values are measured per day, with one set of such measurements forming one datum. All measurements are real-valued. The goal is to implement a fuzzy classification system that, given this dataset of past measurements and without the benefit of an expert in the field at hand, will classify the plant's status and produce human comprehensible explanations of the monitoring results.

The thirty eight features account for the following five aspects of the water treatment plant's operation (see Figure 2 for an illustration of this): input to plant; input to primary settler; input to secondary settler; output from plant; and overall plant performance. The operational state of the plant is represented by a boolean categorisation representing the detection of a fault. The point is to draw the operator's attention to an impending fault.

### 3.2   Fuzzy Classifier

In this experimental study, to obtain a system that will entail classification of the plant's operating status, the fuzzy induction algorithm first reported in [3] is used. This is

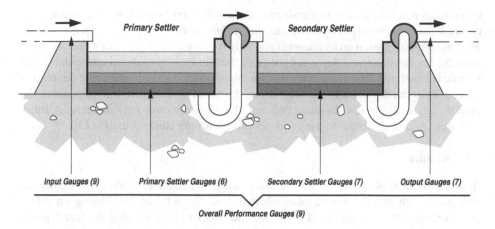

Input Gauges (9)          Primary Settler Gauges (6)          Secondary Settler Gauges (7)          Output Gauges (7)

Overall Performance Gauges (9)

**Fig. 2.** Schematic diagram of the water treatment plant, indicating the number of measurements sampled at various points

adopted simply due to the availability of its software implementation; any other fuzzy rule induction method may be utilised as an alternative for classifier building. The resulting classification system is represented in a set of fuzzy production rules. For the sake of completeness, an outline of the induction algorithm employed is given below.

The algorithm generates a hyperplane of candidate fuzzy rules by fuzzifying the entire training dataset using all permutations of the input features. Thus, a system with $M$ inputs, each of which has a domain fuzzified by $f_j$ fuzzy sets ($1 \leq j \leq M$), the hyperplane is fuzzified into $\prod_{j=1}^{M} f_j$ $M$-dimensional clusters, each representing one vector of rule preconditions. Each cluster $\underline{p} = \langle D^1, D^2, \ldots, D^M \rangle$ may lead to a fuzzy rule, provided that the given dataset supports it.

To obtain a measure of what classification applies to a cluster, fuzzy min-max composition is used. The input feature pattern of each example object is fuzzified according to the fuzzy sets $\{\mu_{D^1}, \mu_{D^2}, \ldots, \mu_{D^M}\}$ that make up cluster $\underline{p}$. For each object $\underline{x} = \langle x_1, x_2, \ldots, x_M \rangle$, the following $t$-norm of it, with respect to cluster $\underline{p}$ and classification $c$, is calculated:

$$T_c^{\underline{p}} \underline{x} = \min \left( \mu_{D^1}(x_1), \mu_{D^2}(x_2), \ldots, \mu_{D^M}(x_M) \right) \tag{9}$$

Furthermore, the maximum of all $t$-norms with respect to $\underline{p}$ and $c$ is then calculated and this is dubbed an $s$-norm:

$$S_c^{\underline{p}} = \max \left\{ T_c^{\underline{p}} \underline{x} \mid \underline{x} \in C_c \right\} \tag{10}$$

where $C_c$ is the set of all examples that can be classified as $c$. This is iterated over all possible classifications to provide a full indication of how well each cluster applies to each classification.

A cluster generates at most one classification rule. The rule's preconditions are the cluster's $M$ corresponding fuzzy sets connected conjunctively. The conclusion is the classification attached to the cluster. Since there may be $s$-norms for more than one classification, it is necessary to decide on one classification for each of the clusters. Such contradictions are resolved by using the *uncertainty margin*, $\varepsilon$ ($0 \leq \varepsilon < 1$). An $s$-norm assigns its classification on its cluster if and only if it is greater by at least $\varepsilon$ than all other $s$-norms for that cluster. If this is not the case, the cluster is considered undecidable and no rule is generated. The uncertainty margin introduces a trade-off in the rule generation process between the size and the accuracy of the resulting classification. In general, the higher $\varepsilon$ is, the less rules are generated, but classification error may increase. A fuller treatment of this algorithm in use for descriptive learning can be found in [3].

### 3.3   Results

Running the RSAR algorithm on the Water Treatment dataset provided a significant reduction, with merely two features selected from the total of 38. Testing on previously unseen data resulted in a classification accuracy of 97.1%, using the fuzzy model generated by the above-mentioned rule induction method.

A comparison against a widely recognised benchmark method should help in establishing the success of the system. C4.5 [10] is a widely accepted and powerful algorithm that provides a good benchmark [5] for learning from data. The decision trees it generates allow for rapid and efficient interpretation. Yet, C4.5's decision tree for the present problem involves a total of three attributes from the dataset, as opposed to two chosen by the RSAR algorithm. In terms of classification performance, C4.5 obtains a compatible accuracy of around 96.8%.

Note that training a fuzzy system on all 38 features would be computationally prohibitive with the adopted learning algorithm. As stated previously, the benefits do not limit themselves to the learning phase; they extend to the runtime use of the learned classifier. By reducing the dimensionality of the data, the dimensionality of the rule-set is also reduced. This results in fewer measured features, which is very important for dynamic systems where observables are often restricted. This in turn leads to fewer connections to instrumentation and faster system responses in emergencies. Both of which are important to the problem domain.

The most important benefit of using RSAR is, however, derived from its conjunctive use with the linguistically expressive fuzzy system. With the learned rules, it can provide explanations of its reasoning to the operator. This leads to increased trust in the system, as its alarms can be understood meaningfully. A classification system consisting of rules involving 38 features, even though they are all directly measurable and hence individually interpretable, is very difficult to understand, whilst one involving only two features is very easy to interpret.

## 4   Application II: Medical Image Analysis

Comparing normal and abnormal blood vessel structures, via the analysis of cell images, plays an important role in pathology and medicine [12]. This forms the focus of this application, analysing medical images by the use of a neural network based image classifier that is supported by RSAR.

## 4.1 Problem Case

Central to this analysis is the capture of the underlying features of the cell images. Many feature extraction methods are available to yield various kinds of characteristic descriptions of a given image. However, little knowledge is available as to what features may be most helpful to provide the discrimination power between normal and abnormal cells and between their types, while it is computationally impractical to generate many features and then to perform classification based on these features for rapid diagnosis. Generating a good number of features and selecting from them the most informative ones off-line, and then using those selected on-line is the usual way to avoid this difficulty. Importantly, the features produced ought to have an embedded meaning and such meaning should not be altered during the selection process. Therefore, this problem presents a challenging case to test the potential of RSAR.

The samples of subcutaneous blood vessels used in this work were taken from patients suffering critical limb ischaemia immediately after leg amputation. The level of amputation was always selected to be in a non-ischaemic area. The vessel segments obtained from this area represented internal proximal (normal) arteries, whilst the distal portion of the limb represented ischaemic (abnormal) ones. Images were collected using an inverted microscope, producing an image database of 318 cell images, each sized $512 \times 512$ pixels with grey levels ranging from 0 to 255. Examples of the three types of cell image taken from non-ischaemic, and those from ischaemic, resistance arteries are shown in Figure 3. Note that many of these images seem rather similar to the eye. It is therefore a difficult task for visual inspection and classification.

## 4.2 Neural Network Classifier

In this work, each image classifier is implemented using a traditional multi-layer feedforward artificial neural network (MFNN). To capture and represent many possible and essential characteristics of a given image, fractal models [4] are used. Note that, although these particular techniques are herein adopted to perform their respective task, the work described does not rely on them, but is generally applicable when other classification and feature extraction methods are employed.

An MFNN-based classifier accomplishes classification by mapping input feature patterns onto their underlying image classes. The design of each MFNN classifier used for the present work is specified as follows. The number of nodes in its input layer is set to that of the dimensionality of the given feature set (before or after feature reduction), and the number of nodes within its output layer is set to the number of underlying classes of interest. The internal structure of the network is designed to be flexible and may contain one or two hidden layers.

The training of the classifier is essential to its runtime performance, and is here carried out using the back-propagation algorithm [11]. For this, feature patterns that represent different images, coupled with their respective underlying image class indices, are selected as the training data, with the input features being normalised into the range of 0 to 1. Here, each feature pattern consists of 9 fractal features (including 5 isotropic fractals measured on the top five finest resolutions and 4 directional fractals [12]) and the mean and standard deviation (STD), with their reference numbers listed in Table 1.

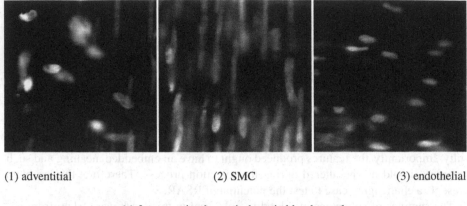

(1) adventitial                    (2) SMC                    (3) endothelial

(a) from proximal, non-ischaemic blood vessels

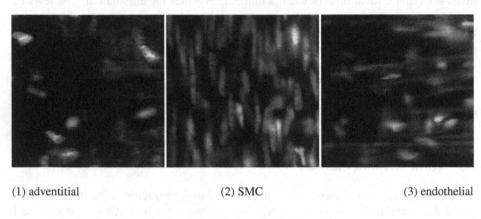

(1) adventitial                    (2) SMC                    (3) endothelial

(b) from distal, ischaemic blood vessels

**Fig. 3.** Section cell images, where the first, second and third columns respectively show adventitial, smooth muscle and endothelial cells in proximal non-ischaemic and distal ischaemic subcutaneous blood vessels, taken from a human lower limb

Note that when applying the trained classifier, only those features selected during the learning phase are required to be extracted and that no discretisation is needed but real-valued features are directly fed to the classifier.

### 4.3 Results

Eighty-five images selected from the image database are used for training and the remaining 233 images are employed for testing. For simplicity, only MFNNs with one hidden layer are considered.

Table 2 lists the results of using RSAR and the original full set of features. The error rate of using the five selected features is lower than that of using the full feature set. This improvement of performance is obtained by a structurally much simpler network of 10

**Table 1.** Features and their reference number

| Feature No. | Feature Meaning | Feature No. | Feature Meaning |
|---|---|---|---|
| 1 | 0° direction | 7 | 3rd finest resolution |
| 2 | 45° direction | 8 | 4th finest resolution |
| 3 | 90° direction | 9 | 5th finest resolution |
| 4 | 135° direction | 10 | Mean |
| 5 | Finest resolution | 11 | STD |
| 6 | 2nd finest resolution | | |

**Table 2.** Results of using rough-selected and the original full set of features

| Method | Dimensionality | Features | Structure | Error |
|---|---|---|---|---|
| Rough | 5 | 1,4,9,10,11 | $5 \times 10 + 10 \times 6$ | 7.55% |
| Original | 11 | 1,2,3,4,5,6,7,8,9,10,11 | $11 \times 24 + 24 \times 6$ | 9.44% |

**Table 3.** Results of using rough and PCA-selected features

| Method | Dimensionality | Features | Structure | Error |
|---|---|---|---|---|
| **Rough** | **5** | **1,4,9,10,11** | $\mathbf{5 \times 10 + 10 \times 6}$ | **7.7%** |
| PCA | 1 | 1 | $1 \times 12 + 12 \times 6$ | 57.1% |
| | 2 | 1,2 | $2 \times 12 + 12 \times 6$ | 32.2% |
| | 3 | 1,2,3 | $3 \times 12 + 12 \times 6$ | 31.3% |
| | 4 | 1,2,3,4 | $4 \times 24 + 24 \times 6$ | 28.8% |
| | **5** | **1,2,3,4,5** | $\mathbf{5 \times 20 + 20 \times 6}$ | **18.9%** |
| | 6 | 1,2,3,4,5,6 | $6 \times 18 + 18 \times 6$ | 15.4% |
| | 7 | 1,2,3,4,5,6,7 | $7 \times 24 + 24 \times 6$ | 11.6% |
| | 8 | 1,2,3,4,5,6,7,8 | $8 \times 24 + 24 \times 6$ | 13.7% |
| | 9 | 1,2,3,4,5,6,7,8,9 | $9 \times 12 + 12 \times 6$ | 9.9% |
| | 10 | 1,2,3,4,5,6,7,8,9,10 | $10 \times 20 + 20 \times 6$ | 7.3% |
| | 11 | 1,2,3,4,5,6,7,8,9,10,11 | $11 \times 8 + 8 \times 6$ | 7.3% |

hidden nodes, as opposed to the classifier that requires 24 hidden nodes to achieve the optimal learning. This is indicative of the power of RSAR in helping reduce not only redundant feature measures but also the noise associated with such measurement. Also, the classifier using those five RSAR-selected features considerably outperforms those using five randomly selected features, with the average error of the latter reaching 19.1%.

Again, a comparison against a widely recognised benchmark method should help reflect the success of the system. For this, the results of rough feature selection are systematically compared to those obtained via the use of Principal Component Analysis (PCA) [1], as summarised in Table 3. Note that PCA is perhaps the most adopted dimensionality reduction technique. Although efficient, it irreversibly destroys the underlying semantics of the feature set. Therefore, in this table, for the results of using PCA, feature number $i, i \in \{1, 2, ..., 11\}$, stands for the $i$th principal component, i.e. the transformed feature that is corresponding to the $i$th largest variance.

The advantages of using RSAR are clear. Of the same dimensionality (i.e., 5), the classifier using the features selected by the rough set approach has a substantially higher classification accuracy, and this is achieved via a considerably simpler neural network. When increasing the dimensionality of principal features, the error rate generally gets reduced, but the classifier generally underperforms until almost the full set of principal features is used. The overall structural complexity of all these classifiers are more complex than that of the classifier using the five RSAR-selected features. In addition, the use of those classifiers that use PCA-selected features would require many more feature measurements to achieve comparable classification results.

## 5   Conclusion

It is well-known that the applicability of most intelligent classification approaches is limited by the curse of dimensionality, which imposes a ceiling on the complexity of the application domain. This paper has demonstrated an effective approach to semantics-preserving dimensionality reduction by exploiting the basic ideas of rough set theory. Such a feature selection tool makes learned classifiers much more transparent and comprehensible to humans, who have inherent trouble understanding high-dimensionality domains, in addition to being able to lessen the obstacles of the dimensionality ceiling.

In summary, Rough Set Attribute Reduction (RSAR) selects the most information rich attributes in a dataset, without transforming the data, all the while attempting to minimise information loss as regards the classification task at hand. When employed by an intelligent classification system (be it a fuzzy system or neural network), by simplifying the problem domain, RSAR helps enhance the transparency and maintain the accuracy of the classifier. With relatively simple system structures, the examination of the quality of the results inferred by the use of such classifiers is made easy. This has been demonstrated in applications to two rather different problem domains, with very promising results.

Although RSAR has been used as a dataset pre-processor with much success, it is reliant upon a crisp dataset. Important information (for choosing the optimal features) may be lost as a result of required boolean discretisation of the underlying numerical features. Further advances have recently been made in proposing a feature selection technique that employs a hybrid variant of rough sets, the fuzzy-rough sets [2], to avoid this information loss [15]. Whilst this is out of the scope of this paper, it is interesting to point out that initial experimental results, of applying this improved version to the problem of industrial plant monitoring, have shown that fuzzy-rough feature selection is more powerful than many conventional approaches, including entropy-based, PCA-based and random-based methods.

## Acknowledgments

The author is very grateful to many of his colleagues, especially to Alexios Chouchoulas, Richard Jensen and Changjing Shang, for their contribution in the work, whilst taking full responsibility for the views expressed in this paper.

# References

1. Devijver, P. and Kittler, J. Pattern Recognition: a Statistical Approach. Prentice Hall, 1982.
2. Dudois, D. and Prade, H. Putting rough sets and fuzzy sets together. In: R. Slowinski (Ed.), Intelligent Decision Support. Kluwer Academic Publishing, pages 203–232, 1992.
3. Lozowski, A. Cholewo, T. and Zurada, J. Crisp rule extraction from perceptron network classifiers. In *Proceedings of International Conference on Neural Networks*, volume Plenary, Panel and Special Sessions, pages 94–99, 1996.
4. Mandelbrot, B. The Fractal Geometry of Nature. San Francisco: Freeman, 1982.
5. Mitchell, T. *Machine Learning*. McGraw-Hill (1997).
6. Orlowska, E. Incomplete Information: Rough Set Analysis. Springer Verlag, 1997.
7. Pawlak, Z. *Rough Sets: Theoretical Aspects of Reasoning About Data*. Kluwer Academic Publishers, Dordrecht (1991).
8. Peters, J. and Skowron, A. A rough set approach to knowledge discovery. *International Journal of Intelligent Systems* **17** (2002) 109-112.
9. Pawlak, Z. and Skowron, A. Rough set rudiments. *Bulletin of International Rough Set Society* **3(4)** (2000) 43-47.
10. Quinlan, J. R. *C4.5: Programs for Machine Learning*. The Morgan Kaufmann Series in Machine Learning. Morgan Kaufmann Publishers (1993).
11. Rumelhant, D. Hinton, E. and Williams, R. Learning internal representations by error propagating. In: E. Rumelhant and J. McClelland (Eds.), Parallel Distributed Processing. MIT Press, 1986.
12. Shang, C. and Shen, Q. Rough feature selection for neural network based image classification. *International Journal of Image and Graphics* **2** (2002) 541-555.
13. Shen, Q. Semantics-preserving dimensionality reduction in intelligent modelling. In: . Lawry, J. Shanahan and A. Ralescu (Eds.), Modelling with Words. Springer, 2003.
14. Shen, Q. and Chouchoulas, A. A fuzzy-rough approach for generating classification rules. *Pattern Recognition*, **35** (2002) 341–354.
15. Shen, Q. and Jensen, R. Selecting informative features with fuzzy-rough sets and its application for complex systems monitoring. *Pattern Recognition*, **37** (2004) 1351–1363.

# Granulation as a Privacy Protection Mechanism[*]

Da-Wei Wang, Churn-Jung Liau, and Tsan-sheng Hsu

Institute of Information Science
Academia Sinica, Taipei 115, Taiwan
{wdw,liaucj,tshsu}@iis.sinica.edu.tw

**Abstract.** How to achieve a balance between data publication and privacy protection has been an important issue in information security for several years. When microdata is released to users, attributes that clearly identify individuals are usually removed. Nevertheless, it is still possible to link released data with some public or easy-to-access databases to obtain confidential information. To safeguard privacy, numerous techniques, such as generalization, suppression, and microaggregation, have been proposed to modify the to-be-released data. In this paper, we propose attribute-oriented granulation as a data protection mechanism that can integrate both generalization and microaggregation into a uniform framework. We address the computational issue of searching for the most specific granulation that satisfies confidentiality requirements. A breadth-first search algorithm with basic pruning strategies is presented and its properties are investigated. The properties can be used to improve the efficiency of our algorithm. We also define some quantitative measures of data quality and security, and apply evolutionary computation techniques to find the optimal granulation for privacy protection.

## 1 Introduction

Privacy protection is one of the main concerns in the field of data security. In recent years, statistical disclosure control [2] has become increasingly important due to the requirements of data security. One of the major issues in disclosure control is the database linking problem. Generally speaking, the problem is how to prevent users[1] obtaining confidential information about an individual[2] by linking to some public or easy-to-access database with data they can obtain legally from a data center.

Though the protection of privacy is very important, over-restriction of access to a database may render the data useless. Therefore, the main challenge is how to achieve a balance between privacy protection and data availability. One

---

[*] A preliminary version of this paper was published in [1]. This work was partially supported by the Taiwan Information Security Center (TWISC) and NSC (Taiwan). NSC Grants: 95-2221-E-001-019 (D.W. Wang), 95-2221-E-001-029-MY3 (C.J. Liau), and 95-2221-E-001-004 (T-s. Hsu).

[1] In this paper, a user refers to anyone receiving data and having the potential to breach the privacy of individuals.

[2] An individual refers to a person whose privacy should be protected.

J.F. Peters et al. (Eds.): Transactions on Rough Sets VII, LNCS 4400, pp. 256–273, 2007.

possibility is to modify the data before it is released by generalizing the values of some data cells to a coarser level of precision. To do this, we can partition the domain of attributes according to a certain level of precision, and generalize the data from the finest to the coarsest level until the privacy requirement is met. This kind of operation is called *attribute-oriented granulation* (AOG). In this paper, we investigate the application of AOG to privacy protection. It is shown that AOG can integrate generalization[3,4,5,6,7] and microaggregation[8] into a uniform framework. To address the computational issue of searching for the most specific granulation that satisfies confidentiality requirements, a breadth-first search algorithm with basic pruning strategies is presented and its properties are investigated. The properties can be used to improve the efficiency of the basic algorithm. We also define the quantitative measures of data quality and security and apply evolutionary computation (EC) techniques to find the optimal granulation for privacy protection.

The remainder of the paper is organized as follows: In Section 2, we use an example to illustrate the concept of AOG, and formally introduce the AOG operation. The logical security of AOG and its computational aspects are explored in Section 3. We also present several properties of AOG that are used to improve the search algorithm. In Section 4, we discuss the security and quality of AOG and apply an EC approach to the search for an optimal AOG. We then present our conclusions in Section 6.

## 2 Attribute-Oriented Granulation

### 2.1 A Running Example

In this paper, we investigate the privacy protection problem that may arise when a data table [9] is released. The data in many application domains, such as medical records, financial transaction records, employee information, and so on, can be organized as data tables. A data table consists of a set of records, each of which corresponds to an individual and has some attributes.

The attributes of a data table can be divided into three sets [10,11]. The first consists of *identifiers* that can be used to identify to whom a data record belongs. Therefore, these attributes are always masked off in response to a query. Let us equate a set of identifiers with a set of individuals. Throughout this paper, a set of individuals (or identifiers) is denoted by $U$. Second, we have a set of *quasi-identifiers*, the values of which are known to the public. For example, in [12], it is pointed out that some attributes like birth-date, gender, ethnicity, etc. are included in some public databases, such as those that contain census data or voter registration lists. These attributes, if not appropriately processed, may be used to re-identify an individual's record in a data table, thus causing a privacy violation. The last kind of attribute is the *confidential attribute*, the values of which we have to protect. It is often the case that an asymmetry exists between the values of a confidential attribute. For example, if the attribute is a HIV test result, then the revelation of a '+' value may cause a serious invasion of privacy, whereas it does not matter to know that an individual has a '−' status. In this

| ID | D.O.B. | ZIP | Height | Income | Health |
|----|--------|-----|--------|--------|--------|
| $u_1$ | 24/09/56 | 24126 | 161 | 400K | 1 |
| $u_2$ | 06/09/56 | 24129 | 167 | 300K | 1 |
| $u_3$ | 30/09/56 | 24133 | 163 | 300K | 1 |
| $u_4$ | 23/03/56 | 10427 | 160 | 300K | 0 |
| $u_5$ | 18/03/56 | 10431 | 165 | 100K | 2 |
| $u_6$ | 05/03/56 | 10466 | 168 | 100K | 2 |
| $u_7$ | 20/04/55 | 26015 | 175 | 400K | 2 |
| $u_8$ | 18/04/55 | 26032 | 170 | 300K | 1 |
| $u_9$ | 09/04/55 | 26617 | 173 | 100K | 0 |
| $u_{10}$ | 01/04/55 | 26628 | 171 | 400K | 0 |
| $u_{11}$ | 23/04/55 | 26328 | 176 | 400K | 0 |

**Fig. 1.** A data table

| | | | | | |
|----|-------|-------|------------|------|---|
| 1 | 09/56 | 24*** | [160,170) | 400K | 1 |
| 2 | 09/56 | 24*** | [160,170) | 300K | 1 |
| 3 | 09/56 | 24*** | [160,170) | 300K | 1 |
| 4 | 03/56 | 10*** | [160,170) | 300K | 0 |
| 5 | 03/56 | 10*** | [160,170) | 100K | 2 |
| 6 | 03/56 | 10*** | [160,170) | 100K | 2 |
| 7 | 04/55 | 26*** | [170,180) | 400K | 2 |
| 8 | 04/55 | 26*** | [170,180) | 300K | 1 |
| 9 | 04/55 | 26*** | [170,180) | 100K | 0 |
| 10 | 04/55 | 26*** | [170,180) | 400K | 0 |
| 11 | 04/55 | 26*** | [170,180) | 400K | 0 |

**Fig. 2.** A generalized data table

paper, let $T$ denote a data table for a set of individuals $U$, and $t_{ij}$ denote the value of an attribute $j$ of an individual $u_i$.

We use the data table in Figure 1 as our running example[7]. In the table, $U = \{u_1, \cdots, u_{11}\}$ is a set of individuals (or identifiers); the quasi-identifiers are date of birth, zip code, and height; and the confidential attributes are income and health status. The values of "Health" are denoted by "normal"(0), "slightly ill"(1), and "seriously ill"(2) respectively.

In [11,5,12], the notion of *bin size* is proposed to resolve the database linkage problem. A *bin* is defined as an equivalence class based on the quasi-identifiers, and the bin's size is its cardinality. To be deemed secure, a table must satisfy the condition that the size of any bin is sufficiently large. The security criterion is called $k$-anonymity if each bin is required to contain at least $k$ individuals. Though, in general, the chance of a user obtaining confidential information is smaller if the bin size is larger, it is well-known that controlling the bin size alone is not sufficient to stop inference attacks [13]. To fully protect privacy, we must consider some alternative criteria to complement the bin size.

One technique of protecting privacy is to release the data in a coarser granularity. For example, the date of birth may be given as only the year and month, or only the first two digits of the ZIP code may be given. In addition, "Height" can be expressed as a range, instead of a precise value. A concrete generalization of the data table in Figure 1 is given in Figure 2. The first column denotes the serial numbers of the released data records.

From the generalized data table, we observe that the bin containing $u_1, u_2$, and $u_3$ is size 3. However, since the health status attribute of the rows in this bin has the value 1, the recipient of the table can infer that $u_1, u_2$, and $u_3$ are all slightly ill, though he does not know which of them has an income of 400K.

## 2.2  AOG Operations

In this section, we formally define the modification operation that can be applied to a data table to enhance privacy protection. As the operation is based on partitioning the domain of attributes according to different granular scales, it is called attribute-oriented granulation (AOG). We first recall the basic definition of a partition. Let $V$ be a domain of values for some attribute; then, a *partition* $\pi$ of $V$ is a set $\{s_1, s_2, \ldots, s_k\}$ of mutually disjoint subsets of $V$ such that $\cup_{i=1}^{k} s_i = V$. Each $s_i$ is called an equivalence class of the partition, and we use $\pi(v)$ to denote the equivalence class containing $v$. Let $\pi_1$ and $\pi_2$ be two partitions of $V$. Then, $\pi_1$ is a *refinement* of $\pi_2$, written as $\pi_1 \preceq \pi_2$ if, for $s \in \pi_1$ and $t \in \pi_2$, either $s \subseteq t$ or $s \cap t = \emptyset$. Let $\pi_1 \prec \pi_2$ denote $\pi_1 \preceq \pi_2$ and $\pi_1 \neq \pi_2$. For a given set $V$, we use $\perp$ and $\top$ (possibly with indices) to denote the finest partition $\{\{v\} \mid v \in V\}$ and the coarsest partition $\{V\}$ respectively.

Let us assume that the set of quasi-identifiers is $\{1, 2, \ldots, m\}$ and denote the domain of attribute $i$ by $V_i$ for $1 \le i \le m$. Then, an AOG operation is specified by a tuple $(\pi_1, \pi_2, \ldots, \pi_m)$, where for $1 \le i \le m$, $\pi_i$ is a partition of $V_i$. Let $\tau_1 = (\pi_1, \pi_2, \ldots, \pi_m)$ and $\tau_2 = (\pi'_1, \pi'_2, \ldots, \pi'_m)$ be two AOG operations. Then, $\tau_1$ is *at least as specific as* $\tau_2$, denoted by $\tau_1 \preceq \tau_2$, if for $1 \le i \le m$, $\pi_i \preceq \pi'_i$; and $\tau_1$ is *more specific than* $\tau_2$, denoted by $\tau_1 \prec \tau_2$, if $\tau_1 \preceq \tau_2$ and $\tau_1 \neq \tau_2$.

Since the number of possible partitions of a domain may be prohibitively large, we sometimes focus on a subset of *admissible partitions*. Let us define $\Pi_i$ as the set of admissible partitions of $V_i$ such that $\perp_i$ and $\top_i \in \Pi_i$ for $1 \le i \le m$; then, the set of *admissible AOGs* is $\Pi = \Pi_1 \times \Pi_2 \times \cdots \times \Pi_m$. $\tau_2$ is called a direct successor of $\tau_1$ in $\Pi$ if $\tau_1 \prec \tau_2$ and there does not exist any $\tau \in \Pi$ such that $\tau_1 \prec \tau \prec \tau_2$.

## 2.3  The Running Example

Figure 3 shows a set of admissible partitions for our running example, where the partitions for the dates of birth and zip codes are obvious, and the partitions for height are defined as

$$I_1 = \{\cdots, \{160\}, \{161\}, \cdots, \{174\}, \{175\}, \cdots\},$$
$$I_5 = \{\cdots, [160, 165), [165, 170), [170, 175), \cdots\},$$
$$I_{10} = \{\cdots, [160, 170), [170, 180), \cdots\},$$
$$I_{20} = \{\cdots, [160, 180), [180, 200), \cdots\}.$$

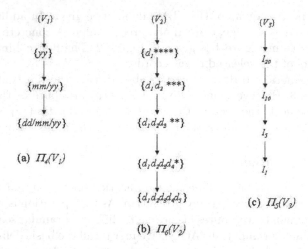

**Fig. 3.** Admissible partitions for the quasi-identifiers in our running example

## 3   Logical Security

### 3.1   Security of AOG

To decide whether an AOG is secure, we use Pawlak's decision logic (DL, [9]) to describe confidential information. The set of *atomic sentences* for DL is $\mathcal{P} = \{(j, v) \mid j \in J, v \in V_j\}$, where $J$ is the set of confidential attributes. The intuitive meaning of the atomic sentence $(j, v)$ is that an individual's attribute $j$ has value $v$. The set of sentences is the smallest set containing $\mathcal{P}$ that is closed on the Boolean connectives $\neg$, $\wedge$, and $\vee$. If $\alpha \subseteq V_j$, we abbreviate $\vee_{v \in \alpha}(j, v)$ as $(j, \alpha)$. We assume that the information an individual $u$ wants to keep confidential is represented by a set of DL sentences, $CON(u)$. As usual, the sentences are evaluated inductively with respect to the data table $T$ and each individual in $U$ as follows:

1. $u_i \models (j, v)$ iff $t_{ij} = v$.
2. $u \models \neg\varphi$ iff $u \not\models \varphi$.
3. $u \models \varphi \wedge \psi$ iff $u \models \varphi$ and $u \models \psi$.
4. $u \models \varphi \vee \psi$ iff $u \models \varphi$ or $u \models \psi$.

The meaning set of a sentence $\varphi$, $\|\varphi\|_T = \{u \in U \mid u \models \varphi\}$, is the set of individuals that satisfies $\varphi$ in the data table $T$. The subscript $T$ is usually omitted when it is clear from the context.

Let $\pi$ be a partition of the domain of an attribute $k$; then, the $\pi$-*indiscernibility relation* with respect to the data table $T$, denoted by $ind_T(\pi)$, is an equivalence relation on $U$ defined by $(u_i, u_j) \in ind_T(\pi) \Leftrightarrow \pi(t_{ik}) = \pi(t_{jk})$. Again, the subscript $T$ is usually omitted for convenience. Let $\tau = (\pi_1, \pi_2, \ldots, \pi_m)$ be an AOG operation; then, the $\tau$-*indiscernibility relation* with respect to the data table $T$ is defined as

$$ind(\tau) = \cap_{1 \leq k \leq m} ind(\pi_k).$$

An AOG operation, $\tau = (\pi_1, \pi_2, \ldots, \pi_m)$, determines how the data is modified before it is released. The requirement is that, for any $u_i, u_j \in U$ and attribute $k$, $\pi_k(t_{ik}) = \pi_k(t_{jk})$ iff $t_{ik}$ and $t_{jk}$ are replaced by the same value in the modified data table. For example, the generalization method in [4,11] replaces a table entry $t_{ik}$ with $\pi_k(t_{ik})$, whereas the microaggregation method in [8] replaces it with some statistics, such as the mean, median, or mode of the multiset[3] $\{t_{jk} \mid (u_i, u_j) \in ind(\pi_k)\}$. Thus, the AOG method subsumes both generalization and microaggregation. In this paper, we do not specify any particular modification method for the AOG operation. We simply use $\tau(T)$ to denote the table derived by modifying the data table $T$ with $\tau$.

Given the $\tau$-indiscernibility relation, the standard definition of the lower approximation in rough set theory is used to define the logical security of an AOG. The lower approximation for any set $X \subseteq U$ is defined as

$$ \underline{ind(\tau)}X = \{u \mid \forall(u, u') \in ind(\tau), u' \in X\}. $$

The AOG operation $\tau$ is *logically secure* (or simply secure) for $u$ if $u \notin \underline{ind(\tau)}\|\varphi\|$ for any $\varphi \in CON(u)$, and secure for $U$ if it is secure for all $u \in U$. Once the data table to released has been modified by an AOG $\tau$, the user can not distinguish the records of two individuals who are indiscernible in the relation $ind(\tau)$. Therefore, even though the user knows the values of all the quasi-identifiers of an individual, as well as how the values are modified, he can not deduce that the individual satisfies a confidential property $\varphi$, provided that $\tau$ is secure.

## 3.2   The Running Example

Let us consider an AOG $\tau = (\pi_1, \pi_2, \pi_3) = (mm/yy, d_1, d_2 * **, I_{10})$. Then

$$ ind(\tau) = ind(\pi_1) = ind(\pi_2) = \{\{u_1, u_2, u_3\}, \{u_4, u_5, u_6\}, \{u_7, u_8, u_9, u_{10}, u_{11}\}\}, $$

$$ ind(\pi_3) = \{\{u_1, u_2, u_3, u_4, u_5, u_6\}, \{u_7, u_8, u_9, u_{10}, u_{11}\}\}. $$

By using the generalization method to modify the data table in Figure 1, we obtain the generalized table in Figure 2. On the other hand, if the microaggregation method is used to modify the data table, and the arithmetical mean and median are taken as the statistical operators of the continuous and ordinal attributes respectively, then we can obtain the modified data table in Figure 4. To understand how the table is derived, let us consider the individual $u_1$. First, for the continuous attribute height, $[u_1]_{ind(\pi_3)} = \{u_1, u_2, u_3, u_4, u_5, u_6\}$; thus, $t_{13}$ is replaced by the arithmetical mean of the multiset $\{161, 167, 163, 160, 165, 168\}$, which is equal to 164. Second, for the ordinal attributes, date of birth and zip code, $[u_1]_{ind(\pi_1)} = [u_1]_{ind(\pi_2)} = \{u_1, u_2, u_3\}$; thus, $t_{11}$ and $t_{12}$ are replaced by the median of $\{24/09/56, 06/09/56, 30/09/56\}$ and $\{24126, 24129, 24133\}$, which are $24/09/56$ and $24129$, respectively.

Note that the tables produced by generalization and microaggregation are structurally isomorphic[14]. It is shown in [15] that isomorphic tables have the

---

[3] A multiset is a set that allows the multiple occurrence of its elements.

| 1 | 24/09/56 | 24129 | 164 | 400K | 1 |
|---|---|---|---|---|---|
| 2 | 24/09/56 | 24129 | 164 | 300K | 1 |
| 3 | 24/09/56 | 24129 | 164 | 300K | 1 |
| 4 | 18/03/56 | 10431 | 164 | 300K | 0 |
| 5 | 18/03/56 | 10431 | 164 | 100K | 2 |
| 6 | 18/03/56 | 10431 | 164 | 100K | 2 |
| 7 | 18/04/55 | 26617 | 173 | 400K | 2 |
| 8 | 18/04/55 | 26617 | 173 | 300K | 1 |
| 9 | 18/04/55 | 26617 | 173 | 100K | 0 |
| 10 | 18/04/55 | 26617 | 173 | 400K | 0 |
| 11 | 18/04/55 | 26617 | 173 | 400K | 0 |

**Fig. 4.** Our running example modified by the microaggregation technique

same granular data model, defined as $(U, Q)$, such that $Q$ is a set of equivalence relations induced by the attributes.

Now, if $\varphi = (\text{Health}, 2)$ is a confidential sentence, then $ind(\tau) \| \varphi \| = \emptyset$, since $\| \varphi \| = \{u_5, u_6, u_7\}$. Thus, $\tau$ is secure for $U$ if $CON(u) = \{(\text{Health}, 2)\}$. On the other hand, if $\varphi' = (\text{Health}, 1) \in CON(u_1)$, then $\tau$ is insecure for $u_1$, since $ind(\tau) \| \varphi' \| = \{u_1, u_2, u_3\}$. The result is matched by our intuition.

## 3.3   The Basic Search Algorithm

Since the goal of privacy protection is to find a secure and maximally informative AOG operation, we can achieve it by a bottom-up search of all possible AOGs. The algorithm proposed in this section is based on a breadth-first search through the set of admissible AOGs using basic pruning strategies. For simplicity, we present the basic algorithm in this section, and discuss improvements that make it more efficient in the next section. Our previous experiments show that the performance of the basic algorithm is acceptable in non-realtime environments [6].

Although it is sufficient to find *a* secure and maximally informative AOG for a data table, for the sake of flexibility, our search algorithm is designed to find *all* secure and maximally informative AOGs for a given data table. We start from the most specific AOG $(\perp_1, \cdots, \perp_m)$ and test its security according to our definition. If this operation is secure, we stop searching. Otherwise, we have to climb the search tree according to the partial order $\preceq$ between AOG operations. Each new AOG must be tested to evaluate its security. If it is secure, then all AOG operations above it can be pruned, since our purpose is to find the maximally informative (i.e., $\preceq$-minimal) AOGs. Thus, the pruning operation substantially reduces the number of AOGs that must be visited. In the search algorithm presented in Figure 5, the function GET-FROM-QUEUE returns the first element of a queue, whereas the procedure PUT-INTO-QUEUE adds an AOG to the end of a queue. These operations are standard and can be found in algorithm textbooks. Also, we record the status of each $\tau$ in a Boolean array $F$, where $F(\tau) = 1$ means that it is not necessary to check $\tau$ any further.

---

**Procedure** SEARCH($\Pi, T, CON$)

1. Initialize a Boolean array $F[\tau] := 0$ for all $\tau \in \Pi$;
2. Initialize a queue of AOG operations $Q \leftarrow \{(\perp_1, \perp_2, \cdots, \perp_m)\}$;
3. **while** $Q \neq \emptyset$ **do**
    **begin**
    **repeat** $\tau \leftarrow$ GET-FROM-QUEUE($Q$) **until** $F[\tau] = 0 \vee Q = \emptyset$;
    **if** $F[\tau] = 1 \wedge Q = \emptyset$ **then** exit;
    $F[\tau] \leftarrow 1$;
    **if** SECURITY($\tau, T, CON$)
        **then begin**
        Output($\tau$);
        $F[\tau'] \leftarrow 1$ for all $\tau'$ such that $\tau \preceq \tau'$
        **end**
        **else for** each direct successor $\tau'$ of $\tau$ **do**
        **if** $F[\tau'] = 0$ **then** PUT-INTO-QUEUE($Q, \tau'$)
    **end**

---

**Fig. 5.** The search algorithm for AOG

---

**Function** SECURITY($\tau, T, CON$)

1. Find $ind(\tau)$ by sorting $\tau(T)$;
2. Initialize Boolean $SF \leftarrow 1$;
3. **for** each $u \in U$ **do**
    **begin**
    (a) $US[u] \leftarrow 0$;
    (b) **for** each $\varphi \in CON[u]$ **do**
        **begin**
        $KN(u, \varphi) \leftarrow 1$;
        **for** each $u' \in [u]_{ind(\tau)}$ **do** $KN(u, \varphi) \leftarrow KN(u, \varphi) \wedge (u' \models \varphi)$;
        $US(u) \leftarrow US(u) \vee KN(u, \varphi)$
        **end**;
    (c) $SF \leftarrow SF \wedge \neg US[u]$
    **end**

---

**Fig. 6.** The security test function for AOG

The SECURITY function takes an AOG $\tau$, a data table $T$, and the confidential data function $CON$ as its arguments and returns 1 if $\tau$ is secure with respect to $T$ according to the confidential requirement specified by $CON$; otherwise, it returns 0. The SECURITY function is presented in Figure 6. By sorting $\tau(T)$ according to its quasi-identifiers, we can partition $U$ into $ind(\tau)$-equivalence classes. Then, we use a Boolean variable, $SF$, and two Boolean arrays, $US$ and $KN$, indexed by $U$ and $U \times \mathcal{L}_0$ respectively, to compute the output, where $\mathcal{L}_0$ denotes the set of confidential sentences. Here $SF$, which is initialized to 1, denotes the security of $\tau$, whereas $US[u] = 1$ means that $\tau$ is not secure for $u$; hence, the final security level is computed by repeat conjunction of $SF$ with $\neg US[u]$ for all $u \in U$. The array $KN$ denotes the user's knowledge about individuals, so $KN(u, \varphi) = 1$ means the user knows that $u$ satisfies $\varphi$, i.e., $u \in \overline{ind(\tau)\|\varphi\|}$. $KN(u, \varphi)$ is computed by repeat conjunction of its initial value 1 with $u' \models \varphi$ for all $u' \in [u]_{ind(\tau)}$, where $u' \models \varphi$ means that $\varphi$ is satisfied by $u'$. Furthermore, $\tau$ is not secure for $u$ if for some $\varphi \in CON[u]$, $KN(u, \varphi) = 1$; consequently, $US(u)$ is computed by repeat disjunction of its initial value 0 with $KN(u, \varphi)$ for all $\varphi \in CON[u]$.

The complexity of the SECURITY function can be analyzed as follows. First, Step 1, the sorting step, needs $O(n \log n)$ time using standard algorithms, where $n$ is the cardinality of $U$. Let us assume the evaluation $u' \models \varphi$ can be performed in constant-bounded time; then, the total execution time of Step 3 is

$$\sum_{u \in U} |CON[u]| \cdot |[u]_{ind(\tau)}|.$$

Assuming the size of each $CON[u]$ is bounded above by a constant $C$, the total execution time of Step 3 is at most

$$C \cdot \sum_{u \in U} |[u]_{ind(\tau)}|,$$

which is in $O(n^2)$ time, since $|[u]_{ind(\tau)}| \leq n$ for all $u \in U$. The $O(n^2)$ bound is quite loose, because $|[u]_{ind(\tau)}|$ may be much smaller than $n$. Furthermore, in the special case where all individuals have the same set of confidential data (or at least in the case where, for all $u_1, u_2 \in U$, $(u_1, u_2) \in ind(\tau)$ implies $CON[u_1] = CON[u_2]$), Step 3(b) is only executed once for each individual corresponding to a different $ind(\tau)$-equivalence class, which reduces its computation time to $O(n)$. Therefore, the total time complexity of the security test procedure is $O(n^2)$ in general, and $O(n \log n)$ in special cases.

## 3.4   Computational Improvement

As noted earlier, our algorithm for finding maximally informative AOGs is based on a breadth-first search with basic pruning strategies. Recently, a more efficient algorithm for full-domain $k$-anonymity, called Incognito, has been proposed [16]. It employs more advanced pruning strategies based on the generalization, rollup, and subset properties.

The generalization and subset properties still hold if $k$-anonymity is replaced by our security criterion, whereas the rollup property can be easily extended to our framework if the frequency set used in [16] is replaced by the characteristic functions of confidential sentences. Therefore, Incognito can be easily adapted to find all maximally informative AOGs for a data table. In the following, we show that the generalization, rollup, and subset properties hold in our framework.

First, the generalization property is an obvious fact that is used in our basic search algorithm.

**Property 1 (Generalization property).** *If $\tau_1 \preceq \tau_2$ and $\tau_1$ is secure, then $\tau_2$ is also secure.*

Second, to demonstrate the subset property, we have to define an AOG for an arbitrary subset of quasi-identifiers. So far, we have only defined an AOG for the set of *all* quasi-identifiers. Let $J$ be a subset of $\{1, 2, \cdots, m\}$, then an AOG for $J$ is specified by an $m$-tuple $(\pi_1, \pi_2, \ldots, \pi_m)$ such that $\pi_i = \top_i$ iff $i \notin J$. If $J_1 \subseteq J_2$, and $\tau_1 = (\pi_1, \pi_2, \ldots, \pi_m)$ and $\tau_2 = (\pi'_1, \pi'_2, \ldots, \pi'_m)$ are the respective AOGs for $J_1$ and $J_2$, we say that $\tau_1$ is a restriction of $\tau_2$, denoted by $\tau_1 = \tau_2|J_1$, if $\pi_i = \pi'_i$ for $i \in J_1$. It is obvious that $\tau_2 \preceq \tau_1$ if $\tau_1$ is a restriction of $\tau_2$. The subset property is therefore a corollary of the generalization property.

**Property 2 (Subset property).** *Any restriction of a secure AOG is also secure.*

Third, the rollup property must be adapted to our framework. We note that it is necessary to count the records with each unique combination of values of quasi-identifiers in order to check $k$-anonymity. The rollup property is used to execute the count efficiently. To check our security criterion, we do not have to count the number of records. Instead, we only need to check whether a confidential sentence is falsified for individuals with a combination of quasi-identifiers values. Thus, we define a characteristic function as a mapping from each equivalence class of $\tau$ to the subset of confidential sentences falsified by some individuals in the class. Then, the rollup property can be reformulated as follows.

**Property 3 (Rollup property).** *If $\tau_1 \succeq \tau_2$, then we can generate each set of falsified sentences for the characteristic function of $\tau_2$ by a set union from the characteristic function of $\tau_1$.*

## 4 Security and Data Quality

### 4.1 Security Measure

The security criterion defined in the preceding section is purely qualitative. Thus, even though the security condition is satisfied, there is still a sufficiently high probability that the user could infer an individual's confidential information. To assess the security of a protection mechanism more precisely, a number of quantitative criteria have been proposed [6,17]. One criterion that measures how much confidential information is leaked is called the *average benefit criterion*, because it was originally used to assess the benefit a user derives when he receives released data. It is especially appropriate for AOG operations and can also be used to measure risk, since the lower the average benefit, the less an individual's privacy can be breached.

To define such a risk measure, we examine the difference between a user's a priori and a posteriori knowledge. Consider a data table containing an $ind(\tau)$-equivalence class, where 99 percent of the individuals in that class have the same confidential value for one specific attribute. It is tempting to conclude that personal privacy could be violated easily. However, if this distribution is close to the prior distribution of the attribute value of the entire population, release of the above-mentioned data would not be a threat to personal privacy, since a user could not learn much about the distribution by database linking. It is therefore important to consider the original distribution of attribute values in a database.

We now propose an information-theoretic approach that measures information gain after receiving $\tau(T)$. The user's a priori knowledge about $\varphi$ can be modeled by the prior probability $Pr(\varphi)$, which is the statistical probability of $\varphi$ for the whole population. If the set $U$ is sufficiently representative of the whole population, then

$$Pr(\varphi) = \frac{|\{x \mid x \models \varphi\}|}{|U|}.$$

On the other hand, the user's a posteriori knowledge about whether $u$ satisfies $\varphi$ is the percentage of individuals satisfying $\varphi$ in the $ind(\tau)$-equivalence class $[u]_{ind(\tau)}$, written as

$$Pr_\tau(\varphi|u) = \frac{|\{x \mid x \in [u]_{ind(\tau)} \wedge x \models \varphi\}|}{|[u]_{ind(\tau)}|}.$$

Note that $Pr_\tau(\varphi|u)$ is the rough membership [18] of $u$ in $\|\varphi\|$. Let $dm(u, \varphi)$ be a positive real number denoting the potential damage to an individual $u$ if his/her confidential information $\varphi$ is breached. We assume the damage values of the individuals are normalized so that $\sum_{\varphi \in CON(u)} dm(u, \varphi) = 1$ for each $u \in U$. Thus, the risk to $u$ due to the release of $\tau(T)$, denoted by $ri(\tau, u)$, is

$$\sum_{\varphi \in CON(u)} dm(\varphi) \cdot \max(\frac{\log Pr(\varphi) - \log Pr_\tau(\varphi|u)}{\log Pr(\varphi)}, 0),$$

and the security measure of $\tau$ is defined as

$$sf(\tau) = 1 - \frac{\sum_{u \in U} ri(\tau, u)}{|U|}.$$

## 4.2   Quality Measure

Privacy protection mechanisms inevitably reduce the quality of released data. We should therefore assess how data quality is affected by AOG operations. Since such operations are based on the partition of quasi-identifier domains, we can use Shannon's entropy to measure data quality. First, the entropy of a partition $\pi$ of a domain $V$ is defined as

$$h(\pi) = \sum_{s \in \pi} -\frac{|s|}{|V|} \cdot \log \frac{|s|}{|V|}.$$

Second, we consider the significance of the quasi-identifiers. Let $w_i \in [0, 1]$ denote the importance of the quasi-identifiers in data utilization. We also assume that

$\sum_{1 \le i \le m} w_i = 1$. In Section 3, we only considered the case where all quasi-identifiers are equally important, i.e., $w_i = 1/m$ for $1 \le i \le m$. Thus, the quality measure defined in this section is more flexible. Finally, the quality of an AOG, $\tau = (\pi_1, \pi_2, \cdots, \pi_m)$, is defined as

$$ql(\tau) = \sum_{1 \le i \le m} w_i \cdot \frac{h(\pi_i)}{\log(|V_i|)},$$

where $V_i$ is the domain of the quasi-identifier $i$.

## 4.3    The Search for Optimal AOGs

Once we can quantitatively measure the security and quality of released data, the search for the optimal AOG for privacy protection becomes an optimization problem. In other words, we have to find $\tau$ in the set of admissible AOGs that maximizes the objective function $sf(\tau) \cdot ql(\tau)$. There are numerous techniques for solving such problems. Here, we use the EC approach to find the optimal AOG.

The EC approach is a class of nature-inspired methodologies that can solve hard problems. By this approach, a population of possible solutions is initially given. Then, three basic mechanisms of evolution, i.e., *reproduction*, *mutation*, and *selection*, are applied to the population of solutions to produce the next generation of the population. The process is repeated until satisfactory solutions are found, or a pre-determined number of iterations is reached. A basic scheme of the EC algorithm is presented in Figure 7. The algorithm is adapted from the approach introduced in [19].

The initial population of the algorithm is a randomly selected subset of admissible AOGs. At every step $t$, also called a generation, each AOG in the population $P(t)$ is evaluated according to some predefined fitness function. Then, a subset of AOGs is selected from $P(t)$ according to the result of the evaluation. The selected subset, known as the *mating pool*, is denoted by $P'(t)$. Next, reproduction and mutation operations are applied to AOGs in $P'(t)$ to produce a

---

**Procedure** EC

1. Initialize a population $P(0) \subseteq \Pi$
2. **while** not done **do**
   **begin**
   EVALUATE $P(t)$;
   $P'(t) \leftarrow$ SELECT$[P(t)]$;
   $P''(t) \leftarrow$ GENETIC-OP$[P'(t)]$;
   $P(t+1) \leftarrow$ INTRO$[P''(t), P(t)]$;
   $t \leftarrow t+1$
   **end**

**Fig. 7.** An EC algorithm for AOG

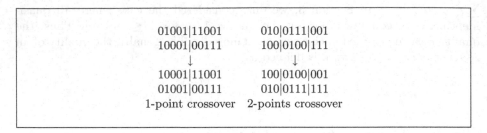

**Fig. 8.** Typical reproduction operations in GA

new population $P''(t)$. The AOGs in $P''(t)$ are offspring of those in $P'(t)$. Finally, $P''(t)$, together with $P(t)$, is introduced into the next-generation of the population $P(t+1)$; usually $P(t)$ is simply replaced by $P''(t)$ to form $P(t+1)$.

A concrete implementation of the skeleton in Figure 7 can be achieved by the standard genetic algorithm (GA). In the GA implementation, we assume that each admissible set of partitions, $\Pi_i$, is identified by a set of integers $\{0, 1, \cdots, |\Pi_i| - 1\}$; therefore, each partition in $\Pi_i$ can be encoded as a binary string of length $\lceil \log |\Pi_i| \rceil$, and each AOG can be encoded as a binary string of length $\sum_{1 \leq i \leq m} \lceil \log |\Pi_i| \rceil$. The fitness function of GA is simply the objective function $sf(\cdot)\overline{ql}(\cdot)$. There are a number of ways to perform the selection. The most popular is the roulette wheel method, where each AOG is selected with a probability proportional to its fitness. The typical reproduction operation for GA is *crossover*, which is performed with a fixed probability, called the *crossover rate*, between two selected AOGs. Figure 8 shows two kinds of crossover operation. The mutation operation is performed by flipping bits at random with some small probability, i.e., the mutation rate. Note that the crossover and mutation operations may produce illegal codes that do not correspond to any AOG, so post-processing is necessary to adjust the codes to legal AOGs.

As an example, we use the admissible partitions in Figure 3. We need an 8-bit string to encode an AOG (2 bits for the date of birth, 3 bits for the zip code, and 3 bits for the height). Thus, for example, (01011010) denotes the AOG $(\{mm/yy\}, d_1 d_2 * **, I_{10})$. If a crossover operation

$$010|11010$$
$$001|00101$$
$$\downarrow$$
$$010|00101$$
$$001|11010$$

is carried out, the resultant codes correspond to (1,0,5) and (0,7,2). However, these are not legal encodings of any AOG, since 5 is not a legal code for height and 7 does not correspond to any partition of zip codes. To transform them into legal encodings, we can change 5 to 5mod5=0 and 7 to 7mod6=1; therefore, the offspring of the crossover operation should be 01000000 and 00001010.

# 5  Related Works

As mentioned in Section 3.1, the granulation approach subsumes two important data protection techniques, generalization[7,11,5,12] and microaggregation[8]. Moreover, rough set theory has been applied to privacy protection previously [20,6,21]. In this section, we further discuss several works related to our approach.

The main concept of logical security models a user's knowledge based on indiscernibility. Traditionally, epistemic logic has been used to represent such knowledge. The relationship between epistemic operators and rough set approximation has been studied extensively[22,23]. Epistemic logic has also been applied to the analysis of security [24,25,26]. The security logic (SL) developed in [24] is a permission-based approach that specifies the knowledge a user is allowed to have, which contrasts with our prohibition-based approach based on the set of confidential sentences. The logic of security (LS) proposed in [25,26] is applied to the analysis of dynamic systems with multiple subjects, where each subject is permitted to know different levels of confidential information according to his role. SL and LS can be applied to the analysis of general security problems; however, our framework is specifically tailored for the database linking problem.

While we are concerned with the issue of *attribute disclosure*, many previous works have addressed the issue of *identity disclosure*. Attribute disclosure occurs when some characteristic of an individual can be inferred more accurately because of the released data, whereas identity disclosure means that an individual can be uniquely identified. The issue of identity disclosure in the database linking context has been addressed in [27,17,11,5,12]. In those works, the main goal of privacy protection is to maintain the anonymity of data records, i.e., to prevent the user from knowing which data record belongs to a specific individual. The $k$-anonymity criterion mentioned earlier is designed to prevent identity disclosure. However, it has been observed that $k$-anonymity is not sufficient for attribute disclosure control, so a logical criterion has been formulated to remedy the problem[4]. A similar problem, called *homogeneity attack*, is also observed in [28]. In this case, the $l$-diversity criterion is proposed to prevent such attacks.

The protection of confidential information has been widely studied in the contexts of disclosure control [29], inference control [30,13,31], access control [32], and data mining [33,34,35,36]. The works most closely related to our approach are those on disclosure control, which modifies data to prevent users from recognizing individual identities in the data or discovering private information about the individuals. Various techniques have been applied in disclosure control. In addition to the granulation approach, whereby released data is made less precise than the original data, other techniques, such as data perturbation [37] or lying [38,39], distort the data to be released. Data perturbation adds noise to the released data, while ensuring that some statistical properties of the whole data set are preserved; whereas lying distorts the truth, i.e., the negation of the correct answer to the user's query to prevent the user from inferring confidential information.

Another important aspect of disclosure control is the assessment of disclosure risk and data quality. A variety of measures for assessing disclosure risk and information loss have been proposed in [40,41,42,17,43,44,28]. Some information measures associated with data tables may also be useful in such assessments. Measures of interest include: Shannon's entropy [45], Kolmogorov's complexity [46], and uncertainty-based information measures [47]. Based on the assessment of disclosure risk and data quality, we can achieve a balance between data availability and privacy protection.

In contrast to our framework for the database linking context, some models have been proposed for dealing with the confidentiality problem in more general contexts [48,49,50,24,51]. Also, complementary to the approach proposed in this paper, some probabilistic or decision-theoretic approaches to data security have been proposed in [20,42,52,53,6,44].

## 6    Conclusion

Granular computing (GrC) is an emerging computing paradigm developed from Pawlak's rough set theory. In recent years, it has had a strong impact on many application domains. In this paper, we apply GrC techniques to privacy protection in the context of data release. Granulation of the domains of quasi-identifiers makes it possible to release microdata without invading individuals' privacy. An attribute-oriented technique is employed to modify the to-be-released data. To achieve a balance between the quality of the released data and privacy protection, we present a basic search algorithm to find the maximally specific AOGs that satisfy the security requirements. We also discuss the properties that can be utilized to improve the efficiency of the algorithm. Then, we define quantitative measures to assess the security and quality of an AOG, and show that EC techniques can be employed to find the optimal granulation for privacy protection.

To demonstrate the performance of the proposed approach, further theoretical analysis and experimental verification of the proposed optimization algorithm are needed. Moreover, to improve the optimization algorithm, more criteria of data quality and security measures could be considered. In the longer term, we will explore the possibility of applying other GrC techniques, such as reduct computation and dependency analysis, to resolve practical data security problems.

## References

1. Wang, D.W., Liau, C.J., Hsu, T.-s: Attribute-oriented granulation for privacy protection. In: Proceedings of the 2nd IEEE International Conference on Granular Computing. (2006) 726–731
2. Willenborg, L., de Waal, T.: Statistical Disclosure Control in Practice. Springer (1996)
3. Chiang, Y.C., Hsu, T.-s., Kuo, S., Liau, C.J., Wang, D.W.: Preserving confidentiality when sharing medical database with the Cellsecu system. International Journal of Medical Informatics **71** (2003) 17–23

4. Hsu, T.-s, Liau, C.J., Wang, D.W.: A logical model for privacy protection. In: Proceedings of the 4th International Conference on Information Security. LNCS 2200, Springer-Verlag (2001) 110–124

5. Sweeney, L.: Achieving $k$-anonymity privacy protection using generalization and suppression. International Journal on Uncertainty, Fuzziness and Knowledge-based Systems **10** (2002) 571–588

6. Wang, D.W., Liau, C.J., Hsu, T.-s: Medical privacy protection based on granular computing. Artificial Intelligence in Medicine **32** (2004) 137–149

7. Wang, D.W., Liau, C.J., Hsu, T.-s: An epistemic framework for privacy protection in database linking. Data and Knowledge Engineering (2007)

8. Domingo-Ferrer, J., Torra, V.: Ordinal, continuous and heterogeneous $k$-anonymity through microaggregation. Data Mining and Knowledge Discovery **11** (2005) 195–212

9. Pawlak, Z.: Rough Sets–Theoretical Aspects of Reasoning about Data. Kluwer Academic Publishers (1991)

10. Dalenius, T.: Finding a needle in a haystack - or identifying anonymous census records. Journal of Official Statistics **2** (1986) 329–336

11. Samarati, P.: Protecting respondents' identities in microdata release. IEEE Transactions on Knowledge and Data Engineering **13** (2001) 1010–1027

12. Sweeney, L.: $k$-anonymity: a model for protecting privacy. International Journal on Uncertainty, Fuzziness and Knowledge-based Systems **10** (2002) 557–570

13. Denning, D.: Cryptography and Data Security. Addison-Wesley Publishing Company (1982)

14. Grzymala-Busse, J.: Algebraic properties of knowledge representation systems. In: Proceedings of the ACM SIGART International Symposium on Methodologies for Intelligent Systems, ACM Press (1986) 432–440

15. Lin, T.Y.: Mining associations by linear inequalities. In: Proceedings of the 4th International Conference on Data Mining, IEEE Press (2004) 154–161

16. LeFevre, K., DeWitt, D., Ramakrishnan, R.: Incognito: efficient full-domain $k$-anonymity. In: Proceedings of the 24th ACM SIGMOD International Conference on Management of Data. (2005) 49–60

17. Iyengar, V.: Transforming data to satisfy privacy constraints. In: Proceedings of the 8th ACM SIGKDD International Conference on Knowledge Discovery and Data Mmining. (2002) 279–288

18. Pawlak, Z.: Rough sets and fuzzy sets. Fuzzy Sets and Systems **17** (1985) 119–123

19. Pena-Reyes, C., Sipper, M.: Evolutionary computation in medicine: An overview. Artificial Intelligence in Medicine **19** (2000) 1–23

20. Chiang, Y.T., Chiang, Y.C., Hsu, T.-s, Liau C.J., Wang D.W.: How much privacy? - a system to safe guard personal privacy while releasing database. In: Proceedings of the 3rd International Conference on Rough Sets and Current Trends in Computing. LNCS 2475, Springer-Verlag (2002) 226–233

21. Wang, D.W., Liau, C.J., Hsu, T.-s, Chen, J.K.P.: Value versus damage of information release: A data privacy perspective. International Journal of Approximate Reasoning **43** (2006) 179–201

22. Orłowska, E.: Logic for reasoning about knowledge. Zeitschrift f. Math. Logik und Grundlagen der Math **35** (1989) 559–572

23. Orłowska, E.: Kripke semantics for knowledge representation logics. Studia Logica **XLIX** (1990) 255–272

24. Glasgow, J., MacEwen, G., Panangaden, P.: A logic for reasoning about security. ACM Transactions on Computer Systems **10** (1992) 226–264

25. Bieber, P., Cuppens, F.: A definition of secure dependencies using the logic of security. In: Proc. of the 4th IEEE Computer Security Foundations Workshop. (1991) 2–11
26. Cuppens, F.: A logical formalization of secrecy. In: Proc. of the 6th IEEE Computer Security Foundations Workshop. (1993) 53–62
27. Dawson, S., di Vimercati, S.D.C., Lincoln, P., Samarati, P.: Maximizing sharing of protected information. Journal of Computer and System Sciences **64** (2002) 496–541
28. Machanavajjhala, A., Gehrke, J., Kifer, D., Venkitasubramaniam, M.: *l*-diversity: privacy beyond *k*-anonymity. In: Proceedings of The 22nd International Conference on Data Engineering. (2006)
29. Bethlehem, J., Keller, W., Pannekoek, J.: Disclosure control of microdata. Journal of the American Statistical Association **85** (1990) 38–45
30. Brodsky, A., Farkas, C., Jajodia, S.: Secure databases: Constraints, inference channels, and monitoring disclosures. IEEE Transactions on Knowledge and Data Engineering **12** (2000) 900–919
31. Morgenstern, M.: Controlling logical inference in multilevel database systems. In: Proc. of the IEEE Symposium on Security and Privacy. (1988) 245–255
32. Bonatti, P., Damiani, E., di Vimercati, S.D.C., Samarati, P.: An access control model for data archives. In: Proceedings of the 16th International Conference on Information Security: Trusted Information: The New Decade Challenge. (2001)
33. Agrawal, D., Aggarwal, C.: On the design and quantification of privacy preserving data mining algorithms. In: Proceedings of the 12th ACM SIGACT-SIGMOD-SIGART Symposium on Principles of Database Systems. (2001) 247–255
34. Clifton, C., Kantarcıoğlu, M., Vaidya, J.: Privacy-preserving data mining. In Chu, W., Lin, T.Y., eds.: Foundations and Advances in Data Mining. Springer-Verlag (2005) 313–344
35. Saygin, Y., Verykios, V., Clifton, C.: Using unknowns to prevent the discovery of association rules. SIGMOD Record **30** (2001) 45–54
36. Srikant, R.: Privacy preserving data mining: challenges and opportunities. In: Proceedings of the 6th Pacific-Asia Conference on Knowledge Discovery and Data Mining. LNCS 2336, Springer-Verlag (2002) 13
37. Muralidhar, K., Sarathy, R.: Security of random data perturbation methods. ACM Transactions on Database Systems **24** (1999) 487–493
38. Biskup, J., Bonatti, P.: Confidentiality policies and their enforcement for controlled query evaluation. In: Proceedings of the 2nd European Symposium on Research in Computer Security. LNCS 2502, Springer-Verlag (2002) 39–55
39. Bonatti, P., Kraus, S., Subrahmanian, V.: Foundations of secure deductive databases. IEEE Transactions on Knowledge and Data Engineering **7** (1995) 406–422
40. Damiani, E., di Vimercati, S.D.C., Jajodia, S., Paraboschi, S., Samarati, P.: Balancing confidentiality and efficiency in untrusted relational dbmss. In: Proceedings of the 10th ACM Conference on Computer and Communication Security. (2003) 93–102
41. Domingo-Ferrer, J.: Advances in inference control in statistical databases: An overview. In: Inference Control in Statistical Databases: From Theory to Practice. LNCS 2316, Springer-Verlag (2002) 1–7
42. Hsu, T.-s, Liau, C.J., Wang, D.W., Chen, J.: Quantifying privacy leakage through answering database queries. In: Proceedings of the 5th International Conference on Information Security. LNCS 2433, Springer-Verlag (2002) 162–175

43. Truta, T., Fotouhi, F., Barth-Jones, D.: Privacy and confidentiality management for the microaggregation disclosure control method: disclosure risk and information loss measures. In: Proceeding of the ACM Workshop on Privacy in the Electronic Society. (2003) 21–30
44. Wang, D.W., Liau, C.J., Hsu, T.-s., Chen, J.K.P.: On the damage and compensation of privacy leakage. In: Proceedings of the 18th Annual IFIP WG 11.3 Working Conference on Data and Applications Security, Kluwer Academic Publisher (2004) 311–324
45. Shannon, C.: The mathematical theory of communication. The Bell System Technical Journal **27** (1948) 379–423,623–656
46. Li, M., Vitanyi, P.: An introduction to Kolmogorov Complexity and its Applications. Springer-Verlag (1993)
47. Klir, G., Wierman, M.: Uncertainty-Based Information : Elements of Generalized Information Theory. Physica-Verlag (1998)
48. Cholvy, L., Cuppens, F.: Analysing consistency of security policies. In: Proc. of the IEEE Symposium on Security and Privacy. (1997) 103–112
49. Cuppens, F., Demolombe, R.: A deontic logic for reasoning about confidentiality. In Brown, M., Carmo, J., eds.: Deontic logic, agency, and normative systems: ΔEON'96, Third International Workshop on Deontic Logic in Computer Science. (1996) 66–79
50. Cuppens, F., Demolombe, R.: A modal logical framework for security policies. In Ras, Z., Skowron, A., eds.: Proc. of the 10th International Symposium on Methodologies for Intelligent Systems. LNAI 1325, Springer-Verlag (1997) 579–589
51. Syverson, P., Stubblebine, S.: Group principals and the formalization of anonymity. In: Proc. of the 1999 World Congress on Formal Methods. LNCS 1708 (1999) 814–833
52. Gray III, J.G., Syverson, P.: A logical approach to multilevel security of probabilistic systems. Distributed Computing **11** (1998) 73–90
53. Syverson, P., Gray III, J.G.: The epistemic representation of information flow security in probabilistic systems. In: Proc. of the 8th IEEE Computer Security Foundations Workshop. (1995) 152–166

# A Note on Definability and Approximations

Yiyu Yao

Department of Computer Science
University of Regina,
Regina, Saskatchewan, Canada S4S 0A2
yyao@cs.uregina.ca

**Abstract.** Definability and approximations are two important notions of the theory of rough sets. In many studies, one is used to define the other. There is a lack of an explicit interpretation of the physical meaning of definability. In this paper, the definability is used as a more primitive notion, interpreted in terms of formulas of a logic language. A set is definable if there is a formula that defines the set, i.e., the set consists of all those elements satisfying the formula. As a derived notion, the lower and upper approximations of a set are two definable sets that approximate the set from below and above, respectively. This formulation may be more natural, bringing new insights into our understanding of rough set approximations.

## 1 Introduction

There exist at least two types of approaches for the development of rough sets, namely, the constructive and algebraic (axiomatic) methods [20,23]. Constructive methods concern various ways to build constructively a pair of lower and upper approximations from more familiar notions, such as information tables [10,11,12,13], equivalence relations (or equivalently partitions) [10,12], binary relations [24], generalized approximation spaces [13], and coverings [26]. Algebraic methods treat the lower and upper approximations as a pair of unary set-theoretic operators that are defined by certain axioms [8,25,26]. Many authors studied various algebras from rough sets [1]. Both types of approaches are useful for rough set theory.

A commonly used constructive method is to define first an equivalence relation from an information table, and then to define a pair of approximations using the equivalence classes induced by the equivalence relation. With this formulation, the notion of definability has been introduced in two ways through equivalence classes and approximations, respectively. The equivalence classes of the equivalence relation are called elementary or basic sets defined by a set of attributes. A set is said to be definable if it is the union of some equivalence classes [3,5,10,11,19]. Alternatively, some authors considered the definability of a set based on its approximations. A set is said to be definable if its lower and upper approximations are the same, and undefinable otherwise [2,9]. The two definitions of definability are equivalent in the sense that the family of definable sets consists exactly of the empty set, the equivalence classes and unions

J.F. Peters et al. (Eds.): Transactions on Rough Sets VII, LNCS 4400, pp. 274–282, 2007.

of equivalence classes [2,10]. They are also equivalent to the ones defined using either the lower or the upper approximations [15].

A difficulty with the existing definitions is that the physical meaning of definability is not entirely clear. On the other hand, the notion of a definable set has been well studied in mathematical logics [6,7], where logic formulas are used to characterize definability. It seems useful to investigate connections of definability in rough set theory and definability in logic. One may also adopt a more intuitive notion of definability from logics into rough set theory. Along this line, initial studies have been made by some authors. Pawlak *et al.* [11] explained the definability of the union of some equivalence classes in terms of a logic condition corresponding to a conjunctive normal form. Buszkowski [2] showed that the definability of rough set theory can be interpreted in terms of propositional definability of a set.

Based on the above mentioned studies, we further examine the notions of definability and approximations. We use definable sets as a primitive notion. The definability of sets is explicitly defined in terms of logic formulas. Once it is established that some sets are not definable, namely, undefinable, their approximations through definable sets come naturally. Instead of defining two types of definability and showing their equivalence as done by Buszkowski [2], we treat approximations as a derived notion constructed from the family of definable sets.

Although the results of the paper are not new, a re-examination and clarification would lead to a better and deeper understanding of rough set approximations. By reinterpreting the existing results, we arrive at a more natural formulation of the theory. The new interpretation not only provides a different point of view, but also allows us to relate rough set theory to other theories. For example, it has been observed that rough set analysis and formal concept analysis are complementary to each other based on two different families of definable sets [22].

## 2  Definability in Information Tables

In the classical view, every concept is understood as a unit of thought that consists of two parts, the intension and the extension of the concept [16,17,18]. The intension (comprehension) of a concept consists of all intrinsic properties or attributes that are valid for all those objects to which the concept applies. The extension of a concept is the set of objects or entities which are instances of the concept. All objects in the extension have the same properties that characterize the concept. In other words, the intension of a concept is an abstract description of common features or properties shared by elements in the extension, and the extension consists of concrete examples of the concept. A concept is thus described jointly by its intension and extension. Such a view of concepts is very useful for rule induction based on rough set theory [9,14].

In order to make the notions of intensions and extensions more concrete, we consider a simple knowledge presentation scheme called information tables. By introducing a logic language in an information table, we can formally define the intension of a concept by a logic formula. We say that a concept is definable if its extension can be precisely defined by a logic formula. In this case, the extension of the concept is called a definable set. It should be pointed out that such a simple view of concepts, though concrete and intuitive appealing, is very restrictive and may not be completely accurate. Nevertheless, it is sufficient for the present investigation on definability and approximations.

## 2.1    Information Tables

Consider a simple knowledge representation scheme in which a finite set of objects is described by using a finite set of attributes. Formally, it can be defined by an information table $M$ expressed as the tuple:

$$M = (U, At, \{V_a | a \in At\}, \{I_a | a \in At\}),  \qquad (1)$$

where $U$ is a finite nonempty set of objects, $At$ is a finite nonempty set of attributes, $V_a$ is a nonempty set of values for an attribute $a \in At$, and $I_a : U \longrightarrow V_a$ is an information function. Furthermore, it is assumed that the mapping $I_a$ is single-valued. In this case, the value of an object $x \in U$ on an attribute $a \in At$ is denoted by $I_a(x)$. In general, for a subset of attributes $A \subseteq At$, we use $I_A(x)$ to denote the vector of values of $x$ on $A$.

A fundamental concept of rough set theory is equivalence relations defined by subsets of attributes.

**Definition 1.** *For a subset of attributes $A \subseteq At$, we can define an equivalence relation $E(A)$ as follows:*

$$xE(A)y \iff \forall a \in A(I_a(x) = I_a(y))$$
$$\iff I_A(x) = I_A(y).  \qquad (2)$$

*That is, $E(A)$ is reflexive, symmetric, and transitive.*

The relation $E(A)$ is commonly known as the indiscernibility relation. If $xE(A)y$, we cannot differentiate $x$ and $y$ based only on attributes in $A$. The equivalence relation $E(A)$ induces a partition of the universe and is denoted by $U/E(A)$. From $U/E(A)$, we can construct an $\sigma$-algebra, $\sigma(U/E(A))$, which contains the empty set $\emptyset$, equivalence classes of $E(A)$, and is closed under set intersection, union and complement. The partition $U/E(A)$ is a base of $\sigma(U/E(A))$.

## 2.2    A Logic Language

In order to formally define intensions of concepts, we adopt the decision logic language $\mathcal{L}$ used by Orlowska [9] and Pawlak [10] for analyzing an information table. Formulas of $\mathcal{L}$ are constructed recursively based on a set of atomic formulas corresponding to some basic concepts. An atomic formula is given by a descriptor $(a = v)$, where $a \in At$ and $v \in V_a$. For each atomic formula $(a = v)$, an object $x$ satisfies it if $I_a(x) = v$, written $x \models (a = v)$. Otherwise, it does not satisfy $(a = v)$ and is written $\neg x \models (a = v)$. From atomic formulas, we can construct other formulas by applying the logic connectives $\neg$, $\wedge$, $\vee$, $\rightarrow$, and $\leftrightarrow$. The satisfiability of any formula is defined recursively as follows:

(1).   $x \models \neg\phi$ iff not $x \models \phi$,

(2).   $x \models \phi \wedge \psi$ iff $x \models \phi$ and $x \models \psi$,

(3).   $x \models \phi \vee \psi$ iff $x \models \phi$ or $x \models \psi$,

(4).   $x \models \phi \rightarrow \psi$ iff $x \models \neg\phi \vee \psi$,

(5).   $x \models \phi \leftrightarrow \psi$ iff $x \models \phi \rightarrow \psi$ and $x \models \psi \rightarrow \phi$.

The language $\mathcal{L}$ can be used to reason about intensions. Each formula represents an intension of a concept. For two formulas $\phi$ and $\psi$, we say that $\phi$ is more specific than $\psi$, and $\psi$ is more general than $\phi$, if and only if $\models \phi \rightarrow \psi$, namely, $\psi$ logically follows from $\phi$. In other words, the formula $\phi \rightarrow \psi$ is satisfied by all objects with respect to any universe $U$ and any information function $I_a$. If $\phi$ is more specific than $\psi$, we write $\phi \preceq \psi$, and call $\phi$ a sub-concept of $\psi$, and $\psi$ a super-concept of $\phi$.

If $\phi$ is a formula, the set $m(\phi)$ defined by:

$$m(\phi) = \{x \in U \mid x \models \phi\}, \tag{3}$$

is called the meaning of the formula $\phi$ in an information table $M$. The meaning of a formula $\phi$ is indeed the set of all objects having the properties expressed by the formula $\phi$. In other words, $\phi$ can be viewed as the description of the set of objects $m(\phi)$. Thus, a connection between formulas and subsets of $U$ is established. The following properties hold [10]:

(a).   $m(\neg\phi) = -m(\phi)$,

(b).   $m(\phi \wedge \psi) = m(\phi) \cap m(\psi)$,

(c).   $m(\phi \vee \psi) = m(\phi) \cup m(\psi)$,

(d).   $m(\phi \rightarrow \psi) = -m(\phi) \cup m(\psi)$,

(e).   $m(\phi \equiv \psi) = (m(\phi) \cap m(\psi)) \cup (-m(\phi) \cap -m(\psi))$.

With the introduction of language $\mathcal{L}$, we have a formal description of concepts. A concept in an information table $M$ is a pair $(\phi, m(\phi))$, where $\phi \in \mathcal{L}$. More specifically, $\phi$ is a description of $m(\phi)$ in $M$, the intension of concept $(\phi, m(\phi))$, and $m(\phi)$ is the set of objects satisfying $\phi$, the extension of concept $(\phi, m(\phi))$.

In many applications of rough set theory, one considers only a subset of attributes $A \subseteq At$. In other words, only attributes from $A$ are used in forming formulas of the logic language. We will use $\mathcal{L}(A)$ to denote the language defined using only attributes from $A$. All the discussions so far still hold if we replace $\mathcal{L}$ by $\mathcal{L}(A)$.

## 2.3   Definability of Sets and Concepts

Given a formula as the intension of a concept, we can easily find its extension through the meaning function $m$. On the other hand, given an arbitrary subset

$X \subseteq U$ as extension of a concept, the task of finding the corresponding intension is not so easy. Several issues have to be considered. The attributes $At$ may not be sufficient for us to define a formula so that its meaning is $X$. Even if such a formula exists, it may not be unique. The first problem leads to the study of definability and the second problem requires a consideration of a restricted language in which only certain logic connectives can be used [21].

Consider a subset of attributes $A \subseteq At$ and the corresponding language $\mathcal{L}(A)$. The definability of a subset of objects can be defined formally.

**Definition 2.** *A subset $X \subseteq U$ is definable by a set of attributes $A \subseteq At$ in an information table $M = (U, At, \{V_a | a \in At\}, \{I_a | a \in At\})$ if and only if there exists a formula $\phi$ in the language $\mathcal{L}(A)$ so that,*

$$X = m(\phi). \tag{4}$$

*Otherwise, it is undefinable.*

This definition is consistent with the notion of definable set in mathematical logic [6,7]. That is, a set is definable if one can find a logic formula that defines the elements of the set. Since a logic formula in $\mathcal{L}(A)$ has a concrete physical interpretation, we therefore associate its meaning set with a concrete interpretation. This point has in fact been made implicitly by many authors [10,11].

According to the definition, the family of all definable sets is given by:

$$\text{Def}(U, \mathcal{L}(A)) = \{m(\phi) \mid \phi \in \mathcal{L}(A)\}. \tag{5}$$

Similarly, the family of concepts that can be defined by the language $\mathcal{L}(A)$ is given by:

$$\text{DefCon}(U, \mathcal{L}(A)) = \{(\phi, m(\phi)) \mid \phi \in \mathcal{L}(A)\}. \tag{6}$$

It should be noted that definability depends on the set of attributes $A$.

With the introduction of language $\mathcal{L}(A)$, we can arrive at an equivalent definition of the equivalence relation.

**Lemma 1.** *Suppose $A \subseteq At$ is a subset of attribute. Let $E(A)$ be the equivalence relation defined by $A$. The following condition holds: for $x, y \in U$,*

$$xE(A)y \text{ if and only if } x \models \phi \Longleftrightarrow y \models \phi \text{ for all } \phi \text{ in the language } \mathcal{L}(A). \tag{7}$$

The result of the lemma can be easily shown by the equivalence of the condition in equation (2) of Definition 1 and the condition in equation (7). That is, two objects $x$ and $y$ satisfy exactly the same set of formulas in $\mathcal{L}(A)$ if and only if they have the same values on all attributes in $A$.

The new definition of the equivalence of objects has been considered by Hobbs [4]. According to Hobbs, two objects are considered to be equivalent, if we cannot distinguish them by all available predicates in a first-order logic theory. One can easily re-express the logic language in the form of a first-order logic theory. The definition of Hobbs is more general in the sense that the set of predicates does not have to be defined with respect to an information table. This offers a new avenue for generalizing rough set theory.

In terms of language $\mathcal{L}(A)$, two objects are considered to be equivalent if they satisfy exactly the same set of formulas in $\mathcal{L}(A)$. With this interpretation, the following lemma follows immediately.

**Lemma 2.** *Suppose $X \subseteq U$ is a definable set with reference to a language $\mathcal{L}(A)$. For two elements $x, y \in U$ with $xE(A)y$, $x \in X$ if and only if $y \in X$.*

According to the lemma, for any equivalence class $[x]_{E(A)}$ of $E(A)$, a definable set either contains $[x]_{E(A)}$ or is disjoint with $[x]_{E(A)}$. That is, a definable set is the union of some equivalence classes. This immediately leads to the main result of the paper.

**Theorem 1.** *The family of definable sets with reference to a language $\mathcal{L}(A)$ is exactly the $\sigma$-algebra $\sigma(U/E(A))$. That is,*

$$\text{Def}(U, \mathcal{L}(A)) = \sigma(U/E(A)). \tag{8}$$

Although the discussion produces the same result of earlier studies that the union of some equivalence classes is a definable set, there is a subtle difference. In many studies, the union of some equivalence classes is simply called a definable set without giving an explicit interpretation. The logic based explicit interpretation examined in this paper not only justifies the earlier result but also provides insights into definability.

## 3   Rough Set Approximations

The dual notion of definable sets is undefinable sets. For an undefinable set, it is impossible to construct a formula with the set as its meaning set. In order to characterize an undefinable set, one may approximate it from below and above by two definable sets. The family of definable sets is a subsystem of the power set. We can use the subsystem-based definition of rough set approximations.

**Definition 3.** *For a subset of objects $X \subseteq U$, we define a pair of lower and upper approximations as:*

$$\underline{apr}(X) = \bigcup \{Y \mid Y \in \text{Def}(U, \mathcal{L}(A)), Y \subseteq X\},$$
$$\overline{apr}(X) = \bigcap \{Y \mid Y \in \text{Def}(U, \mathcal{L}(A)), X \subseteq Y\}. \tag{9}$$

*This is, $\underline{apr}(X)$ is the largest definable set contained in $X$, and $\overline{apr}(X)$ smallest definable set containing $X$.*

The definition is well defined, for the family of definable sets is closed under set intersection, union and complement. By definition, a definable set has the same lower and upper approximation.

**Theorem 2.** *A set of objects $X \subseteq U$ is a definable set if and only if the following condition holds:*

$$apr(X) = \overline{apr}(X). \tag{10}$$

The theorem easily follows from the fact that in general $apr(X) \subseteq X \subseteq \overline{apr}(X)$ and both $apr(X)$ and $\overline{apr}(X)$ are definable sets.

According to our reformulation, approximations are a derived notion from definability. Approximations are due to the fact that certain sets are not definable. Since definable sets have clear interpretations in terms of their intensions (i.e., logic formulas), the lower and upper approximations have clear interpretations. The modeling of undefinability through definability seems to capture the central ideas of rough set theory [11]. In other words, one can only approximately say something about an undefinable set and the corresponding concept, based on definable sets.

## 4    Concluding Remarks

In addition to providing many useful methodologies and tools, rough set theory offers a new philosophical view for dealing with uncertainty characterized by indiscernibility. In order to appreciate this view, it is necessary to examine the fundamental notions, of which definability and approximations are examples.

In this paper, we examine these two basic notions. By treating definability as a primitive notion, we define a definable set by a logic formula of a logic language in an information table. It is shown that the family of definable sets indeed coincides with the $\sigma$-algebra constructed from the partition of an equivalence relation. Rough set approximations are formulated as a derived notion from definable sets. Specifically, the lower and upper approximations are two definable sets that approach a set from below and above.

The paper makes three contributions. First, it reformulates the existing results in an attempt to have a more coherent framework. Second, it reinterprets the existing results in order to gain a better understanding of the theory. Third, it formally makes ideas that have been developed explicit. Through such an investigation, we hope to gain more insights into the theory of rough sets.

## Acknowledgment

This study is partially supported by NSERC Canada. The author thanks Dr. Mohua Banerjee for her valuable discussion, and Drs. Andrzej Skowron and James Peters for their kind support when he is preparing the manuscript.

## References

1. Banerjee, M. and Chakraborty, M.K. Algebras from rough sets, in: Pal, S.K., Polkowski, L. and Skowron, A. (Eds.), *Rough-Neural Computing*, Springer, Berlin, 157-184, 2004.
2. Buszkowski, W. Approximation spaces and definability for incomplete information systems, *Rough Sets and Current Trends in Computing (RSCTC'98), LNAI 1424*, 115-122, 1998.

3. Grzymala-Busse, J.W. Incomplete data and generalization of indiscernibility relation, definability, and approximations, *Rough Sets, Fuzzy Sets, Data Mining, and Granular Computing(RSFDGrC'05), LNAI 3641*, 244-253, 2005.
4. Hobbs, J.R. Granularity, *Proceedings of the 9th International Joint Conference on Artificial Intelligence*, 432-435, 1985.
5. Järvinen, J. and Kortelainen, J. A note on definability in rough set theory, in: De Baets, B., De Caluwe, R. De Tré, G., Fodor, J., Kacprzyk, J. and Zadrożny, S. (Eds.), *Current Issues in Data and Knowledge Engineering*, Akademicka Oficyna Wydawnicza EXIT, Warsaw, 272-277, 2004.
6. Kreisel, G. and Krivine, J.L. *Elements of Mathematical Logic*, North-Holland, Amsterdam, 1971.
7. Kunen, K. *Set Theory: An Introduction to Independence Proofs*, North-Holland, Amsterdam, 1980.
8. Mi, J.S. and Zhang, W.X. An axiomatic characterization of a fuzzy generalization of rough sets, *Information Sciences*, **160**, 235-249, 2004.
9. Orlowska, E. Logical aspects of learning concepts, *International Journal of Approximate Reasoning*, **2**, 349-364, 1988.
10. Pawlak, Z. *Rough Sets, Theoretical Aspects of Reasoning about Data*, Kluwer Academic Publishers, Dordrecht, 1991.
11. Pawlak, Z., Grzymala-Busse, J., Slowinski, R. and Ziarko, W. Rough sets, *Communications of the ACM*, **38**, 89-95, 1995.
12. Pawlak, Z. and Skowron, A. Rudiments of rough sets, *Information Sciences*, **177**, 3-27, 2007.
13. Pawlak, Z. and Skowron, A. Rough sets: some extensions, *Information Sciences*, **177**, 28-40, 2007.
14. Pawlak, Z. and Skowron, A. Rough sets and Boolean reasoning, *Information Sciences*, **177**, 41-73, 2007.
15. Pomykala, J.A. On definability in the nondeterministic information system, *Bulletin of the Polish Academy of Sciences: Mathematics*, **36**, 193-210, 1987.
16. Smith, E.E. Concepts and induction, in: Posner, M.I. (Ed.), *Foundations of Cognitive Science*, The MIT Press, Cambridge, Massachusetts, 501-526, 1989.
17. Sowa, J.F. *Conceptual Structures, Information Processing in Mind and Machine*, Addison-Wesley, Reading, Massachusetts, 1984.
18. Van Mechelen, I., Hampton, J., Michalski, R.S. and Theuns, P. (Eds.) *Categories and Concepts, Theoretical Views and Inductive Data Analysis*, Academic Press, New York, 1993.
19. Wasilewska, A. Definable ses in knowledge representation systems, *Bulletin of the Polish Academy of Sciences: Mathematics*, **35**, 629-635, 1987.
20. Wu, W.Z. and Zhang, W.X. Constructive and axiomatic approaches of fuzzy approximation operators, *Information Sciences*, **159**, 233-254, 2004.
21. Yao, J.T., Yao, Y.Y. and Zhao, Y. Foundations of classification, in: Lin, T.Y., Ohsuga, S., Liau, C.J. and Hu, X. (Eds.), *Foundations and Novel Approaches in Data Mining*, Springer, Berlin, 75-97, 2006.
22. Yao, Y.Y. A comparative study of formal concept analysis and rough set theory in data analysis, *Rough Sets and Current Trends in Computing (RSCTC'04), LNAI 3066*, 59-68, 2004.
23. Yao, Y.Y. Constructive and algebraic methods of the theory of rough sets, *Information Sciences*, **109**, 21-47, 1998.

24. Yao, Y.Y. Relational interpretations of neighborhood operators and rough set approximation operators, *Information Sciences*, **111**, 239-259, 1998.
25. Zhu, W. Topological approaches to covering rough sets, *Information Sciences*, in press.
26. Zhu, W. and Wang, F.Y. Reduction and axiomization of covering generalized rough sets, *Information Sciences*, **152**, 217-230, 2003.

# Feature Selection in Audiovisual Emotion Recognition Based on Rough Set Theory*

Yong Yang[1,2], Guoyin Wang[2,**], Peijun Chen[1], Jian Zhou[1], and Kun He[2]

[1] School of Information Science and Technology,
Southwest Jiaotong University,
Chengdou, 610031, P.R. China
[2] Institute of Computer Science & Technology,
Chongqing University of Posts and Telecommunications,
Chongqing, 400065, P.R. China
yangyong@cqupt.edu.cn
Wanggy@ieee.org

**Abstract.** Affective computing is becoming an important research area in intelligent computing technology. Furthermore, emotion recognition is one of the hot topics in affective computing. It is usually studied based on facial and audio information with technologies such as ANN, fuzzy set, SVM, HMM, etc. Many different facial and acoustic features are considered in emotion recognition by researchers. The question which features are important for emotion recognition is discussed in this paper. Rough set based reduction algorithms are taken as a method for feature selection in a proposed emotion recognition system. Our simulation experiment results show that rough set theory is effective in emotion recognition. Some useful features for audiovisual emotion recognition are discovered.

**Keywords:** Affective computing, Emotion recognition, Pattern recognition, Rough set, Feature selection.

## 1 Introduction

It is always a dream that computers can simulate and communicate with a human, or have emotions that human have. A lot of research works have been done in this field in recent years. Affective computing is one of them. Affective computing is computing that relates to, arises from, or deliberately influences emotion, which is firstly proposed by Picard at MIT in 1997 [1]. Affective computing consists of recognition, expressing, modeling, communicating and responding to

---

* This paper is partially supported by National Natural Science Foundation of China under Grant No.60373111, Program for New Century Excellent Talents in University (NCET), Natural Science Foundation of Chongqing under Grant No.2005BA2003, Science & Technology Research Program of Chongqing Education Commission under Grant No.040505.
** Corresponding author.

J.F. Peters et al. (Eds.): Transactions on Rough Sets VII, LNCS 4400, pp. 283–294, 2007.

emotion [2]. Emotion recognition is one of the most fundamental and important modules in affective computing. It is always based on facial and audio information, which is in accordance with people's recognition to emotion. Its applications have reached almost every aspect of our daily life, for example, health care, children education, game software design, and human-computer interaction [1][3].

Nowadays, emotion recognition is usually studied using ANN, Fuzzy set, SVM, HMM and the recognition rate is 64% to 98% [1][4]. In these research works, many different features are used for recognition, the question which one is important for emotion recognition among the features is not answered, yet. On the other hand, Rough Set is a valid theory for data mining. The most advantage of Rough Set is its great ability of attribute reduction. It has been successfully used in many domains such as machine learning, pattern recognition, intelligent data analyzing and control algorithm acquiring, etc. In some research studies [5][6][7][8], rough set has been used in emotion recognition through audio and visual information, and high recognition rate are achieved. In this paper, based on the previous work by the authors in [5][6][7][8], some important features for emotion recognition are identified based on rough set attribute reduction algorithms. Based on the selected subset of features, a higher recognition rate is resulted.

The rest of this paper is organized as follows. In Section 2, some related works are reviewed. In Section 3, basic concepts and methods of Rough Set theory are introduced. In Section 4, the framework of an emotion recognition system based on rough set theory (ERSBRS) is proposed. Simulation experiments are done by means of ERSBRS in Section 5. Finally, conclusion and future works are discussed in Section 6.

## 2    Review of Related Works

### 2.1    Recognition of Emotion from Both Speech and Video Images

Combining audio and visual cues has been studied in recent years for emotion recognition. For emotional expression recognition, the coupling is not so tight. Few research works have been done to utilize both modalities for recognizing emotions.

De Silva proposed a rule-based method for singular classification of audiovisual input data into one of the six emotion categories: happiness, sadness, fear, anger, surprise, and dislike. The audio and visual material were processed separately. They used optical flow for detecting the displacement and velocity of some key facial features (e.g., corners of the mouth, inner corners of the eye brows). From the audio signal, pitch and pitch contours were estimated. A nearest neighbor method was used to classify the extracted facial features and the HMM technique was used to classify the estimated acoustic features into one of the emotion categories. Per subject, the results of the classification were plotted in two graphs and based upon these graphs the rules for emotion classification of the audiovisual input material were defined. A correct recognition rate of 72% was achieved [9].

Chen and Huang proposed a set of methods for singular classification of input audiovisual data into one of the basic emotion categories: happiness, sadness, disgust, fear, anger, and surprise. They collected data from five subjects which displayed 6 basic emotions 6 times by producing the appropriate facial expression. Each of these single-emotion sequences started and ended with a neutral expression. Considering the fact that in the recorded data a pure facial expression occurred right before or after the sentence spoken with the appropriate vocal emotion, the authors applied a single-modal classification method in a sequential manner. An average recognition rate of 79% was achieved in their experiment [10].

Cheng-Yao Chen et al. took speech and video images as the input of an emotion recognition system. 27 facial features and 8 acoustic features were selected. Based on the method of SVM, an average recognition ratio of 84% was achieved [11].

From the related works above, we can find that different researchers took different features as input for emotion recognition. Based on these features, few of them took feature selection as a module of their emotion recognition systems. They did not pay much attention to finding and selecting important facial and acoustic features in developing emotion recognition systems.

## 2.2 Emotion Recognition Based on Rough Set from Speech and Video Images

Yong Yang et al. tried to use rough set as an approach for emotion recognition. Firstly, Rough set was applied for emotion recognition from facial images, 8 features were selected and a correct recognition ratio of 81 % was achieved [5]. Then, the method was also used for both audio and facial images. As a result, 10 features were selected and a correct recognition rate of 77.8% was achieved [8]. Yong Yang et al. also tried to use the method of rough set + SVM for emotion recognition from speech and video images separately. That is, rough set was taken as a module of feature selection and SVM was used as a classifier in emotion recognition. 13 audio features were selected and a correct recognition rate of 74.75% was achieved [6], 10 facial features were selected and a correct recognition rate of 81.96% was achieved [7]. In the previous work, we found that experiment results were not stable when rough set was taken as feature selection and classifier. Usually, a higher recognition rate could be achieved when rough set was taken as feature selection and SVM as a classifier.

## 2.3 Feature Selection

Data preprocessing is an indispensable step in effective data analysis. It prepares data for data mining and machine learning, which aims to turn data into knowledge. Feature selection is a preprocessing technique commonly used on high dimensional data. Feature selection studies how to select a subset or list of attributes or variables that are used to construct models describing data. Its purposes include reducing dimensionality, removing irrelevant and redundant features, reducing the amount of data needed for learning, improving algorithms' predictive accuracy, and increasing the constructed models' comprehensibility. Feature selection is different from feature extraction. It just selects a subset of raw features while feature extraction may generate some new features from the raw features and data. Feature selection methods are particularly useful in interdisciplinary collaborations because the selected features retain the original meanings domain experts are familiar with. The rapid developments in computer science and engineering allow for data collection at an unprecedented speed and present new challenges to feature selection. Feature selection methods attempt to explore data's intrinsic properties by employing statistics or information theory.

Following supervised-learning terminology, feature selection methods are always categorized as filter or wrapper approaches. Filter methods take advantage of some intrinsic property of the data to select features without applying the clustering algorithm. The basic components are the feature search method and the feature selection criterion, such as FOCUS[12], RELIEF[13] and PRESET[14]. On the other hand, wrapper methods apply the unsupervised-learning algorithm to each candidate feature subset and then evaluate the feature subset by criterion functions that use the clustering result. Wrapper methods directly incorporate the clustering algorithm's bias in search and selection. The basic components are the feature search method, the clustering algorithm, and the feature selection criterion.

## 3    Rough Set Theory

Rough Set (RS) is a valid mathematical theory to deal with imprecise, uncertain, and vague information. It has been applied successfully in such fields as machine learning, data mining, pattern recognition, intelligent data analyzing and control algorithm acquiring, etc, since it was proposed by Z. Pawlak in 1980s [15][16]. Rough set was also used in pattern recognition successfully [17][18][19].

The most advantage of RS is its great ability to compute the reductions of information systems. There are a lot of research works on attribute reduction, such as [20][21][22][23][24].

### 3.1    Basic Concepts

For illustration, Some basic concepts of rough set are introduced here.

**Def.1** A decision table information system is a formal representation of a data set to be analyzed. It is defined as a pair $s = (U, R, V, f)$, where $U$ is a finite set of objects and $R = C \cup D$ is a finite set of attributes, $C$ is the condition attribute set and $D = \{d\}$ is the decision attribute set. With every attribute $a \in R$, set of its values $V_a$ is associated. Each attribute $a$ determines function $f_a : U \to V_a$.

**Def.2** For a subset of attributes $B \subseteq A$, the indiscernibility relation is defined by $Ind(B) = \{(x, y) \in U \times U : a(x) = a(y), \forall a \in B\}$.

**Def.3** The lower approximation $B_-(X)$ and the upper $B^-(X)$ approximation of a set of objects $X \subseteq U$ with reference to a set of attributes $B \subseteq A$ may be defined in terms of the classes in the equivalence relation, as follows:

$$B_-(X) = \bigcup\{E \in U/Ind(B)|E \subseteq X\}, \; B^-(X) = \bigcup\{E \in U/Ind(B)|E \cap X \neq \Phi\}$$

is called as the $B_-$lower and upper $B^-$approximation, respectively. The $B_-$lower and $B_-$upper approximation can also be defined as follows:

$$B_-(X) = \{x \in U|[x]_B \subseteq X,\ B^-(X) = \{x \in U|[x]_B \cap X \neq \Phi\}$$

where $[x]_B \in U/Ind(B)$ is the equivalence class of object induced by the set of attributes $B \subseteq A$.

**Def.4** $POS_P(Q) = \bigcup_{x \in U/Ind(B)} P_-(X)$ is the $P$ positive region of $Q$, where $P$ and $Q$ are both attribute sets of an information system.

**Def.5** A reduction of $P$ in an information system is a set of attributes $S \subseteq P$ such that all attributes $a \in P - S$ are dispensable, all attributes $a \in S$ are indispensable and $POS_S(Q) = POS_P(Q)$. We use the term $RED_Q(P)$ to denote the family of reductions of $p$. $CORE_Q(P) = \bigcap RED_Q(P)$is called as the $Q$-core of attribute set $P$.

**Def.6** Correlation degree of condition attributes set $C$ and decision attributes set $D$ is defined as $K_\beta(C, D) = \frac{card(POS_C(D) \cup NEG_C(D))}{card(U)}$.

## 3.2   Attribute Reduction Algorithm for Feature Selection

X. H. Hu and N. Cercone proposed an attribute reduction algorithm for feature selection in [24]. They intended to keep the attribute reduction algorithm bias as small as possible and find a subset attributes which can generate good results by applying a suite of data mining algorithms. An algorithm which can find a relevant feature subset and eliminate unnecessary attributes effectively was proposed.

The authors believe traditional techniques make use of feature merits based on the information theories, statistics correlation between each feature and the class, or the significant values based on rough set theory. All these measures only consider the single feature's effect on the class distinguishability. However, in general, one feature does not distinguish classes by itself; it does so in combination with other features. Therefore, it is desirable to obtain the feature's correlation to the class in the context of other features. Based on the early works, a feature merit measure to rank the feature in the context of other features is proposed. The idea is to assign the contextual merit value based on the component distance of feature weighted by the degree of difference between instances in different classes. The feature merit measure consists of two parts: Weighted Feature Difference ($WFD$) based on the precise unequality of the values of the predictor fields (attributes) and the Value Difference ($VD$) by taking into account the difference of values to evaluate the distinguishability of the two instances. Based on $WFD$ and $VD$, the contextual merit $CM_k$ of a feature $C_k$ is defined. Based on the above concepts, X. H. Hu and N. Cercone developed the following rough set based filter feature selection algorithm.

Alg 1: rough set based filter feature selection algorithm.
Input: conditional attribute set $C$ and decision attribute set $D$
Output: A selected condition attribute set $REDU$
For(C,D) do

Compute $K_\beta(C, D)$ for conditional attribute set $C$ and decision attribute set $D$
$REDU = C$
End
While $K_\beta(C, D) = K_\beta(REDU, D)$ Do
    Compute the contextual merits $CM$ for all attributes of $REDU$;
    Sort the set of attributes $REDU$ based on contextual merits;
    Choose an attribute with the least merit value and
    $REDU = REDU - \{a_j\}$
End
Return $REDU$.

A selected condition attribute set, which strongly rely on the decision attribute set, can result from this algorithm.

## 4 Emotion Recognition System

An audiovisual emotion recognition system composed of such modules as input, pre-processing, feature extraction, feature selection, classification model and output shown in Fig. 1 is developed for the purpose of this paper.

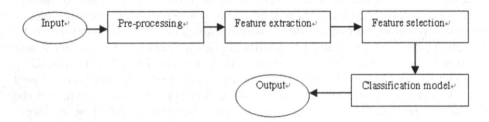

**Fig. 1.** Audiovisual Emotion Recognition System

### 4.1 Input

Emotion recognition is always performed through video, facial image and voice. Although a tactile computer-sensing modality is more natural and it has been proven that affective arousal has a range of somatic and physiological correlations in psychophysiology, only few works aimed at automatic analysis of affective physiological signals have been done in the last years, such as the work introduced by Picard [3]. The lack of interest in this research topic might be due to the lack of interest by research sponsors and there are complicated theoretical and practical open complicated problems in this field, such as the application of haptic technology might have a profound impact on the users' fatigue if it is done improperly; currently available wearable sensors of physiological reactions imply wiring subjects, which are usually experienced as uncomfortable; skin sensors are very fragile and their measuring accuracy is easily affected.

In our emotion recognition system, facial images and audio information of human are taken into consideration.

## 4.2   Pre-processing

Features for emotion recognition always can't be extracted immediately from raw input data, since there may be interference on input data, such as illumination, accessories, side face, etc. There are different methods for different interference, for examples, filter can be used for filtrate noise around, Gamma Intensity Correction (GIC) [25] can be used to normalize the overall image intensity at the given illumination level.

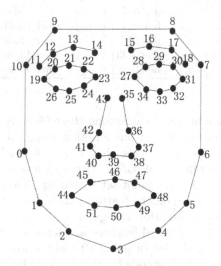

**Fig. 2.** 52 feature points

## 4.3   Feature Extraction

### 4.3.1 Facial Feature Extraction

Human facial expression is expressed by the shape and position of facial components such as eyebrows, eyes, mouth, nose, etc. The geometric facial features present the shape and locations of facial components. The facial feature points are extracted to form feature vector that represents the face geometry. Accordingly, we locate feature points firstly, and then calculate the geometric feature. There are many methods for feature points locating, such as Active Contour Model (Snake) [26], Active Shape Models (ASM) [27], Active Appearance Models (AAM) [28], etc. AAM have been successfully used for deformable objects locating. In this paper, AAM is adopted to locate feature points. On the other hand, the MPEG-4 standard is a popular standard for feature point selection. It extends Facial Action Coding System (FACS) to derive Facial Definition Parameters (FDP) and Facial Animation Parameters (FAP). FAP has been used widely in facial animation for its good performance on compression in recent years. Besides, the FDP and low level FAP constitute a concise representation of a face. They are adequate for basic emotion recognition because of the varieties of expressive parameter. In FAP, 66 low level parameters are defined to

**Table 1.** definition of distance feature

| feature | description | feature | description | feature | description |
|---------|-------------|---------|-------------|---------|-------------|
| $d_0$ | dis(11,19) | $d_{11}$ | dis(39,44) | $d_{22}$ | dis(44,48)/2 |
| $d_1$ | dis(18,31) | $d_{12}$ | dis(39,48) | $d_{23}$ | dis(45,51) |
| $d_2$ | dis(21,25) | $d_{13}$ | dis(44,48) | $d_{24}$ | dis(47,49) |
| $d_3$ | dis(20,26) | $d_{14}$ | dis(46,50) | $d_{25}$ | dis(14,23) |
| $d_4$ | dis(22,24) | $d_{15}$ | dis(39,3) | $d_{26}$ | dis(15,27) |
| $d_5$ | dis(29,33) | $d_{16}$ | dis(21,A) | $d_{27}$ | dis(19,23)/2 |
| $d_6$ | dis(28,34) | $d_{17}$ | dis(A,25) | $d_{28}$ | dis(27,31)/2 |
| $d_7$ | dis(30,32) | $d_{18}$ | hei(A,44) | $d_{29}$ | (wid(19,23)+wid(27,31))/2 |
| $d_8$ | dis(39,46) | $d_{19}$ | dis(29,B) | $d_{30}$ | (hei(11,39)+hei(18,39))/2 |
| $d_9$ | dis(23,44) | $d_{20}$ | dis(B,33) | $d_{31}$ | (hei(14,39)+hei(15,39))/2 |
| $d_{10}$ | dis(27,48) | $d_{21}$ | hei(B,48) | $d_{32}$ | (hei(44,39)+hei(48,39))/2 |

describe the motion of a human face. Among these 66 parameters, 52 parameters are chosen to represent emotion in our emotion recognition system. The other 14 parameters aren't selected because they have not much impact on emotion research, such as the parameter of *pull_l_ear*, which is the horizontal displacement of left ear. Thus, a feature point set including 52 feature points is defined in the image sequence. Based on the feature points, feature distance can be calculated as the features for emotion recognition. Based on the feature distance defined in [29][30][31][32], 33 facial features are defined in the paper.

In Table 1, A is the midpoint of point 19 and 23, and B is the midpoint of point 27 and 31. The $d$, which stands for the distance between point 23 and 27, is changeless for all kinds of expression. dis(i, j) denotes the normalized Euclid distance between point i and j, see Equation (1); hei(i, j) denotes the normalized horizontal distance between point i and j, see Equation (2); wid(i, j) denotes the normalized vertical distance between i and j, see Equation (3).

$$dis(i,j) = \sqrt{(x_i - x_j)^2 + (y_i - y_j)^2}/d, \qquad (1)$$

$$hei(i,j) = |y_i - y_j|/d, \qquad (2)$$

$$wid(i,j) = |x_i - x_j|/d. \qquad (3)$$

Where $x_i$ and $x_j$ denote the $x$-axis of points i, j; $y_i$ and $y_j$ denote the $y$-axis of points i, j.

### 4.3.2 Audio Feature Extraction
Based on the previous work by the authors of [6], 37 secondary (statistical) speech features are taken as audio features used in speech emotion recognition. All features referenced are listed in Table 2.

### 4.4   Feature Selection
Feature selection module is the key part of our audiovisual emotion recognition system. The purpose of this module is to select valuable features for emotion

**Table 2.** Speech features

| NO. | Feature Description | NO. | Feature Description |
|---|---|---|---|
| 1 | Maximum value of energy | 20 | Mean duration of rising slopes (pitch) |
| 2 | Mean value of energy | 21 | Maximum value of falling slopes (pitch) |
| 3 | Median value of energy | 22 | Mean value of falling slopes (pitch) |
| 4 | Variance of energy | 23 | Maximum duration of falling slopes(pitch) |
| 5 | Maximum value of rising slopes (energy)) | 24 | Mean duration of falling slopes (pitch) |
| 6 | Mean value of rising slopes(energy) | 25 | Maximum value of the first formant |
| 7 | Maximum duration of rising slopes(energy) | 26 | Mean value of the first formant |
| 8 | Mean duration of rising slopes(energy) | 27 | Median value of the first formant |
| 9 | Maximum value of falling slopes(energy) | 28 | Variance of the first formant |
| 10 | Mean value of falling slopes(energy) | 29 | Maximum value of the second formant |
| 11 | Maximum duration of falling slopes(energy) | 30 | Mean value of the second formant |
| 12 | Mean duration of falling slopes(energy) | 31 | Median value of the second formant |
| 13 | Maximum value of pitch contour | 32 | Variance of the second formant |
| 14 | Mean value of pitch contour | 33 | Maximum value of the third formant |
| 15 | Median value of pitch contour | 34 | Mean value of the third formant |
| 16 | Variance of pitch contour | 35 | Median value of the third formant |
| 17 | Maximum value of rising slopes (pitch) | 36 | Variance of the third formant |
| 18 | Mean value of rising slopes(pitch) | 37 | Speech rate |
| 19 | Maximum duration of rising slopes(pitch) | | |

recognition from all features. In this module, the number of features for emotion recognition is decreased. Thus, the complexity and cost for the following classification procedure are reduced. It could improve the efficiency of the whole system.

In our emotion recognition system, feature selection module is based on the rough set theory. The attribute reduction algorithm adopted to select features is introduced in section 3, Alg. 1, and results of experiment show that the algorithm is effective.

### 4.5 Classification Module

The classification techniques used in existing emotion recognition systems include: template-based classification, rule-based classification, ANN-based classification, HMM-based classification, Bayesian classification, SVM based classification, etc [2][3][4].

In our emotion recognition system, SVM are taken as the classifier.

## 5  Experiment and Result

We set up a database of 500 video samples including joy, anger, sadness, disgust, surprise and fear. In each video sequence, a frame with the maximum intense of emotion is picked. Among the 500 samples, 226 samples are taken as train set and other 226 samples are taken as test set. Total 70 feature including 33 facial features listed in Table 1 and 37 audio features listed in Table 2 are extracted. Then, a decision information system is generated. With the reduction algorithm of Alg.1, 10 features are selected, which are *Mean value of energy, Median value of energy, Mean value of rising slopes(pitch), Mean value of falling slopes (pitch), Maximum value of the second formant, Median value of the third formant, Speech rate*, $d_{13}$, $d_{14}$ and $d_{21}$. These 10 features are taken as the input of the classifier of SVM.

A comparative experiment using total 70 features and 10 features listed above as the input of SVM is performed. The experiment results are shown in Table 3.

**Table 3.** Experiment results

|         | Rough Set+SVM | | | | | | SVM | | | | | |
|---------|-----|-------|-----|---------|---------|------|-----|-------|-----|---------|---------|------|
|         | Joy | Anger | Sad | Disgust | Suprise | Fear | Joy | Anger | Sad | Disgust | Suprise | Fear |
| Joy     | 68  | 1     | 0   | 1       | 8       | 1    | 76  | 1     | 0   | 0       | 3       | 3    |
| Anger   | 5   | 83    | 0   | 0       | 1       | 3    | 2   | 85    | 0   | 0       | 1       | 0    |
| Sad     | 0   | 0     | 69  | 5       | 0       | 4    | 0   | 0     | 70  | 1       | 0       | 1    |
| Disgust | 2   | 0     | 2   | 59      | 4       | 2    | 1   | 0     | 1   | 62      | 0       | 1    |
| Suprise | 6   | 2     | 0   | 1       | 54      | 3    | 2   | 0     | 0   | 3       | 62      | 2    |
| Fear    | 1   | 0     | 3   | 6       | 1       | 57   | 1   | 0     | 3   | 6       | 2       | 63   |
| average | 86.283186% | | | | | | 92.477876% | | | | | |

In these 10 features selected by rough set based attribute reduction algorithm, there are 7 acoustic features and 3 facial features.

Among seven acoustic features, *Mean value of energy* and *Median value of energy* are energy features, *Mean value of rising slope* and *Mean value of falling slopes* are pitch features, *Maximum value of the second formant* and *Median value of the third formant* are formant features. The last one is *speed rate* feature.

The 3 facial features, $d_{13}$, $d_{14}$ and $d_{21}$, are the width of mouth, height of mouth and the distance between pupil and lip corner. Although both $d_{13}$ and $d_{22}$ are features of width of mouth, both $d_{14}$ and $d_{24}$ are features of height of mouth, but only $d_{13}$ and $d_{14}$ are reserved as the output of rough set attribute reduction algorithm, which means that $d_{22}$ and $d_{24}$ are redundant to emotion recognition.

## 6  Conclusion and Future Works

In this paper, based on rough set theory, an audiovisual emotion recognition system is proposed. Based on rough set reduction, some important features for audiovisual emotion recognition are discovered according to our simulation experiment results. Depending on these features, an average recognition rate of 86.24% is achieved using SVM. In the future, relationship between facial and audio information for emotion recognition will be further studied. Effective reduction algorithm for audiovisual feature selection for emotion recognition will be also studied.

# References

1. R. W. Picard. Affective Computing. Cambridge: MIT Press, 1997.
2. R. W. Picard. Affective Computing: Challenges. International Journal of Human-Computer Studies, 2003, 59(1): 55-64.
3. R. W. Picard, E. Vyzas, and J. Healey. Toward Machine Emotional Intelligence: Analysis of Affective Physiological State. IEEE Transactions on Pattern Analysis and Machine Intelligence, 2001, 23(10): 1175-1191.
4. M. Pantci, L. J. M. Rothkrantz. Toward an Affect-Sensitive Multimodal Human-Computer Interaction. Proceedings of The IEEE, 2003,91(9): 1370-1390.
5. Y. Yang, G. Y. Wang, P. J. Chen and J. Zhou. An Emotion Recognition System Based on Rough Set Theory. Proceedings of Active Media Technology 2006, pages 293-297, 2006.
6. J. Zhou, G. Y. Wang, Y. Yang and P. J. Chen. Speech Emotion Recognition Based on Rough Set and SVM. Proceeding of Fifth IEEE International Conference on Cognitive Informatics, Pages 53-61, 2006.
7. P. J. Chen, G. Y. Wang, Y. Yang and J. Zhou. Facial Expression Recognition Based on Rough Set Theory and SVM. Proceedings of First International Conference on Rough Sets and Knowledge Technology, pages 772-777, 2006.
8. Y. Yang, G. Y. Wang, P. J. Chen and J. Zhou. An Audiovisual Emotion Recognition System Based on Rough Set Theory. Proceedings of 2006 International Conference on Artificial Intelligence, pages 690-693, 2006.
9. L. C. De Silva, P. C. Ng. Bimodal emotion recognition. In Proc. Automatic Face and Gesture Recog-nition, pages 332-335, 2000.
10. L. S. Chen, T. S. Huang. Emotional expressions in audiovisual human computer interaction. In Proc. International Conference on Multimedia and Expo (ICME), pages 423-426, 2000.
11. C. Y. Chen, Y. K. Huang and P. Cook. Visual/Acoustic Emotion Recognition. Multimedia and Expo, 2005. ICME 2005. IEEE International Conference on, pages 1468- 1471,2005.
12. H. Almuallim, T.G. Dietterich; Learning boolean concepts in the presence of many irrelevant features. Artificial Intelligence, 69(1-2):279-305, 1994.
13. K. Kira, L. A. Rendell; The feature selection problem: Traditional methods and a new algorithm, Proceedings of the Ninth National conference on Artificial Intelligence, 1992: 129-134.
14. M. Modrzejewski; Feature Selection Using Rough Stes Theory, Proceedings of the European Conference on Machine Learning, 1993: 213-226.
15. Z. Pawlak. On Rough Sets. Bulletin of the EATCS, 1984,24: 94-108.
16. Z. Pawlak. Rough Classification. International Journal of Man-Machine Studies, 1984,20(5): 469-483.
17. A. Skowron, S. k. Pal. Rough sets, pattern recognition, and data mining. Pattern Recognition Letters, 2003, 24(6): 829-933.

18. Z. M. Wojcik, Z. Pawlak. The rough sets utilization for linguistic pattern recognition. Bulletin of the Polish Academy of Sciences. Technical sciences , 1986, 34: 285-314.
19. K. Cyran, A. Mrozek. Rough sets in hybrid methods for pattern recognition. International journal of intelligent systems, 2000, 15: 919-938.
20. G. Y. Wang. Rough Reduction in Algebra View and Information View. International Journal of Intelli-gent System, 2003, 18(6): 679-688.
21. N. Zhong, J. Z. Dong and S. Ohsuga. Using Rough sets with Heuristics for Feature Selection. Journal of Intelligent Information Systems, 2001, 16(3): 199-214.
22. R. W. Swiniarski, A. Skowron. Rough set methods in feature selection and recognition. Pattern Recognition Letters 24(6): 833-849 (2003).
23. A. Skowron, L. Polkowski. Decision Algorithms: A Survey of Rough Set - Theoretic Methods. Fundamenta Informaticae. 30(3/4): 345-358 (1997)
24. X. H. Hu, N. Cercone. Learning Maximal Generalized Decision Rules via Discretization, Generalization and Rough set Feature Selection. Proceedings of 9th International Conference on Tools with Artificial Intelligence (ICTAI '97), pages 548 - 556, 1997.
25. S. G. Shan, W. Gao, B. Cao and D. Zhao. Illumination Normalization for Robust Face Recognition against Varying Lighting Conditions, IEEE International Workshop on Analysis and Modeling of Faces and Gestures, pp157-164, Nice, France, 2003.10
26. M. Kass, A.Witkin and D. Terzopoulos. Snakes: Active Contour Models. International Journal of Computer Vision. pages 321-331, 1988.
27. T. F. Cootes and C. J. Taylor. Active Shape Models - Smart Snakes. British Machine Vision Conference. pages 266-275, 1992.
28. T. F. Cootes and G. J. Edwards, C. J.Taylor. Active Appearance Models. In:5th European Conference on Computer Vision. pages 484-498,1998.
29. M. Pantic, L.J.M. Rothkrantz. Expert system for automatic analysis of facial expressions. Image and Vision Computing. (2000) 881-905.
30. Y. L. Tian, R. M. Bolle. Automatic Detecting Neutral Face for Face Authentication and Facial Expression Analysis. AAAI-03 Spring Symposium on Intelligent Multimedia Knowledge Management. 3 (2003) 24-26.
31. H. Seyedarabi, A. Aghagolzadeh and S. Khanmohammadi. Recognition of Six Basic Facial Expressions by Feature-Points Tracking using RBF Neural Network and Fuzzy Inference System. 2004 IEEE International Conference on Volume 2. (2004) 1219-1222.
32. S. Liu, Z. L. Ying. Facial expression recognition based on fusing local and global feature. Journal of Computer Applications. 3 (2005) 4-6.

# Novel Classification and Segmentation Techniques with Application to Remotely Sensed Images

B. Uma Shankar

Machine Intelligence Unit, Indian Statistical Institute, Kolkata 700 108, India
uma@isical.ac.in

**Abstract.** The article deals with some new results of investigation, both theoretical and experimental, in the area of image classification and segmentation of remotely sensed images. The article has mainly four parts. Supervised classification is considered in the first part. The remaining three parts address the problem of unsupervised classification (segmentation). The effectiveness of an active support vector classifier that requires reduced number of additional labeled data for improved learning is demonstrated in the first part. Usefulness of various fuzzy thresholding techniques for segmentation of remote sensing images is demonstrated in the second part. A quantitative index of measuring the quality of classification/ segmentation in terms of homogeneity of regions is introduced in this regard. Rough entropy (in granular computing framework) of images is defined and used for segmentation in the third part. In the fourth part a homogeneous region in an image is defined as a union of homogeneous line segments for image segmentation. Here Hough transform is used to generate these line segments. Comparative study is also made with related techniques.

**Keywords:** Active learning, Support vector machine, Fuzzy sets, Fuzzy entropy, Fuzzy correlation, Rough sets, Rough entropy, Granular computing, Soft-computing, Hough transform, Remotely sensed images.

## 1 Introduction

There are several types of images, namely, light intensity (visual) image, range image (depth image), nuclear magnetic resonance image (commonly known as magnetic resonance image (MRI)), thermal image and so on. Light intensity (LI) images, the most common type of image we encounter in our daily experience, represent the variation of light intensity on the scene. Range image (RI), on the other hand, is a map of depth information at different points on the scene. In digital light intensity image the intensity is quantized, while in the case of range image the depth value is digitized. Magnetic resonance images represent the intensity variation of radio waves generated by biological systems when exposed to radio frequency pulses. Biological bodies (human/animals) are built up of atoms and molecules. Some of the nuclei behave like tiny magnets [1], commonly

J.F. Peters et al. (Eds.): Transactions on Rough Sets VII, LNCS 4400, pp. 295–380, 2007.

known as spins. Therefore, if a patient (or any living being) is placed in a strong
magnetic field, the magnetic nuclei tend to align with the applied magnetic
field. For MRI the patient is subject to a radio frequency pulse. As a result of
this the magnetic nuclei pass into a high energy state, and then immediately
relieve themselves of this stress by emitting radio waves through a process called
relaxation. This radio wave is recorded to form the MRI. In digital MRI, the
intensity of the radio wave is digitized with respect to both intensity and spatial
coordinates. Thus in general, any image can be described by a two-dimension
function $f'(x,y)$, where $(x,y)$ denotes the spatial coordinate and $f'(x,y)$ the
feature value at $(x,y)$. Depending on the type of image, the feature value could
be light intensity, depth, intensity of radio wave or temperature. A digital image,
on the other hand, is a two-dimensional discrete function $f(x,y)$ which has
been digitized both in spatial coordinates and magnitude of feature value. We
shall view a digital image as a two-dimensional matrix whose row and column
indices identify a point, called a pixel, in the image and the corresponding matrix
element value identifies the feature intensity level. Throughout our discussion a
digital image will be represented as

$$F_{P \times Q} = [f(x,y)]_{P \times Q} \tag{1}$$

where $P \times Q$ is the size of the image and $f(x,y) \in G_L = \{0, 1, \ldots, L-1\}$,
the set of discrete levels of the feature value. Since the techniques we are going
to discuss in this article are developed for ordinary intensity images, in our
subsequent discussion, we shall usually refer to $f(x,y)$ as gray value (although
it could be depth or temperature or intensity of radio wave).

Segmentation is first essential and important step of low level vision[2,3,4,5].
Its application area varies from the detection of cancerous cells to the identification of an airport from remote sensing data, etc. In all these areas, the quality
of the final output depends largely on the quality of the segmented output. Segmentation is a process of partitioning image into some non-intersecting regions
such that each regions is homogeneous and union of no two adjacent regions is
homogeneous. Formally, it can be defined [6] as follows: if $F$ is the set of all
pixels and $P(\ )$ is a uniformity (homogeneity) predicate defined on groups of
connected pixels, then segmentation is a partitioning of the set $F$ into a set of
connected subsets or regions $(S_1, S_2, \ldots, S_n)$ such that

$$\bigcup_{i=1}^{n} S_i = F \text{ with } S_i \bigcap S_j = \emptyset, \ i \neq j. \tag{2}$$

The uniformity predicate $P(S_i)$=true for all regions $(S_i)$ and $P(S_i \cup S_j)$=false,
when $S_i$ is adjacent to $S_j$. Note that this definition is applicable to all types of
images we have described. For light intensity images the uniformity predicate
measures the uniformity of light intensity, while for range images it could be the
uniformity of surfaces. Algorithm developed for one class of image (say ordinary
intensity image) may not always be applied to other classes of images (MRI/RI).
This is particulary true when the algorithm uses a specific image formation model.

For example, some visual image segmentation algorithms are based on the assumption that the gray level function $f(x, y)$ can be modeled as a product of an illumination component and a reflectance component [7]. On the other hand, in Pal and Pal [8] the gray level distributions have been modeled as Poisson distributions, based on the theory of formation of visual images. Such methods [7,8] should not be applied to MRI/RIs. However, most of the segmentation methods developed for one class of images can be easily applied/extended to another class of images. For example, the variable order surface fitting method [9], although developed for range images can be applied for other images that can be modeled as a noisy version of piece-wise smooth surface.

The present article deals with some new methods of images segmentation for light intensity (LI) images. These methods have been extensively demonstrated on remote sensing image data. Here modern tools like active support vector machine, and rough and granular computing are used besides Hough transform based region segmentation, and fuzzy set theoretic thresholding techniques. Before we describe the scope of the article (in Section 1.6), we provide a brief review on image segmentation in Section 1.1, and characteristics of remote sensing images Section 1.4 and various methods used for their segmentation in Section 1.5.

## 1.1   Image Segmentation Techniques

Segmentation subdivides an image into its constituent regions or objects. The level to which the subdivision is carried depends on the problem being solved. That is segmentation should stop when the objects of interest in an application have been isolated. For example, in the automated inspection of electronic assemblies, interest lies in analyzing images of the products with the objective of determining the presence or absence of specific anomalies, such as missing components or broken connection paths. There is no point in carrying segmentation after the level of detail required to identify those elements is achieved.

Segmentation of nontrivial images is one of the most difficult tasks in image processing. Segmentation accuracy determines the eventual success or failure of computerized analysis procedures. For this reason, considerable care should be taken to improve the probability of rugged segmentation. In some situations, such as industrial inspection applications, at least some measure of control over the environment is possible at times. The experienced image processing system designer invariably pays considerable attention to such opportunities. In other applications, such as autonomous target acquisition, the system designer has no control of the environment. Then the usual approach is to focus on selecting the types of sensors most likely to enhance the objects of interest while diminishing the contribution of irrelevant image details. A good example is the use of infrared imaging by the military to detect objects with strong heat signatures, such as equipment and troops in motion [2].

Image (light intensity) segmentation algorithms generally are based on one of two basic properties of intensity values: discontinuity and similarity. In the first category, the approach is to partition an image based on abrupt changes in intensity, such as edges in an image. The principal approaches in the second category are based on partitioning an image into region that are similar according to set of predefined criteria. Thresholding, pixel classification, region growing, and region splitting and merging are examples of methods in this category [2]. These are described in the following sections.

**Gray Level Thresholding.** Thresholding is one of the old, simple and popular techniques for image segmentation. Thresholding can be done based on global information (e.g., gray level histogram of the entire image) or it can be done using local information (e.g., co-occurrence matrix) of the image. Taxt *et al.* [10] refer to the local and global information based techniques as contextual and non-contextual methods, respectively. Under each of these schemes (contextual/non-contextual) if only one threshold is used for the entire image then it is called global thresholding. On the other hand, when the image is partitioned into several subregions and a threshold is determined for each of the subregions, it is referred to as local thresholding [10]. Some authors [11,12,13] call these local thresholding methods adaptive thresholding schemes. Thresholding techniques can also be classified as bi-level thresholding and multi-level thresholding. In bi-level thresholding the image is partitioned into two regions – object (black) and background (white). When the image is composed of several objects with different surface characteristics (for a light intensity image, objects with different coefficient of reflection, for a range image there can be objects with different depths and so on) one needs several thresholds for segmentation. This is known as multi-level thresholding. In such a situation we try to get a set of thresholds $(T_1, T_2, ..., T_k)$ such that all pixels with $f(x, y) \in [T_i, T_{i+1})$, $i = 0, 1, \ldots, k$, constitute the $i$th region type ($T_0$ and $T_{k+1}$ are taken as $0$ and $L-1$, respectively). Note that thresholding can also be viewed as a classification problem. For example, bi-level segmentation is equivalent to classifying the pixels into two classes: object and background. Mardia and Hainsworth [14] have shown that the main idea behind the iterative thresholding schemes of Ridler and Calvard [15] and Lloyd [16] can be defined as special cases of the classical Bayes' discrimination rule.

If the image is composed of regions with different gray level ranges, i.e., the regions are distinct, the histogram of the image usually shows different peaks, each corresponding to one region and adjacent peaks are likely to be separated by a valley. For example, if the image has a distinct object on a background, the gray level histogram is likely to be bimodal with a deep valley. In this case, the bottom of the valley (T) is taken as the threshold for object background separation. Therefore, when the histogram has a (or a set of) deep valley(s), selection of threshold(s) becomes easy because it becomes a problem of detecting valleys. However, normally the situation is not like this and threshold selection is not a trivial job. There are various methods available for this. For example, Otsu [17] maximized a measure of class separability. He maximized the ratio of the between class variance to the local variance to obtain thresholds. Nakagawa and Rosenfeld [12] assumed that the object and background populations are distributed normally with distinct means and standard deviations. Under this assumption they selected the threshold by minimizing the total misclassification error. This method is computationally involved. Kittler and Illingworth [18], under the same assumption of normal mixture, suggested a computationally less involved method. Their method optimizes a criterion function related to average pixel classification error rate that finds out an approximate minimum error threshold. Pal and Bhandari [19] optimized the same criterion function but assumed Poisson distributions to model the gray level histogram.

Pun [20] assumed that an image is the outcome of an L symbol source. He maximized an upper bound of the total a posteriors entropy of the partitioned image for the purpose of selecting the threshold. Kapur *et al.* [21] on the other hand, assumed two probability distributions, one for the object area and the other for the background area. They then, maximized the total entropy of the partitioned image in order to arrive at the threshold level. Wong and Sahoo [22] maximized the a posterior entropy of a partitioned image subject to a constraint on the uniformity measure of Levine and Nazif [23] and a shape measure. They maximized the a posterior entropy over min $(S_1, S_2)$ and max $(S_1, S_2)$ to get the threshold for segmentation; where $S_1$ and $S_2$ are the threshold levels at which the uniformity and the shape measure attain the maximum values, respectively. Pal and Pal [8] modeled the image as a mixture of two Poisson distributions and developed several parametric methods for segmentation. The assumption of the Poisson distribution has been justified based on the theory of image formation. These algorithms maximize either entropy or minimize the $\chi^2$ statistic. Though these methods use only the histogram, they produce good results due to the incorporation of the image formation model.

All these methods have a common drawback, they take into account only the histogram information (ignoring the spatial details). As a result, such an algorithm may fail to detect thresholds if these are not properly reflected as valleys in the histogram, which is normally the case. There are many thresholding schemes that use spatial information, instead of histogram information. For example, the busyness measure of Weszka and Rosenfeld [24] is dependent on the co-occurrence of adjacent pixels in an image. They minimized the busyness measure in order to arrive at the threshold for segmentation. Deravi and Pal [25] minimized the conditional probability of transition across the boundary between two regions. This method also uses the local information contained in the co-occurrence matrix of the image. However, finally all these methods threshold the histogram, but since they make use of the spatial details, they result in a more meaningful segmentation than the methods which use only the histogram information. Based on the co-occurrence matrix, Chanda *et al.* [26] have given an average contrast measure for segmentation. Pal and Pal [27] proposed measures of contrast between regions and homogeneity of regions using the brightness perception aspect of the human psycho-visual system, and applied them to segmentation. They also defined [28] the higher order entropy and conditional entropy of an image giving measures of homogeneity and contrast, respectively. These measures are finally applied to develop object extraction algorithms. A concept similar to the second order local entropy of Pal and Pal [28] has been used by Abutaleb [29] for segmentation. The gray value of a pixel and the average of its neighboring pixels have been used there for the computation of the co-occurrence matrix. As a result the boundary of the segmented object usually becomes blurred.

The philosophy behind gray level thresholding, "pixels with gray level $< T$ fall into one region and the remaining pixels belong to another region", may not be true on many occasions, particularly, when the image is noisy or the background is uneven and illumination is poor. In such cases the objects will still be lighter or darker than the background, but any fixed threshold level for the entire image will usually fail to separate the objects from the background. This leads one to the methods of adaptive thresholding. In adaptive thresholding [11,12,13] normally the image is partitioned into several non-overlapping blocks of equal area and a threshold for each block is computed independently. Chow and Kaneko [11] used the (sub) histogram of each block to determine local threshold values for the corresponding cell centers. These local thresholds are then interpolated over the entire image to yield a threshold surface. They [11] used only gray level information. Yanowitz and Bruckstein [13] extended this idea to use combined edge and gray level information. They computed the gray level gradient magnitude from a smooth version of the image. The gradient values have then been thresholded and thinned using a local maxima directed thinning process. Locations of these local gradient maxima are taken as boundary pixels between object and background. The corresponding gray levels in the image are taken as local thresholds. The sampled gray levels are then interpolated over the entire image to obtain an adaptive threshold surface.

## Iterative Pixel Classification

***Relaxation.*** Relaxation [5,30,31] is an iterative approach to segmentation in which the classification decision about each pixel can be taken in parallel. Decisions made at neighboring points in the current iteration are then combined to make a decision in the next iteration. There are two types of relaxation: probabilistic and fuzzy. We discuss here the probabilistic relaxation. Suppose a set of pixels $\{f_1, f_2, ..., f_n\}$ is to be classified into $m$ classes $\{C_1, C_2, ..., C_m\}$. For the probabilistic relaxation we assume that for each pair of class assignments $f_i \in C_j$ and $f_h \in C_k$, there exists a quantitative measure of compatibility $C(i, j; h, k)$ of this pair, i.e., the class assignment of pixels is interdependent. It is reasonable to assume that a positive value of $C(i, j; h, k)$ indicates the compatibility of $f_i \in C_j$ and $f_h \in C_k$, while a negative value represents incompatibility and a zero don't care situation. The function $C$ need not be symmetric.

Let $p_{ij}$ represent the probability that $f_i \in C_j, 1 \leq i \leq n$ and $1 \leq j \leq m$, with $0 \leq p_{ij} \leq 1, \sum_j p_{ij} = 1$. Intuitively, if $p_{hk}$ is high and $C(i, j; h, k)$ is positive, we increase $p_{ij}$ since it is compatible with the high probability event $f_h \in C_k$. Similarly, if $p_{hk}$ is high and $C(i, j; h, k)$ is negative, we reduce $p_{ij}$ as it is incompatible with $f_h \in C_k$. On the other hand, if $p_{hk}$ is low or $C(i, j; h, k)$ is nearly zero, $p_{ij}$ is not changed as either $f_h \in C_k$ has a low probability or is irrelevant to $f_i \in C_j$. The fuzzy relaxation based on fuzzy set theory *(to be defined in Section 1.1)* is similar in concept [5,31].

*MRF Based Approaches.* There are many image segmentation methods [32,33,34] which use the spatial interaction models like Markov Random Field (MRF) or Gibbs Random Field (GRF) to model digital images. Geman and Geman [34] have proposed a hierarchical stochastic model for the original image and developed a restoration algorithm, based on stochastic relaxation (SR) and annealing, for computing the maximum a posterior estimate of the original scene given a degraded realization. Due to the use of annealing, the restoration algorithm does not stop at a local maxima but finds the global maximum of the a posterior probability. We mention here that the probabilistic relaxation [31] (also known as relaxation labeling (RL)) and stochastic relaxation, although they share some common features like parallelism and locality, are quite distinct. RL is essentially a non-stochastic (deterministic) process which allows jumps to states (configurations) of lower energy. On the other hand, SR transition to a configuration which increases the energy (decreases the probability) is also allowed. In fact, if the new configuration decreases the energy, the system transits to that state, while if the new configuration increases the energy the system accepts that state with a probability. This helps the system to avoid the local minima. RL usually gets stuck in a local minima. Moreover, in RL there is nothing corresponding to an equilibrium state or even a joint probability law over the configurations. Derin *et al.* [33] extended the one-dimensional Bayes smoothing algorithm of Asker and Derin [32] to two dimensions to get the optimum Bayes estimate for the scene value at every pixel. In order to reduce the computational complexity of the algorithm, the scene is modeled as a special class of MRF models, called Markov mesh random fields which are characterized by causal transition distributions. The processing is done over relatively narrow strips and estimates are obtained at the middle section of the strips. These pieces together with overlapping strips yield a suboptimal estimate of the scene. Without parallel implementation these algorithms become computationally prohibitive.

**Neural Network Based Approaches.** For any artificial vision application, one desires to achieve robustness of the system with respect to random noise and failure of processors. Moreover, a system can (probably) be made artificially intelligent if it is able to emulate some aspects of the human information processing system. Another important requirement is to have the output in real time. Neural network based approaches are attempts to achieve these goals. Neural networks are massively connected networks of elementary processors [35,36]. Architecture and dynamics of some networks are claimed to resemble information processing in biological neurons [35]. The massive connectionist architecture usually makes the system robust while the parallel processing enables the system to produce output in real time. Several authors [37,38,39,40,41,42,43,44] have attempted to segment an image using neural networks. Blanz and Gish [38] used a three-layer feed forward network for image segmentation, where the number of neurons in the input layer depends on the number of input features for each pixel and the number of neurons in the output layer is equal to the number of classes. Babaguchi *et al.* [37] used a multilayer network trained with backpropagation, for thresholding an image. The input to the network is the histogram while the output is the desirable threshold. In this method at the time of learning a large set of sample images with known thresholds which produce visually suitable outputs are required. But for practical applications it is very difficult to get many sample images.

Ghosh *et al.* [40,41] used a massively connected network for extraction of objects in a noisy environment. The maximum a posterior probability estimate of a scene modeled as a GRF and corrupted by additive Gaussian noise has been done using a neural network [40]. The hardware realization of neurons to be used for such a network has also been suggested. This NN (neural networks) based method takes into account the contextual information, because the GRF model considers the spatial interactions among neighboring pixels. Another robust algorithm for the extraction of objects from highly noise corrupted scenes using a Hopfield type neural network has been developed in reference [42]. The energy function of the network has been constructed in such a manner that in a stable state of the net it extracts compact regions from a noisy scene. A multilayer neural network [44] where each neuron in layer $i(i > 1)$ is connected to the corresponding neuron in layer $(i - 1)$ and some of its neighboring neurons (in layer $i - 1$), has been used to segment noisy images. The output status of the neurons in the output layer has been viewed as a fuzzy set *(to be defined in Section 1.1)*. The weight updating rules have been derived to minimize the fuzziness in the system. For this algorithm the architecture of the network enforces the system to consider the contextual information. Moreover, this algorithm integrates the advantages of both fuzzy sets (decision from imprecise/incomplete knowledge) and neural networks (robustness). In reference [39] the image segmentation problem has been formulated as a constraint satisfaction problem (CSP) and a class of constraint satisfaction neural network (CSNN) is proposed. A CSNN consists of a set of objects, a set of labels, a collection of constraint relations and a topological constraint describing the neighborhood relationships among various objects. The CSNN is viewed as a collection of interconnected neurons. The architecture is chosen in such a way that it represents constraints in the CSP. The method is found to be successful on CT (computed tomography) images and MRIs. However, robustness of the algorithm with noisy data has not been investigated. Moreover, a large number of neurons are required even for an image of moderate size.

Kuntimad and Ranganath [43] have describes a method for segmenting digital images using pulse coupled neural networks (PCNN). The pulse coupled neuron (PCN) model used in PCNN is a modification of the cortical neuron model of Eckhorn *et al.* [45]. A single layered laterally connected PCNN is capable of perfectly segmenting digital images even when there is a considerable overlap in the intensity ranges of adjacent regions.

Ghosh and Ghosh [46] used fuzzy logic reasoning into the Neuro-GA (Hopfield type neural network) hybrid framework where GA (Genetic Algorithm) has been used to evolve Hopfield type optimum neural network architecture for object background classification. Each chromosome of the GA represents an architecture. The output status of the neurons at the converged state of the network is viewed as a fuzzy set and measure of fuzziness of this set is taken as a measure of fitness of the chromosome. The best chromosome of the final generation represents the optimum network configuration.

Pal *et al.* [47] described a method using Genetic Algorithms (GAs) to evolve Hopfield type optimum neural network architectures for object extraction problem. Different optimizing functions involving minimization of energy value of the network, maximization of percentage of correct classification of pixels (pcc), minimization of number of connections of the network (noc), and a combination of pcc and noc are considered. The number of connections of the evolved (sub)optimal architectures is seen to be reduced to two-third compared to a fully connected version. They found that the performance of Genetic algorithm is better than that of Simulated Annealing for this problem.

Jiang and Zhou [48] described an image segmentation method based on ensemble of SOM (self-organized map) neural networks. This clusters the pixels in an image according to color and spatial features using each SOM, and then combines the clustering results to give the final segmentation.

**Edge Detection.** Segmentation can also be obtained through detection of edges of various regions, which normally tries to locate points of abrupt changes in gray level intensity values. Since edges are local features, they are determined based on local information. A large variety of methods are available in the literature [3,5,49] for edge finding. Davis [50] classified edge detection techniques into two categories: sequential and parallel. In the sequential technique the decision whether a pixel is an edge pixel or not is dependent on the result of the detector at some previously examined pixels. On the other hand, in the parallel method the decision whether a point is an edge or not is made based on the point under consideration and some of its neighboring points. As a result of this the operator can be applied to every point in the image simultaneously. The performance of a sequential edge detection method is dependent on the choice of an appropriate starting point and how the results of previous points influence the selection and result of the next point.

There are different types of parallel differential operators such as Roberts gradient, Sobel gradient, Prewitt gradient and the Laplacian operator [3,5,49]. These difference operators respond to changes in gray level or average gray level. The gradient operators, not only respond to edges but also to isolated points. For Prewitt's operator the response to the diagonal edge is weak, while for Sobel's operator it is not that weak as it gives greater weights to points lying close to the point $(x, y)$ under consideration. However, both Prewitt's and Sobel's operators possess greater noise immunity. The preceding operators are called the first difference operator. Laplacian, on the other hand, is a second difference operator.

The digital Laplacian being a second order difference operator, has a zero response to linear ramps. It responds strongly to corners, lines, and isolated points. Thus for a noisy picture, unless it has a low contrast, the noise will produce higher Laplacian values than the edges. Moreover, the digital Laplacian is not orientation invariant. A good edge detector, should be a filter with the following two features. First, it should be a differential operator, taking either a first or second spatial derivative of the image. Second, it should be capable of being tuned to act at any desired scale, so that large filters can be used to detect blurry shadow edges, and small ones to detect sharply focused fine details. The second requirement is very useful as intensity changes occur at different scales in an image.

All the edges produced by these operators are, normally, not significant (relevant) edges when viewed by human beings. Therefore, one needs to find out prominent (valid) edges from the output of the edge operators. Kundu and Pal [51] have suggested a method of thresholding to extract the prominent edges based psycho-visual phenomena. Haddon [52] developed a technique to derive a threshold for any edge operator, based on the noise statistics of the image.

**Methods Based on Fuzzy Set Theory.** Application of fuzzy sets [53] to image processing was based on the realization that many of the basic concepts in image analysis, e.g., the concept of an edge or a corner or a boundary or a relation between regions, do not lend themselves well to precise definition. A gray tone image possesses ambiguity within pixels due to the possible multi-valued levels of brightness in the image. This indeterminacy is due to inherent vagueness rather than randomness. Incertitude in an image pattern may be explained in terms of grayness ambiguity or spatial (geometrical) ambiguity or both. Grayness ambiguity means "indefiniteness" in deciding whether a pixel is white or black. Spatial ambiguity refers to "indefiniteness" in the shape and geometry of a region within the image.

We shall mention here a few methods of fuzzy segmentation (based on both gray level thresholding and pixel classification) using global and/or local information of an image space. Note that it is Prewitt [54] who first suggested that the result of segmentation should be fuzzy subsets rather than ordinary subsets.

*Fuzzy Thresholding.* Different histogram thresholding techniques in providing both fuzzy and non-fuzzy segmented versions by minimizing the grayness ambiguity (global entropy, index of fuzziness, index of crispness) and geometrical ambiguity (fuzzy compactness) of an image have been described in [55,56]. These algorithms use different S-type membership functions [53] to define fuzzy "object regions" and then select the one which is associated with the minimum (optimum) value of the aforesaid ambiguity measures. The optimum membership function thus obtained enhances the object from background and denotes the membership values of the pixels for the fuzzy object region. Note that the cross-over point (the point with membership value of 0.5) of the optimum membership function may be considered as a threshold for crisp segmentation. Its extension to multi-level thresholding has also been made. The mathematical framework of the algorithm including the selection of S functions, its bandwidth and bounds has been established by Murthy and Pal [57].

The problem of determining the appropriate membership function in image processing drew the attention of many researchers. Reconsider the problem of gray level thresholding using S functions. If there is a difference in opinion in defining an S function (i.e., instead of a single membership function, we have a set of monotonically non-decreasing functions), the concept of spectral fuzzy sets

[58] can be used to provide soft decisions (a set of thresholds along with their certainty values) by giving due respect to all opinions. In making such a decision, the algorithm minimizes differences in opinions in addition to the ambiguity measures mentioned earlier; thereby managing the uncertainty. The bounds for S-type functions have been defined based on the properties of fuzzy correlation [57] so that any function lying in the bounds would give satisfactory segmentation results. It, therefore, demonstrates the flexibility of fuzzy algorithms. Xie and Bedrosian [59] have also made attempts in determining membership functions for gray level images.

Recently Tobias and Seara [60] used a procedure for histogram thresholding which is not based on the minimization of a criterion function. Instead, the histogram threshold is determined according to the similarity between gray levels. Such a similarity is assessed through a fuzzy measure. The fuzzy framework is used to obtain a mathematical model of such a concept. Because of the used assumption, in which objects and background must occupy non-overlapping regions of the histogram, the applicability of the proposed method is limited to images that satisfy such a requirement.

*Fuzzy Clustering.* The fuzzy c-means (FCM) clustering algorithm [61] has also been used in image segmentation [62,63,64]. The fuzzy c-means algorithm uses an iterative optimization of an objective function based on a weighted similarity measure between the pixels in the image and each of the c-cluster centers. A local extremum of this objective function indicates an optimal clustering of the input data.

Trivedi and Bezdek [64] proposed a fuzzy set theoretic image segmentation algorithm for aerial images. The method is based upon region growing principles using a pyramid data structure. The algorithm is hierarchical in nature. Segmentation of the image at a particular processing level is done by the FCM algorithm. In a multilevel segmentation experiment, level $i$ regions are considered homogeneous when image elements have largest cluster membership values of greater than a prescribed threshold. If the homogeneity test fails, regions are split to form the next level regions which are again subjected to the FCM algorithm. This algorithm is a region splitting algorithm, where the acceptance of a region is determined by fuzzy membership values to different regions. Hall *et al.* [62] segmented magnetic resonance brain images using the unsupervised fuzzy c-means and also by a supervised computational network–a dynamic multi-layered perceptron trained with the cascade correlation learning algorithm. The different aspects of both approaches and their utility for the diagnostic process have been discussed. However, computational complexity of fuzzy c-mean is too high to apply it for real time application of MRI segmentation. One of the advantages in using fuzzy clustering algorithms is that one can dynamically select the appropriate number of clusters depending on the strength of memberships across clusters [65]. Keller and Carpenter [66] used a modified version of FCM for image segmentation. The cluster centers are updated using the FCM formula but new membership values for each point are calculated using an S-type function based on the feature value of each point and the fuzzy means.

## Methods Using Other Tools

*Mathematical Morphology.* The mathematical morphology is used as a tool for extracting image components that are useful in the representation and description of region shape, such as boundaries, skeletons and the convex hull. Morphological techniques are also used for pre or post processing of images, such as morphological filtering, thinning, and pruning. Morphological segmentation methods include boundary extraction via morphological gradients operation, region partitioning based on texture content, and size distribution of particles in an image using granulomentry [2]. Image segmentation using morphological watersheds now became an important research area. Segmentation by watersheds embodies many of the concepts of three approaches used in image processing (i.e., detection of discontinuities, thresholding and region processing) and often produces more stable segmentation results, including continuous segmentation boundaries. This approach also provides a simple framework for incorporating knowledge-based constraints in the segmentation process [2]. The paper by Couprie *et al.* [67] describes an algorithm to compute topological watersheds. Additional research in this line is reported in [68,69].

*Genetic Algorithm.* Genetic algorithm, being an adaptive search techniques, has been used for image segmentation. For example, Bhanu and Fonder [70] described an approach for automatic image segmentation, in which user selected sets of examples and counter-examples supply information about the specific segmentation problem. In their approach, image segmentation is guided by a genetic algorithm which learns the appropriate subset and spatial combination of a collection of discriminating functions, associated with image features. The genetic algorithm encodes discriminating functions into a functional template representation, which can be applied to the input image to produce a candidate segmentation. The performance of each candidate segmentation is evaluated within the genetic algorithm, by a comparison to two physics-based techniques for region growing and edge detection. Through the process of segmentation, evaluation, and recombination, the genetic algorithm optimizes functional template design efficiently. Results are presented on real synthetic aperture radar (SAR) imagery of varying complexity. The first closed-loop image segmentation system is presented in [71] that incorporates genetic and other algorithms to adapt the segmentation process to changes in image characteristics caused by variable environmental conditions, such as time of day, time of year, and weather.

*Wavelet Based Segmentation.* Wavelets are commonly used as a tool for coding and compression. Wavelets and multi-resolution analysis together form a field which is explored for image segmentation [2]. Acharyya *et al.* [72] explained a scheme for segmentation of multitexture images. The methodology involves extraction of texture features using an wavelet decomposition scheme called discrete M-band wavelet packet frame (DMbWPF). This is followed by the process of selection of important features using a neuro-fuzzy algorithm under unsupervised learning. Using selected feature segmentation is performed on the IRS

(Indian Remote Sensing) and SPOT[1] (Satellite Pour l'Observation de la Terre) images using k-means clustering. Some other attempts on segmenting remotely sensed images are explained in Section 1.5.

*Level Set Methods and Image Segmentation.* Image segmentation using level set is another area that has received importance these days. Level set methods are used for implementing curve evolution or diffusion under various forces. These can also be used for surface evolution for volume segmentation in 3-D images. Heiler and Schnörr [73] integrated a model for filter response statistics of natural images into a variational framework for image segmentation. This model drives level sets toward meaningful segmentations of complex textures and natural scenes. Despite its enhanced descriptive power, the approach preserves the efficiency of level set based segmentation since each region comprises two model parameters only. Other methods using level sets are available in [74,75,76].

*Methods in Soft Computing Framework.* A large number of researchers, all over the word, have been engaged in developing soft computing methodologies for designing intelligent information systems for image processing and analysis [77]. The objective of the soft computing is to exploit the tolerance for imprecision, uncertainty, approximate reasoning and partial truth in order to achieve tractability, robustness, low solution cost and close resemblance with human like decision making. Usually, it attempts to find an acceptable solution at low cost by seeking for an approximate solution to a precisely or imprecisely formulated problem [78]. The principal components of soft computing are fuzzy logic, rough sets, neural computing, probabilistic reasoning, genetic algorithm, chaotic systems, belief networks and some parts of learning theory like support vector machine (SVM). Fuzzy logic and rough sets are mainly concerned with providing algorithm for dealing with imprecision and approximate reasoning, and for computing with words. Neural computing provides the machinery for adaptive learning and curve fitting. Probabilistic reasoning is for propagation of belief, and genetic algorithms are for efficient search and optimization. All these partners of soft computing are complementary rather than competitive. Because of this reason it is found frequently to be advantageous to use these components in combination rather than in isolation; thereby developing more intelligent systems in hybrid domain. Segmentation methods using soft computing tools like fuzzy sets, neural networks and genetic algorithms are explained before. Similar methods using these tools in the integrated framework are reported in [46,47,79].

## 1.2  Segmentation of Color Images

Color is a very important perceptual phenomenon related to human response to different wavelengths in the visible electromagnetic spectrum [80,81]. The image is usually described by the distribution of three color components R (red), G (green), B (blue). Color image is often also represented by three psychological qualities hue, saturation and intensity.

---

[1] SPOT Image - Home ( http://www.spotimage.fr/html).

According to Cheng *et al.* [82], which is more recent study on color images segmentation, there are two critical issues for color image segmentation: (1) what segmentation method should be utilized; and (2) what color space should be adopted. At present, color image segmentation methods are generally extended from monochrome segmentation approaches. Several approaches applied to color image are discussed in their article [82], which includes histogram thresholding, region based approaches, edge detection and fuzzy techniques. A combination of these approaches is often utilized for color image segmentation.

Naik and Murthy [83] recently worked on the problem of edge detection in color images. They used this edge detection method for object recognition [84], having the knowledge of the appearance of the objects from different view points. Appearance of each view of the object is encoded using the descriptions of the regions involving multiple segments on the image surface. Description of all such distinct regions are combined to represent each view of the object. Two different methods to locate and find the different colors in such regions involving multiple segments on the surface of the object have been proposed and tested on some standard data sets (like COIL-100, SOIL-47A, SOIL-47B and ALOI color image data sets).

## 1.3   Objective Evaluation of Segmentation Results

We have discussed above several methods of image segmentation. It is known that no method is equally good for all images and all methods are not good for a particular type of images. Here an important problem remains to be discussed, how to make a quantitative evaluation of segmentation results. Such a quantitative measure would be quite useful for vision applications where automatic decisions are required. Also this will help to justify an algorithm. Unfortunately, a human being is the best judge to evaluate the output of any segmentation algorithm. However, some attempts have already been made for the quantitative evaluation. Levine and Nazif [85] used a two dimensional distance measure that quantifies the difference between two segmented images, one proposed by a human being the other by an algorithm. Later on they [23] defined another set of performance parameters such as region uniformity, region contrast, line contrast, etc. These measures have also been used for quantitative evaluation of segmentation algorithms. Lim and Lee [86] attempted to do this by computing the probability of error between the manually segmented image and the segmentation result. Pal and Bhandari [87] used the higher order local entropy as an index to measure the quality of the output. They also suggested the use of symmetric divergence between two probability distributions, one for the output generated by an algorithm and the other for the manually segmented image. The correlation measure [88] between the original image and the segmented one has also been used for the purpose of quantitative evaluation [87]. It is already mentioned that a human being is the ultimate judge to make an evaluation of the result. In [89] it is suggested that one can use a vector of such measures for objective evaluation. For example, if for some segmented image, the correlation, uniformity, and entropy are all high and divergence is low then one can consider the output to be good.

In the survey of Sezgin and Sankur [90], instead of one criteria, they used five different criteria to get better understanding of the performance features of thresholding methods. The five performance criteria are as follows: misclassification error (ME), edge mismatch (EMM), relative foreground area error (RAE), modified Hausdorff distance (MHD), and region nonuniformity (NU). A combination of these performance measures is used in two different ways. The first method is based on the arithmetic average of the normalized scores obtained from the five measures, while the second method is based on the average of the ranks of the images.

There are several reviews available in the literature for image segmentation and evaluation. One may refer to [89,90,91,92,93,94,95,96,97].

## 1.4   Remote Sensing Images: Characteristics

Remotely sensed image data of the earth's surface acquired from either aircraft or spacecraft platforms is readily available in digital format; spatially the data is composed of discrete picture elements or pixels and radiometrically it is quantized into discrete brightness levels. Even the data that is not recorded in digital form initially can be converted into discrete data by the use of digitizing equipments.

The great advantage of having data available digitally is that it can be processed by computer either for machine assisted information extraction or for enhancement before an image product is formed. The latter is used to assist the role of photo-interpretation.

A major characteristic of an image in remote sensing is the wavelength band it represents. Some images are measurements of the spatial distribution of reflected solar radiation in the ultraviolet, visible and near-to-middle infrared range of wavelengths. Others are measurements of the spatial distribution of energy emitted by the earth itself (dominant in the so-called thermal infrared wavelength range); yet others, particularly in the microwave band of wavelengths, measure the relative return from the earth's surface of energy actually transmitted from the vehicle itself. Systems of this last type are referred to as active since the energy source is provided by the remote sensing platform; by comparison systems involving remote sensing measurements that depend upon an external energy source, such as the sun, are called passive [98,99,100]. From a data handling and analysis point of view the properties of image data of significance are the number and location of the spectral measurements (or spectral bands or channels) provided by a particular sensor, the spatial resolution as described by the pixel size in equivalent ground metres, and the radiometric resolution. The last describes the range and discernable number of discrete brightness values and is sometimes referred to alternatively as dynamic range. Frequently the radiometric resolution is expressed in terms of the number of binary digits, or bits, necessary to represent the range of available brightness values. Thus data with 8 bit radiometric resolution has 256 levels of brightness. Together with the frame size of an image, in equivalent ground kilometres (which is determined usually by the size of the recorded image swath), the number of spectral bands', radiometric resolution and spatial resolution determine the data volume provided by a particular sensor and thus establish the amount of data to be processed, at least in principle. As an illustration consider the IRS-1A/1B multi-spectral scanner image taken from LISS-II (Linear Imaging Self Scanner) camera. It has spatial resolution of 36.25m in 4 wavelength bands with 7 bits radiometric resolution. A typical image frame consists of 25.2 million pixels (25.2Mbytes) [101]. Remote sensing system could measure energy emanating from the earth's surface in any sensible range of wavelengths. However technological consideration and selective opacity of the earth's atmosphere, scattering from atmospheric particulate and significance of the data provided, excludes certain wavelengths. The major ranges utilized for earth resources sensing are between about $0.4\mu m$ to $12~\mu m$ (visible and infra-red range) and between about 30mm to 300mm (microwave range normally mentioned in terms of 1GHz to 10GHz). In the atmospheric remote sensing about 20GHz to 60GHz range is used. Significance of these different ranges lies in the interaction mechanism between the electromagnetic radiation and the materials being examined. Each range of wavelength has its own strength in terms of the information it can contribute to the remote sensing process. The purpose of acquiring remote sensing image data is to be able to identify and assess by some means, either surface materials or their spatial properties. The remote sensing in different ranges and different bands provides information, which helps in identifying objects with their signature in different bands. Therefore, we find systems available that are optimized for and operate in particular spectral ranges, and provide data that complements the information received from other sensors and systems, the details can be found in [99].

In the present investigation, we have used data sets which are acquired in the visible and near infra-red bands either in single or multi-spectral form for segmentation and classification of regions. The Remote Sensing images are normally acquired and distributed by private and government organizations through various agencies. In India images are available for users through National Remote Sensing Agency (NRSA)[2]. An image received from distribution agency goes through various types of processing before final interpretation and understanding of the image, normally know as "image analysis". In the process of image analysis one has to segment and/or classify the image into its constituent components. Some of the methods used in the past are discussed in the next section.

## 1.5   Remote Sensing Images: Different Approaches and Methodologies for Classification and Segmentation

Remotely sensed images are normally poorly illuminated, highly dependent on the environmental conditions, and have very low spatial resolution. Most of the times a scene contains too many objects (or regions), and these regions are ill-defined because of both grayness and spatial ambiguities. Moreover, the gray value assigned to a pixel is the average reflectance of different types of ground covers present in the corresponding pixel area. Assigning unique class labels with certainty is thus a problem for remotely sensed images. Fuzzy set theory

[2] The official web-site of NRSA is http://www.nrsa.gov.in

and rough set theory can provide a better way of handling this problem by associating certainty factors with class labels.

There are mainly two approaches to pixel classification in remote sensing. One of them attempts to relate pixel groups with actual earth-surface cover types, e.g., vegetation, soil, urban area, water. These groups of pixels are called information classes. The other approach determines the characteristics of non-overlapping groups of pixels in terms of their spectral band values. The groups of pixels in this case are known as spectral classes [102]. Remote sensing is successful because in many instances these spectral classes coincide with information classes. The former approach, where samples from information classes (training data) are used for learning and then classifying unknown pixels (patterns), is called supervised. On the other hand, the latter approach where at first the spectral classes are found without a priori knowledge on the information classes and then their relationship with the information classes is established using a map and/or ground truth is called unsupervised. In other words, segmentation (and associating the segments with information classes), and classification of remote sensing image data are referred to as unsupervised and supervised classification, respectively.

Below we provide in brief, some of the methods used for segmentation and classification of remote sensing images using both classical approaches and modern tools like fuzzy sets and artificial neural networks.

**Classical Methods.** Maximum likelihood classifier (in the Bayesian paradigm) is commonly used as statistical methods for supervised classification [99,100,102]. This has been mostly used in quantitative analysis of the remotely sensed images for estimation of vegetation, crop, water and soil [102,103,104]. Some of its other applications can be found in [105,106].

One can find several references in pattern recognition, where the results of the newly adopted supervised classifiers are compared with that of the maximum likelihood classifier [107,108,109,110]. The effectiveness of maximum likelihood classification depends upon estimation of the mean vector and the covariance matrix for each spectral class. This in turn is dependent upon having a sufficient number of training pixels for each of those classes. When this is not so, inaccurate estimates of the covariance matrix, leads to poor classification. Therefore, where the number of training samples per class is limited it can be more effective to resort to a classifier that does not make use of covariance information, instead depends only upon the sample points from classes and their means, e.g., minimum distance classifier, decision tree, neural network, fuzzy set theoretic classifier and support vector machine. The minimum distance classifier has been used for classification of remote sensing images by Murthy *et al.* [111] using two bands (green and near infra-red band) of Indian Remote Sensing (IRS) multispectral image. From the bivariate frequency table, seed points are chosen for six classes and minimum distance classifier is applied to segment the image. Minimum distance classification can be performed using distance measures other than Euclidean [112]. Methods of table-look up classification, parallelepiped classification, linear discriminant function can also be found in the literature [99,100].

Classification methods that take into account the labels of neighbors when seeking to determine the most appropriate class for a pixel are said to be context classifiers. A method is developed by Khazenie and Crawford [113] for contextual classification which considers both spatial and temporal correlation for processes which satisfy second-order stationary conditions. According to Li and Peng [114], by incorporating the local statistics of an image, a semi-causal non-stationary autoregressive random field can be applied to a non-stationary image for segmentation. Since this non-stationary random field can provide a better description of the image texture than the stationary one, an image can be better segmented. The proposed technique is applied to extract urban areas from a Landsat image. A Bayesian contextual classification scheme is described in connection with modified M-estimates and a discrete Markov random field model by Jhung and Swain [115]. The spatial dependence of adjacent class labels is characterized based on local transition probabilities in order to use contextual information. Due to the computational load required to estimate class labels in the final stage of optimization and the need to acquire robust spectral attributes derived from the training samples, modified M-estimates are implemented to characterize the joint class-conditional distribution. The experimental results show that the suggested scheme outperforms conventional noncontextual classifiers as well as contextual classifiers which are based on least squares estimates or other spatial interaction models.

Demonstration of probabilistic relaxation for pixel classification can be found in Gong and Howarth [116], Richards et al. [117]. Recently Sun et al. [106] described an information fusion method for the extraction of land-use information. It integrates spectral, spatial and structural information existing in the image. A thematic map was first produced with a maximum-likelihood classification (MLC) applied to the multispectral imagery. "Probabilistic relaxation" (PR) was then performed on the thematic map to refine the classification with neighborhood information. Furthermore, they incorporated edges extracted from the higher resolution panchromatic imagery in the classification. An edge map was generated using operations such as edge detection, edge thresholding and edge thinning. Finally, a modified region-growing approach was used to improve image classification. The result of the method is an improved land-use map, which is characterized with sharp interregional boundaries, reduced number of mixed pixels and more homogeneous regions.

A general model for multi-source classification of remotely sensed data based on Markov random fields (MRF) is described in [118]. A model for fusion of optical images, synthetic aperture radar (SAR) images, and GIS (geographic information systems) ground cover data is presented in detail and tested. The MRF model exploits spatial class dependencies (spatial context) between neighboring pixels in an image, and temporal class dependencies between different images of the same scene. By including the temporal aspect of the data, it is suitable for detection of class changes between the acquisition dates of different images. The performance of the model is investigated by fusing Landsat TM images, multitemporal ERS-1 (Earth Resources Satellite) SAR images, and GIS ground-cover maps for land-use classification and applied on agricultural crop classification.

The use of decision tree (DT) for classification in remote sensing problems is described by Swain and Hauska [119]. The use of classification trees (CT) for land cover mapping is becoming increasingly common. Classification trees, sometimes called decision trees, or CART (Classification and Regression Trees) offer several advantages over classification algorithms traditionally used for land cover mapping. One advantage is the ability to effectively use both categorical and continuous predictor data sets with different measurement scales. Other advantages include the ability to handle nonparametric training and predictor data, good computational efficiency, and an intuitive hierarchical representation of discrimination rules. Classification trees use multiple explanatory variables to predict a single response variable. Pal and Mather [120] in a recent study, using separate test and training data sets from two different geographical areas and two different sensors multispectral Landsat ETM+ (Enhanced Thematic Mapper Plus) and hyperspectral DAIS (The Digital Airborne Imaging Spectrometer), evaluated the performance of univariate and multivariate DTs for land cover classification. Factors considered are: the effects of variations in training data set size and of the dimensionality of the feature space, together with the impact of boosting, attribute selection measures, and pruning.

In the case of unsupervised classification k-means clustering is a most commonly used technique. However, methods like iso-data clustering, agglomerative hierarchical clustering and clustering by histogram peak selection are also found to be useful [99]. Parui et al. [121] used the four bands of IRS images and k-means algorithm for estimating the centroid of the classes and finally segmenting the image into seven classes. Acharyya et al. [72] explained a scheme for segmentation of multitexture images using selected wavelet features and k-means clustering and applied on IRS and SPOT images. The use of k-means algorithm can also be found in [122] in the application of evolutionary algorithm for segmentation.

A method of evaluating the suitability of valleys as threshold has been described by Sahasrabudhe and Dasgupta [123], and applied to satellite image segmentation. Laprade [124] described a split-and-merge technique using F-test and a mean predicate to test the uniformity of regions and applied it to aerial photographs. This method approximates the image intensity surface by planar facets. He applied a least square (fitting) plane to the intensity surface and this procedure is incorporated into a split-and-merge algorithm. Baraldi and Parmiggiani [125] presented a class of single linkage region growing (SLRG) algorithms, in which pairs of neighboring pixels are compared for merging. This is one of the conceptually simplest approaches to image segmentation. Their method has two main properties: (i) it combines single linkage, centroid linkage and hybrid linkage criteria; and (ii) its goal is to detect areas characterized by low contrast in an image. They developed two new SLRG algorithms, and applied to multiband images by exploiting the VDM (vector degree of match) criterion for grouping of two adjacent pixels. This method requires only one user-defined parameter, namely, the VDM Threshold (VDMT), which is a normalized value featuring the local adaptivity.

A general method of statistical clustering by means of expectation maximization (EM) algorithm, along with rough sets, is used by Pal and Mitra [126]. EM provides the statistical model of the data, and handles the associated measurement and representation uncertainties. Rough set theory helps in faster convergence and in avoiding the local minima problem, thereby enhancing the performance of EM. For rough-set-theoretic rule generation, each band is discretized using fuzzy-correlation-based gray-level thresholding. This method is applied on IRS-1A four-band images.

Shah *et al.* [127] described a method which uses an Independent Component Analysis (ICA) based approach for unsupervised classification of hyper-spectral imagery. ICA is employed for a mixture model, it estimates the data density in each class and models class distributions with non-Gaussian structure, formulating the ICA mixture model (ICAMM). Four feature extraction techniques, namely, Principal Component Analysis, Segmented Principal Component Analysis, Orthogonal Subspace Projection and Projection Pursuit have been considered as preprocessing steps for reducing the data dimensionality. The results demonstrate that the ICAMM significantly outperforms the k-means algorithm for land cover classification of hyper-spectral imagery implemented on reduced data sets. Moreover, data sets extracted using Segmented Principal Component Analysis produce the highest classification accuracy.

Band ratios and vegetation indices [100,128] are normally used with multi-spectral imagery for segmentation of vegetated areas (e.g., forest and agricultural regions). The simple vegetation index (SVI) and the normalized difference vegetation index (NDVI) are two frequently used indices, and their mathematical form is the basis for the development of other modified vegetation indices, such as transformed vegetation index (TVI) and soil adjusted vegetation index (SAVI). In a study by Vaiopoloulos *et al.* [129] recently, a different approach based on probability theory is developed in order to evaluate the efficiency of SVI and NDVI, and to suggest two modified vegetation indices, namely, modified simple vegetation index (MSVI) and modified normalized difference vegetation index (MNDVI). These methods are applied on a Landsat-7 Enhanced Thematic Mapper (ETM) image of an island in western Greece. By choosing the proper value for a characteristic parameter in the expression of MSVI and MNDVI, these methods are seen to perform better than SVI and NDVI.

Like NDVI, McFeeters [130] described a normalized difference water index (NDWI) to delineate open water features and enhance their presence in remotely sensed imagery. The NDWI makes use of the reflected near-infrared radiation and visible green light to enhance the presence of such features, while eliminating the presence of soil and terrestrial vegetation features.

**Fuzzy Set Theoretic Methods.** In conventional remote sensing supervised classification, the concept of a pixel originating from more than a class (i.e., mixture pixel) is not taken into consideration in training a classifier and in determining pixels' memberships. This expressive limitation has reduced the classification accuracy and led to the poor extraction of class information. Wang [131] describes a fuzzy supervised classification method in which geographical information is represented as fuzzy sets. The algorithm consists of two major steps: estimation of fuzzy parameters from fuzzy training data, and fuzzy partition of the spectral space. The concept of partial membership of pixels allows component cover classes of mixed pixels to be identified and more accurate statistical parameters to be generated, resulting in a higher classification accuracy. They applied the method for classifying a Landsat MSS (Multispectral Scanner) image.

Melgani *et al.* [132] describes an explicit fuzzy supervised classification method which consists of three steps. The explicit fuzzyfication is the first step where the pixel is transformed into a matrix of membership degrees representing the fuzzy inputs of the process. Then, in the second step, a MIN fuzzy reasoning rule followed by a rescaling operation is applied to deduce the fuzzy outputs, or in other words, the fuzzy classification of the pixel. Finally, a defuzzyfication step is carried out to produce a hard classification. The classification results on Landsat TM data demonstrate the promising performances of the method with low classification time.

In Mandal *et al.* [133] a multivalued recognition system has been used, which is capable of handling various imprecise inputs and in providing multiple choices of classes corresponding to any input pattern. The work describes a method of analyzing Indian Remote Sensing (IRS) satellite imagery for detecting various man-made objects, namely, roads, bridges, airports, seaports, city area and township/industrial areas. The recognition system initially classifies the image pixels into six land cover types by providing multiple choices of classes. In order to identify the targets, some spatial knowledge about them and their inter-relationships have been incorporated on the classified image using some heuristic rules. It is found that use of multiple class choices makes the detection procedures effective.

Pal *et al.* [134] demonstrated an application of the principle of shape estimation with the concept of fuzzy notion to the problem of extracting different regions from satellite imagery. The recognition system, capable of providing output in multiple states, reduces the uncertainty in decisions.

Fuzzy unsupervised methods mainly deals with fuzzy c-means (FCM) clustering and its other variations [61]. A two-stage fuzzy c-means algorithm was applied by Cannon *et al.* [135] on a Landsat-4 image with six-bands to demonstrate the feasibility of the methodology for segmentation. Trivedi and Bezdek [64] used fuzzy c-means clustering for low-level image segmentation. A modification of FCM to make it fast is described by Shankar and Pal [136] in this regard. Use of fuzzy c-means can be found along with genetic algorithm in [137] and along with wavelet features in [138] for segmentation.

An investigation of urban areas detection in satellite images was carried out by Lorette *et al.* [139] using fuzzy c-means and Markovian model of segmentation. At first they analyze the texture through the modelling of the luminance field with eight different chain-based models. Then derived a texture parameter from these models. The effect of the lattice anisotropy is corrected by a renormalization group technique, which comes from statistical physics. This parameter, which takes into account local conditional variances of the image, is compared to classical methods of texture analysis. A modified fuzzy c-means algorithm that includes an entropy term is developed and used here. The advantage of such an algorithm is that the number of classes does not need to be known a priori. Besides, this algorithm provides with further information such as the probability that a given pixel belongs to a given cluster. Finally this information is introduced in a Markovian model of segmentation. The method is applied on SPOT-5 simulated images, SPOT-3 images and ERS-1 radar images.

**Neural Network Based Methods.** Neural network based supervised approaches are also developed for various land cover type classification [140,141,142,143,144]. These have also been used successfully for the classification of synthetic aperture data (SAR) for terrain images by Decatur [143] using a three-layer back-propagation networks. Lee *et al.* [144] used a four-layer back-propagation networks for cloud classification of Landsat MSS data. They compared the neural network classifier with various statistical classifiers.

In the work of Benediktsson *et al.* [141] neural learning procedures and statistical classification methods are applied and compared empirically in classification of multisource remote sensing and geographic data. In [142] Benediktsson *et al.* used a Conjugate - Gradient Neural Networks for classification of multisource high dimension aircraft scanner Remote sensing data.

Bischof *et al.* [108] used the three-layered back propagation neural networks for classification of Landsat TM (Thematic Mapper) data on a pixel-by-pixel basis and the results are compared with that of Gaussian maximum likelihood classification. They showed that neural networks performed better and the textural information can be integrated into the network without explicit definition of the texture measure. Similar investigations for land use activities are reported by using MLP (multilayer perceptron) and learning vector quantization (LVQ) in [109], and by using a neural classifier evolved with genetic algorithm in [145].

Baraldi and Parmiggiani [140] described an artificial neural network (ANN) which performs unsupervised detection of categories from arbitrary sequences of multivalued input patterns and applied for satellite image clustering. Recently, Villmann *et al.* [146,147] studied the application of self-organizing maps (SOMs) for the analysis of very high-dimensional remote sensing spectral images. They concentrated on the issue of faithful topological mapping in order to avoid false interpretations of cluster maps created by a SOM. They described several new extensions of the standard SOM: the growing SOM, magnification control, and generalized relevance learning vector quantization, and demonstrate their effect on both low-dimensional traditional multi-spectral imagery and 200-dimensional hyperspectral imagery.

For the speckle contained in SAR image, SAR image can not be segmented efficiently with traditional methods. In the work of Xue *et al.* [148], an automatic clustering method based on competitive Hopfield NN is used to segment SAR images.

**Some Other Methods.** Here we describe some of the methods developed for segmentation of remote sensing images using knowledge based system, evolutionary computation, support vector machine, wavelets and level set.

The use of knowledge based system for segmentation and classification, can be found in [149,150,151]. Mandal *et al.* [149] used a multi valued recognition system to segment the image into various land cover types by providing multiple choices of classes. Using the spatial knowledge and their inter-relation along with some heuristic rules, they detected man-made objects like roads, bridges, airports seaports, city area and township and industrial areas from the IRS images. Ton *et al.* [150,151] used a knowledge-based approach for Landsat image segmentation. The image segmentation problem is solved by extracting kernel information from the input image to provide an initial interpretation of the image and by using a knowledge-based hierarchical classifier to discriminate between major land-cover types in the study area. The method is designed in such a way that a Landsat image can be segmented and interpreted without any prior image-dependent information. The general spectral land-cover knowledge is constructed from the training land-cover data, and the road information of an image is obtained through a road-detection program.

Ho and Lee [122] designed an efficient evolutionary image segmentation algorithm (EISA). EISA uses a k-means algorithm to split an image into many homogeneous regions, and then uses an intelligent genetic algorithm (IGA) associated with an effective chromosome encoding method to merge the regions automatically such that the objective of the desired segmentation can be effectively achieved, where IGA is superior to conventional genetic algorithms in solving large parameter optimization problems. High performance of EISA is illustrated in terms of both evaluation performance and computation time.

In the article of Pal *et al.* [152], the effectiveness of some genetic algorithm based pattern supervised classifiers has been investigated in the domain of satellite imagery which usually have complex and overlapping class boundaries. Landsat data, SPOT image and IRS image are considered as input. The superiority of these classifiers over k-NN (nearest neighbor) rule, Bayes maximum likelihood classifier and multilayer perceptron for partitioning different landcover types is established. Incorporation of the concept of variable length chromosomes and chromosome discrimination led to superior performance in terms of automatic evolution of the number of hyperplanes for modeling the class boundaries, and the convergence time. This non parametric classifier requires very little a priori information, unlike k-NN rule and MLP (where the performance depends heavily on the value of k and the architecture, respectively), and Bayes maximum likelihood classifier (where assumptions regarding the class distribution functions needs to be made). Bandyopadhyay and Pal [107], used the concept of chromosome differentiation, commonly witnessed in nature as male and female sexes, in genetic algorithms with variable length strings for designing a nonparametric classification methodology. Its significance in partitioning different landcover regions from satellite images, having complex/overlapping class boundaries, is demonstrated. The classifier is able to evolve automatically the appropriate number of hyperplanes efficiently for modeling any kind of class boundaries optimally. Merits of the system over the related ones are established through the use of several quantitative measures.

The article by Liu *et al.* [145] describes the effectiveness of the genetic algorithm evolved neural network classifier and its application to the land cover classification of remotely sensed multispectral imagery. The methodology adopts a real coded GA strategy and hybrids with a back propagation (BP) algorithm. The genetic operators are carefully designed to optimize the neural network, avoiding premature convergence and permutation problems. A SPOT-4 XS imagery is employed to evaluate its accuracy. A more complicate experiment on CBERS (China-Brazil Earth Resources Satellite) data also demonstrates that carefully designed genetic algorithm-based neural networks can outperform gradient descent-based neural networks [142].

Under unsupervised framework, GA has been used for clustering data described as in [137,153]. The problem of classifying an image into different homogeneous regions is viewed as the task of clustering the pixels in the intensity space. In [137] a real-coded variable string length genetic fuzzy clustering with automatic evolution of clusters is used. The cluster centers are encoded in chromosomes, and the Xie-Beni index [154] is used as a measure of the validity of the corresponding partition.

Classification using support vector machine can be found in the Brown *et al.* [155] and Huang *et al.* [156]. Huang *et al.* [156] used support vector machines for land cover classification and assessment. This paper gives an experimental evaluation of its accuracy, stability and training speed in deriving land cover classifications from satellite images. The SVM was compared with three other popular classifiers (e.g., maximum likelihood classifier (MLC), neural network classifier(NNC) and decision tree classifier (DTC)). The impact of kernel configuration on the performance of the SVM, and the effect of the training data and input variables on the four classifiers were also evaluated.

Mixture modeling, which is an increasingly important tool in the remote sensing research community, tries to resolve area information in to subpixel level. The process of mixed-pixel classification is to model the class mixing proportions (percentage ground cover area) rather than estimating the class probability that a signature corresponds to a particular class. Brown *et al.* [155] described a support vector machine (SVM) using linearly separating hyperplane to solve the problem of pixel unmixing (considering non-linear mixture regions), and overlapping of the spectral classes. This method seems to perform better than a traditional method, known as constrained least square linear spectral mixture models (CLS LSMM), that assumes linear mixture regions. The potential of SVM is demonstrated using a labeled area of Landsat TM data set.

Pal and Mather [110] have used multi-class Support vector machines (SVMs) for classification of DAIS hyperspectral remotely sensed data. Their results show that the SVM performs better than maximum likelihood, univariate decision tree and back propagation neural network classifiers, even with small training data sets, and is almost unaffected by the Hughes phenomenon (curse of dimensionality) [157].

There have been several attempts in applying wavelets for segmentation of remotely sensed images. Texture properties of a remotely sensed image provide valuable information for analysis, where different object regions are treated as different texture classes. This feature is utilized by various researchers in their methods. Kundu and Acharyya [158] described an application of M-band wavelets for texture segmentation. Different quasi-homogenous regions in the remotely sensed image are treated to have different texture properties. Based on this assumption a multi-class texture segmentation scheme is developed for segmentation of remote sensing images. A feature extraction method based on M-band wavelet packet frames is described in [72] in this regard. Mecocci et al. [138] described an octave-band wavelet decomposition scheme for texture segmentation combined with fuzzy c-means classifier, whereas Lindsay et al. [159] used the one-dimensional discrete wavelet transform (DWT) based on Daubechies wavelet filter. The aim of the work of Neidermeier et al. [160] is to show how coastline can be derived from Synthetic Aperture Radar (SAR) images by using wavelet and active contour methods.

Application of level set in image segmentation has recently drawn the attraction of researchers. For example, Kervrann and Trubuil [75] described a method for finding the boundaries of level sets of the image segments and applied to aerial images.

**Curvilinear Structure Detection.** The methodologies described so far in Section 1.4, are mostly useful for segmenting or partitioning different regions from remotely sensed images. One may note that there are many important curvilinear features (like roads, canals, rivers) that can be found in remotely sensed images. The segmentation and detection of these structures cover a major area of analysis of remote sensing images. Some of the attempts made so far are explained below. To enhance and detect linear structures in a gray level image, local operations with an additive score are normally used. Parui et al. [161] used a multiplicative score instead of additive score, which gives better results than the additive one. The threshold values are automatically selected in the enhanced image for segmenting out the linear structures.

Hu et al. [162] used a graph-based approach for detection of roads. Zlotnick and Carnine [163] described a method that track roads by searching for anti-parallel edges as starting points for roads tracking and linking. Barzohar and Cooper [164] described an automatic method of extracting main roads in aerial images. The aerial image is partitioned into windows, road extraction starts from the window of high confidence estimates, while road tracing is to perform a dynamic programming to find an optimal global estimate.

Mandal et al. [133] used the concept of multiple choices for the recognition of roadlike structures from IRS images. Geman and Jedynak [165] described a semi-automatic method where, given a start point and start direction, a road is extracted from a panchromatic SPOT satellite image. Gruen and Li [166] described a linear feature extraction method using Active Contour models called snakes. They combined characteristics of snakes and adaptive least squares correlation method. This method might need large computation time on high-resolution images. Some other research efforts can be found in Park and Kim [167] which presents a road extraction algorithm using template matching. But its limitations is that it requires initial seed points on the road central lines and each road segment requires separate seed points. In a recent article, Stoica *et al.* [168] described a method for the extraction of roads from remotely sensed images. Under the assumption that roads form a thin network in the image, they approximated such a network by connected line segments. This probabilistic model uses Gibbs point process framework. The estimation for the network is found by minimizing an energy function. A simulated annealing algorithm, based on a Monte Carlo dynamics (RJMCMC: Reverse-Jump Markov Chain Monte Carlo) for finite point processes is used. Recently a survey on state of the art on automatic road extraction for GIS update from aerial and satellite imagery is provided in [169].

Before we describe the scope of the article in the next section, we reiterate that the methods of segmentation of remote sensing images are based on either pixel classification or gray level thresholding. These images, usually being mutlispectral, make the classification based approach a natural choice. On the other hand, thresholding based techniques are seen to be computationally less expensive, but the thresholded outputs on different bands need to be integrated to arrive at a decision. One may note that in certain remote sensing applications the image is available only in one band, for example, Panchromatic images, for which segmentation based on thresholding seems to be convenient and appropriate.

## 1.6   Scope of the Article

The objective of this article is to present some new results of investigation, both theoretical and experimental, in the area of image classification and segmentation with application to remote sensing images. Both classical and soft computing techniques are used.

The problem of scarcity of labeled pixels, required for segmentation of remotely sensed satellite images in supervised pixel classification framework, is addressed in Section 2 of the article. A support vector machine (SVM) is considered for classifying the pixels into different landcover types. It is initially designed using a small set of labeled points, and subsequently refined by actively querying for the labels of pixels from a pool of unlabeled data. The label of the most interesting/ambiguous unlabeled point is queried at each step. Here, the principle of active learning is exploited to minimize the number of labeled data used by the SVM classifier by several orders. These features are demonstrated on an IRS-1A four band multispectral image. Comparison with related methods is made in terms of the number of data points used, computational time and a classification/segmentation quality measure [170,171]. A new quantitative index for image classification/segmentation using the concept of homogeneity within regions is introduced in this regard [172].

Effectiveness of various fuzzy set theoretic thresholding techniques (based on entropy of fuzzy sets, fuzzy geometrical properties, and fuzzy correlation) on remotely sensed images is demonstrated in Section 3. Results are compared with those of probabilistic thresholding, and fuzzy c-means and hard c-means clustering algorithms, both in terms of index value (quantitatively) and structural details (qualitatively). Fuzzy set theoretic algorithms are seen to be superior to their respective non-fuzzy counter parts. Among all the techniques fuzzy correlation, followed by fuzzy entropy, performed better for extracting the structures. Fuzzy geometry based thresholding algorithms produced a single stable threshold for a wide range of membership variation. Both IRS and SPOT imagery are considered for this investigation [172,173].

The problem of image object extraction using rough sets and granular computing is addressed in Section 4. An image in the framework of rough set theory is defined with respect to a partition on it. For a given partition, the object and background regions are described in terms of upper and lower approximation of rough sets. This represents the associated uncertainty arising from the granularity (in terms of windows) in the image space. This uncertainty is quantified using the concept of object roughness and background roughness, and then minimized over various partitions on the image space using a measure called "rough entropy". Rough entropy of an image is defined based on the concept of image granules. Its maximization results in minimization of the roughness in both object and background regions; thereby determining the threshold of partitioning. Methods of selecting the appropriate granule size and efficient computation of rough entropy are also described [174].

While Sections 3 and 4 concern with the problem of segmenting an image using gray level thresholding into homogeneous regions, Section 5 deals with the said problem in terms of homogeneous line segments. Here we present a technique for extracting homogeneous regions of arbitrary shape and size in a gray level image based on Hough transform. The regions are defined in terms of unions of homogeneous line segments. A line segment in an image is viewed as a collection of pixels having the property of straight line in Euclidean plane and possessing the same homogeneity property. The detection of homogeneous line segments is made directly from gray level images. A definition of "region" in terms of these line segments, with constraints on its length and variance (i.e., gray level variation within the line segment), is provided. The proposed method is able to extract regions irrespective of their shape and size from a gray level image. Its effectiveness is demonstrated on IRS images [175].

Conclusions of the article and future research is described in the Section 6.

# 2 Active Learning, Support Vector Machine and Pixel Classification

The primary problem in supervised pixel classification is the pure availability of labeled data, which can be obtained only from ground truths and by costly manual labeling. Recently, active learning has become a popular paradigm for reducing the data requirement of large scale learning tasks [176,177,178]. Here, instead of learning from 'random samples', the learner has the ability to select its own training data. This is done iteratively, and the output of a step is used to select the examples for the next step. Several active learning strategies exist in practice, e.g., error driven techniques, uncertainty sampling, version space reduction and adaptive resampling.

Support vector machines (SVM) are particularly suited for active learning since a SVM classifier is characterized by a small set of support vectors (SVs) which can be easily updated over successive learning steps. One of the most efficient active SVM learning strategy is to iteratively requests the label of the data point closest to the current separating hyperplane or which violates the margin constraint maximally [178,179,180]. This accelerates the learning drastically compared to random data selection. The above technique is often referred to as active/query SVM. Besides active SVM, another active learning strategy based on version space splitting is presented in [181]. The points which split the current version space into two halves having equal volumes are selected at each step, as they are likely to be the actual support vectors. Three heuristics for approximating the above criterion are described, the simplest among them selects the point closest to the current hyperplane as in [179]. A greedy optimal strategy for active SV learning is also described in [182]. Here, logistic regression is used to compute the class probabilities, which is further used to estimate the expected error after adding an example. The example that minimizes this error is selected as a candidate SV.

The present section describes a pixel classification algorithm [170,171] based on the query SVM algorithm. A conventional SVM is initially designed using a small set of points labeled manually. The SVM is subsequently refined by actively querying for the labels of pixels from a pool of unlabeled data. The most interesting/ambiguous unlabeled point is queried at each step and is labeled by an human expert. It is seen that the above active learning strategy reduces the number of labeled data used by the SVM classifier by several orders compared to conventional SVM, while providing comparable segmentation quality. These features are demonstrated on an IRS-1A four band image. Comparison with related methods is made in terms of the number of data points used, computation time and a classification/segmentation index.

After explaining in brief, the fundamentals of support vector machines in Section 2.1, the active SVM learning algorithm for pixel classification is described in Section 2.2. Classification/segmentation index-$\beta$ is defined in Section 2.3. Experimental results are provided in Section 2.4.

## 2.1  Support Vector Machines

Support vector machines are a general class of learning architecture inspired from statistical learning theory that performs *structural risk minimization* on a nested set structure of separating hyperplanes [183]. Given a training data, the SVM training algorithm obtains the optimal separating hyperplane in terms of generalization error. We describe below the SVM design algorithm for a two class problem. Multiclass extensions can be done by designing a number of one-against-all or one-against-one two class SVMs.

*Algorithm 1*

Suppose we are given a set of examples $(\mathbf{x}_1, y_1), \ldots, (\mathbf{x}_l, y_l), \mathbf{x} \in R^N, y_i \in \{-1, +1\}$. We consider functions of the form $sgn((\mathbf{w} \cdot \mathbf{x}) + b)$, in addition we impose the condition

$$\inf_{i=1,\ldots,l} |(\mathbf{w} \cdot \mathbf{x}_i) + b| = 1. \tag{3}$$

We would like to find a decision function $f_{\mathbf{w},b}$ with the properties $f_{\mathbf{w},b}(x_i) = y_i$; $i = 1, \ldots, l$. If this function exists, condition (3) implies

$$y_i((\mathbf{w} \cdot \mathbf{x}_i) + b) \geq 1, \quad i = 1, \ldots, l. \tag{4}$$

In many practical situations, a separating hyperplane does not exist. To allow for possibilities of violating Equation (4), slack variables are introduced like

$$\xi_i \geq 0, \quad i = 1, \ldots, l \tag{5}$$

to get

$$y_i((\mathbf{w} \cdot \mathbf{x}_i) + b) \geq 1 - \xi_i, \quad i = 1, \ldots, l. \tag{6}$$

The support vector approach for minimizing the generalization error consists of the following:

$$\textbf{Minimize}: \qquad \Phi(\mathbf{w}, \xi) = (\mathbf{w} \cdot \mathbf{w}) + C \sum_{i=1}^{l} \xi_i \tag{7}$$

subject to the constraints (5) and (6).

It can be shown that minimizing the first term in Equation (7), amounts to minimizing the VC-dimension, and minimizing the second term corresponds to minimizing the misclassification [184]. The above minimization problem can be posed as a constrained quadratic programming (QP) problem.

The solution gives rise to a decision function of the form:

$$f(\mathbf{x}) = sgn \left[ \sum_{i=1}^{l} y_i \alpha_i (\mathbf{x} \cdot \mathbf{x_i}) + b \right]. \tag{8}$$

Only a small fraction of the $\alpha_i$ coefficients are non-zero. The corresponding set of $\mathbf{x_i}$ entries are known as *support vectors* and they fully define the decision function. The support vectors are geometrically the points lying near the class boundaries.

The linear SVM was described above. However, nonlinear kernels like polynomial, sigmoidal and radial basis functions (RBF) may also be used. Here, the decision function is of the form:

$$f(\mathbf{x}) = sgn \left[ \sum_{i=1}^{l} y_i \alpha_i \kappa(\mathbf{x}, \mathbf{x_i}) + b \right]. \tag{9}$$

Where $\kappa(\mathbf{x}, \mathbf{x_i})$ is the corresponding nonlinear kernel function. In remote sensing images, classes are usually spherical shaped and the use of spherical RBF kernel is most appropriate. RBF kernels are of the form $\kappa(\mathbf{x_1}, \mathbf{x_2}) = e^{-\gamma|\mathbf{x_1}-\mathbf{x_2}|^2}$, (where $\gamma > 0$). Again, the aforesaid two class SVM can easily be extended for multiclass classification by designing a number of one-against-all, a two class SVMs, e.g., a $k$-class problem is handled with $k$ two class SVMs.

## 2.2    Active Support Vector Learning for Pixel Classification

A limitation of the SVM design algorithm, described above, is the need to solve a quadratic programming (QP) problem involving a dense $l \times l$ matrix, where $l$ is the number of points in the data set. Since most QP routines have quadratic complexity, SVM design requires huge memory and computational time for large data applications. Several approaches exist for circumventing the above short-comings as well as to minimise the number of labeled points required to design the classifier. Many of them exploit the fact that the solution of the SVM prob-lem remains the same if one removes the points that correspond to zero Lagrange multipliers of the QP problem (the non-SV points). The large QP problem can thus be broken down into a series of smaller QP problems, whose ultimate goal is to identify all of the non-zero Lagrange multipliers (SVs) while discarding the zero Lagrange multipliers (non-SVs). At every step, one solves a QP problem that consists of the non-zero Lagrange multiplier points from the previous step, and a number of other points queried. At the final step, the entire set of non-zero Lagrange multipliers has been identified; thereby solving the large QP problem. The active SVM design algorithm used here for pixel classification is based on the aforesaid principle. At each step the most informative point not belonging to the current SV set is queried along with its label; the goal is to minimise the total number of labeled points used by the learning algorithm. The method is described below and illustrated in Fig. 1. The steps need to be repeated $k$ times for a $k$ class problem with data from respective classes.

*Algorithm 2*
Let $\mathbf{x} = [x_1, x_2, \ldots, x_d]$ represent a pixel of a $d$-band multispectral image. Here, $x_i$ is the grey value of the $i$th band for pixel $\mathbf{x}$. Let $A = \{\mathbf{x_1}, \mathbf{x_2}, \ldots, \mathbf{x}_{l_1}\}$ denote the set of pixels for which class labels are known, and $B = \{\mathbf{x_1}, \mathbf{x_2}, \ldots, \mathbf{x}_{l_2}\}$ the set of pixels for which class labels are unknown. Usually, $l_2 >> l_1$. $SV(C)$ denotes the set of support vectors of the set $C$ obtained using the methodology described in Section 2.1. $S_t = \{\mathbf{s_1}, \mathbf{s_2}, \ldots, \mathbf{s_m}\}$ is the support vector set obtained after $t$th iteration, and $< \mathbf{w_t}, b_t >$ is the corresponding separating hypersurface. $Q_t$ is the point actively queried for at step $t$. The learning steps involved are given below:

**Initial Step:** Set $t = 0$ and $S_0 = SV(A)$. Let the parameters of the correspond-ing RBF be $< \mathbf{w_0}, b_0 >$.
**While** *Stopping Criterion* is not satisfied:
   $Q_t = \{\mathbf{x}| \min_{\mathbf{x} \in B} \kappa(\mathbf{w_t}, \mathbf{x})\} + b.$

Request label of $Q_t$.
$S_t = SV(S_t \cup Q_t)$.
$B = B - Q_t$.
$t = t + 1$.
**End While**

The set $S_T$, where $T$ is the iteration at which the algorithm terminates, contains the final SV set representing the classifier.

**Stopping Criterion:** $\min_{\mathbf{x} \in B} \kappa(\mathbf{w}_t \cdot \mathbf{x}) + b > 1$. In other words, training is stopped when none of the unlabeled points lie within the margin of the separating hypersurface.

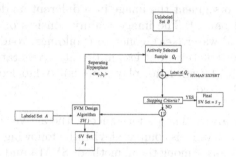

**Fig. 1.** Block Diagram of the active SVM learning algorithm for pixel classification

## 2.3  Quantitative Measure (Index-$\beta$))

Several methods have been used with different parameters in this investigation. Each of these gives rise to a classification (segmentation) of the image space. We intend to evaluate this classification/segmentation results quantitatively using an index (say $\beta$) [172,173].

Let $n_i$ be the number of pixels in the $i$th ($i = 1, 2, ..., k$) region obtained by a segmentation method. Let $X_{ij}$ be the gray value of $j$th pixel ($j = 1, ...., n_i$) in region $i$, and $\bar{X}_i$ the mean of $n_i$ gray values of $i$th region. Then index-$\beta$ is defined as:

$$\beta = \frac{\frac{1}{n} \sum_{i=1}^{k} \sum_{j=1}^{n_i} (X_{ij} - \bar{X})^2}{\sum_{i=1}^{k} \frac{n_i}{n} \times \frac{1}{n_i} \sum_{j=1}^{n_i} (X_{ij} - \bar{X}_i)^2} = \frac{\sum_{i=1}^{k} \sum_{j=1}^{n_i} (X_{ij} - \bar{X})^2}{\sum_{i=1}^{k} \sum_{j=1}^{n_i} (X_{ij} - \bar{X}_i)^2} \quad (10)$$

where $n$ is the size of the image and $\bar{X}$ is the mean gray value of the image. Note that the above measure is nothing but the ratio of the total variation and within-class variation. Since the numerator is constant for an image, $\beta$ value is dependent only on the denominator. The denominator decreases with increase in homogeneity in the regions. Therefore, for a given image and $k$ value, the higher the homogeneity within the regions, the higher would be the $\beta$ value. The value of $\beta$ also increases with $k$. In an extreme case when there is no partition (i.e., the entire image space is being considered as one class), we have $k = 1$ and $\beta = 1$. Otherwise the value of $\beta$ is always greater than 1. Significance of the index-$\beta$ in evaluation of segmentation results is demonstrated in Section 3.4.

## 2.4   Experimental Results

The multispectral IRS-1A image data, used in our experiment, contains observations for the city of Mumbai (Fig. 2). The data contains images of 4 spectral bands, namely blue, green, red and near infra-red. The images contain $512 \times 512$ pixels and each pixel represents a 36.25m $\times$ 36.25m region.

Here the task is to segment the image into different landcover regions, using 4 features (spectral bands). The image mainly consists of 6 classes e.g., clear water (ponds), turbid water (sea), concrete (buildings, roads, airport tarmacs), habitation (concrete structures but less in density), vegetation (crop, forest areas) and open spaces (barren land, playgrounds). A labeled set ($A$) containing 198 points is initially used.

**Algorithms Compared.** The performance of the active support vector learning algorithm ('active SVM') is compared with the following multispectral image segmentation algorithms. Among them, methods SVM 1 and SVM 2 represent extreme conditions on the use of labeled samples. In SVM 1 the labeled set is very small in size but the labels are accurate, while in SVM 2 a large fraction of the entire data constitutes the labeled set, but the labels may be inaccurate. The $k$-means algorithm is a completely unsupervised scheme requiring no class labels.

(i) SVM 1: The conventional support vector machine, using only the initial labeled set as the entire design set.
(ii) $k$-means: The unsupervised $k$-means clustering algorithm.
(iii) SVM 2: The conventional support vector machine, using 10% of the entire set of pixels as the design set. The labels are supplied by the output of the $k$-means algorithm.

**Evaluation Criterion.** The image segmentation algorithms are compared on the basis of the following quantities:

1. Total number of labeled data points used in training ($n_{labeled}$).
2. Training time ($t_{training}$) on a Sun UltraSparc 350 MHz workstation.
3. Quantitative Classification/Segmenatation Index-$\beta$ (Section 2.3).

**Comparative Results.** The performances of different multispectral image segmentation methods are presented in Table 1. Among them, the proposed active SVM learning algorithm provides the best segmentation quality as measured by the index-$\beta$. The SVM 1 algorithm provides the lowest $\beta$ value, which is expected since it uses a very small number of training samples. The unsupervised $k$-means

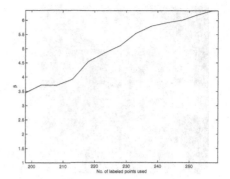

**Fig. 2.** Original band 4, IRS-1A image of Mumbai

**Fig. 3.** Variation of $\beta$ value with the number of labeled data points used by the active SVM algorithm

**Table 1.** Comparative results for the Mumbai IRS image in terms of computation time and index-$\beta$

| Method | $n_{labeled}$ | $t_{training}$ (sec) | $\beta$ |
|---|---|---|---|
| active SVM | 259 | 72.02 + (time for labeling 54 pixels) | 6.35 |
| SVM 1 | 198 | 28.15 | 3.45 |
| $k$-means | 0 | 1054.10 | 2.54 |
| SVM 2 | 26214 | 2.44 $\times 10^5$ | 4.72 |

algorithm also provides much lower $\beta$ value compared to the active SVM algorithm. The SVM 2 algorithm uses the labels generated by the $k$-means algorithm, but provides a relatively small improvement in performance compared to $k$-means. The visual quality of the classified images (Fig. 4) also reinforce these conclusion.

Among the supervised classification algorithms, namely, active SVM, SVM 1 and SVM 2, SVM 1 uses the least number of labeled samples and has minimum training time. However, the active SVM algorithm uses only 54 additional labeled points compared to SVM 1 with a substantial improvement in segmentation quality. This is due to the fact that the additional points queried by active SVM were the most informative ones and contributed to the increase in segmentation quality. On the other hand, SVM 2 uses a large sized labeled set, consisting of randomly chosen points, for training. Since, accurate labels for the large training set used were not available, slightly inaccurate labels were used. The overall effect shows that the performance of the SVM 2 algorithm is poorer compared to active SVM inspite of it requiring a much higher computation time.

The variation in segmentation quality (as measured by index-$\beta$) with the number of labeled samples queried by the active SVM algorithm is shown in Fig. 3. It is seen that the initial SVM designed using the training set of SVM 1 provides a $\beta$ value of 3.45 which subsequently increases as more point are queried to a final value of 6.35.

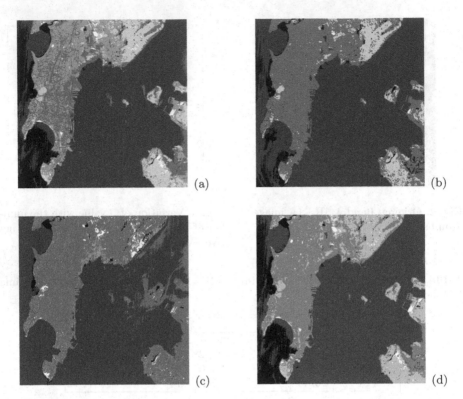

**Fig. 4.** Classified Mumbai image using (a) active SVM, (b) SVM 1, (c) $k$-means, and (d) SVM 2

The main goal of the active learning algorithm is to reduce the requirement of labeled pixels. Hence, an aggressive query strategy is adopted here. However, the aggressive strategy is sensitive to wrong labeling by a human expert, resulting in performance degradation. If in some application, a higher number of labeled pixels, with possibly few wrong labels, are available, a more conservative query strategy will provide better performance.

The next two sections deal with the task of pixel classification (image partitioning) in unsupervised framework. While Section 3 addresses the problem of fuzzy thresholding, Section 4 provides an object extraction method using the principle of rough sets and granular computing.

## 3    Fuzzy Set Theoretic Thresholding

Remotely sensed images are normally poorly illuminated, highly dependent on the environmental conditions, and have very low spatial resolution. Most of the times a scene contains too many objects (or regions), and these regions are ill-defined because of both grayness and spatial ambiguities. Moreover, the gray value assigned to a pixel is the average reflectance of different types of ground covers present in the corresponding pixel area. Assigning unique class labels with certainty is thus a problem for remotely sensed images. Fuzzy set theory provides a way of handling this problem by associating certainty factors with class labels.

Of the two broad approaches, grey level thresholding techniques are usually computationally less expensive as compared to those based on pixel classification. Moreover, in certain remote sensing applications the image is available only in one band, for example, Panchromatic images, for which segmentation based on thresholding seems to be convenient and appropriate. Furthermore, the methods available for image segmentation, like other processing techniques, are problem dependent. When an image is segmented for visual interpretation, it is ultimately up to the viewers to judge its quality for a specific application. The process of evaluation of image quality therefore becomes a subjective one.

In the present investigation we demonstrate the effectiveness of various fuzzy thresholding and clustering techniques along with quantitative evaluation for segmentation of remotely sensed images [172,173]. Their comparison with the respective non-fuzzy techniques is also made both qualitatively and quantitatively. Five different thresholding techniques based on fuzzy and non-fuzzy (probabilistic) entropy, fuzzy geometry and fuzzy correlation, and two clustering (both fuzzy and non-fuzzy) techniques are considered. Some of the algorithms use only global information of input images and the others use local information [21,28,56,58,89,185,186]. The quantitative index used here is based on the concept of homogeneity within a region. Results are demonstrated on the certain bands of both IRS and SPOT satellite images.

Sections 3.1 and 3.2 present an overview of the thresholding methods used, including the basic definitions. In Section 3.3 we present, in brief, fuzzy c-means and hard c-means algorithms and a method to measure the quality of segmented images. Section 3.4 depicts the performance of various algorithms used.

## 3.1  Probabilistic Entropy Based Thresholding

In this section we describe entropy of an image, and a few thresholding algorithms. (Performance of these methods have been compared with those of fuzzy segmentation techniques.)

**Global Entropy of an Image.** Based on the concept of Shannon [187], *entropy of an image* (or its histogram) can be defined as follows. Let $F = [f(p,q)]_{P \times Q}$ be an image of size $P \times Q$, where $f(p,q)$ is the gray value at $(p,q)$; $f(p,q) \in G_L = \{0, 1, ..., i, ..., L-1\}$, the set of gray levels. Let $n_i$ be the frequency of occurrence of the gray level $i$ ($i \in G_L$). Then $\sum_{i=0}^{L-1} n_i = P \times Q = n$ (say). The global entropy of the image is then expressed as

$$H = -\sum_{i=0}^{L-1} p_i \, log_2 \, p_i \; ; \quad p_i = \frac{n_i}{n} . \tag{11}$$

$H$ is called global, as it depends only on the histogram of the image.

The concept of global entropy of an image can be viewed from a different angle also. Instead of considering one probability distribution for the entire image, let us consider two probability distributions, one for the object and the other for the background. The sum of the individual entropy of the object and the background gives the total entropy of the image.

If $S$ is an assumed threshold (i.e., $S$ is the boundary gray value between background and object), then the probability distribution of the gray levels over the background portion of the image (assuming lower gray values correspond to background) is

$$\frac{p_0}{P_S}, \frac{p_1}{P_S}, ...., \frac{p_S}{P_S}, \tag{12}$$

and that of the object portion of the image is

$$\frac{p_{S+1}}{1 - P_S}, \frac{p_{S+2}}{1 - P_S}, ...., \frac{p_{L-1}}{1 - P_S}; \tag{13}$$

where $P_S = \sum_{i=0}^{S} p_i$.

The entropy of the background portion of the image

$$H_{Bg}(S) = -\sum_{i=0}^{S} \frac{p_i}{P_S} log_2(\frac{p_i}{P_S}), \tag{14}$$

and that of the object portion is

$$H_{Obj}(S) = -\sum_{i=S+1}^{L-1} \frac{p_i}{1 - P_S} log_2 \frac{p_i}{1 - P_S}. \tag{15}$$

The total entropy of the image is

$$H_{Tot}(S) = H_{Obj}(S) + H_{Bg}(S). \tag{16}$$

In order to segregate the object regions from the background [21], one needs to maximize $H_{Tot}(S)$ which results in equiprobable gray levels in each region; and thus maximizes the sum of homogeneities in gray levels within object and background. Therefore, the value of $S$ which maximizes $H_{Tot}(S)$ gives the threshold for object and background classification.

**Higher Order Entropy of an Image.** In an image, pixel intensities are not independent of each other. This dependency of pixel intensities can be incorporated by considering sequences of pixels for defining image properties. Entropy of order $r$ ($r = 1, 2, 3...$) of an image was defined [28] based on the concept of sequence of pixels as follows.

Let $p(S_i)$ be the probability of occurrence of a sequence $S_i$ of gray levels of length $r$, where a sequence $S_i$ of length $r$ is defined as a permutation of $r$ gray levels. Let

$$H^{(r)} = -\frac{1}{r} \sum_i p(S_i) \, log_2 \, p(S_i), \tag{17}$$

where the summation is taken over all gray level sequences of length $r$. For different values of $r$ we get various orders of entropy.
If $r = 1$, we get

$$H^{(1)} = -\sum_{i=0}^{L-1} p_i \, log_2 \, p_i, \qquad (18)$$

where $p_i$ is the probability of occurrence of the gray level $i$. Note that Equation (18) is the same as Equation (11) representing the "global entropy" of the image. For $r = 2$

$$H^{(2)} = -\frac{1}{2}\sum_{i=0}^{L-1} p_i(S_i) \, log_2 \, p_i(S_i), \qquad (19)$$

$$= -\frac{1}{2}\sum_{i}\sum_{j} p_{ij} \, log_2 \, p_{ij}, \qquad (20)$$

where $S_i$ is the sequence of gray level of length two and $p_{ij}$ is the probability of co-occurrence of gray levels $i$ and $j$. Thus, $H^{(2)}$ (second order entropy of an image) can be obtained from the co-occurrence matrix (as shown in Fig. 5) of an image.

$H^{(2)}$ takes into account the spatial distribution of gray levels. Therefore, two images, with identical histograms but having different spatial distributions will have the same $H^{(1)}$ value, but different $H^{(2)}$ values. Expressions for higher order entropies for $(r > 2)$ can also be deduced in a similar manner. $H^{(r)}, r \geq 2$, is also called the **local entropy** of order $r$ of an image [28].

**Conditional Entropy:** Suppose an image has two distinct portions, the object $X$ and the background $Y$. Suppose the object consists of the gray levels $\{x_i\}$ and the background contains the gray levels $\{y_i\}$. The conditional entropy of the object $X$ given the background $Y$, i.e., the average amount of information that may be obtained from $X$, given that one has viewed the background $Y$, is defined as

$$H\left(\frac{X}{Y}\right) = -\sum_{x_i \in X}\sum_{y_j \in Y} p\left(\frac{x_i}{y_j}\right) log_2 \, p\left(\frac{x_i}{y_j}\right). \qquad (21)$$

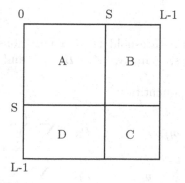

**Fig. 5.** Four quadrants of the co-occurrence matrix

Similarly, the conditional entropy of the background $Y$, given the object $X$, is defined as

$$H\left(\frac{Y}{X}\right) = -\sum_{y_j \in Y}\sum_{x_i \in X} p\left(\frac{y_j}{x_i}\right) log_2\, p\left(\frac{y_j}{x_i}\right). \tag{22}$$

The pixel having gray level $y_j$, in general, can be an $m$th order neighbor of the pixel with gray level $x_i$. Since the estimation of probability of such an occurrence is very difficult, we assume $x_i$ and $y_j$ to be gray levels of adjacent pixel.

The conditional entropy of an image is then defined as [28]

$$H_{Con} = \frac{1}{2}\left[H\left(\frac{X}{Y}\right) + H\left(\frac{Y}{X}\right)\right]. \tag{23}$$

As mentioned before, higher order entropy of an image takes into account the spatial details of it. The way of computing them from the first order (4-neighbors) co-occurrence matrix of the image is described below.

**Strategies for Computing the Probabilities of Co-occurrence.** The frequency of occurrence $(t_{ij})$ of gray level $i$ followed by gray level $j$ is defined as follows:

$$t_{ij} = \sum_{l=1}^{P}\sum_{k=1}^{Q}\delta\;; \tag{24}$$

$$\delta = 1, \quad \text{if}\begin{cases} f(l,k) = i \text{ and } f(l, k+1) = j \\ \qquad\qquad \text{or} \\ f(l,k) = i \text{ and } f(l+1, k) = j; \end{cases}$$

$$\delta = 0, \quad \text{otherwise.}$$

The probability of co-occurrence $(p_{ij})$ of gray levels $i$ and $j$ therefore is

$$p_{ij} = \frac{t_{ij}}{\left(\sum_{g}\sum_{h} t_{g,h}\right)}, \quad 0 \le g, h \le L - 1, \tag{25}$$

where $0 \le p_{ij} \le 1$.

If $S$, $0 \le S \le L - 1$, is a threshold then $S$ partitions the co-occurrence matrix into four quadrants, namely, $A$, $B$, $C$ and $D$ as shown in Fig. 5.

Let us define the following quantities:

$$P_A = \sum_{i=0}^{S}\sum_{j=0}^{S} p_{ij}, \qquad\qquad P_B = \sum_{i=0}^{S}\sum_{j=S+1}^{L-1} p_{ij},$$

$$P_C = \sum_{i=S+1}^{L-1}\sum_{j=S+1}^{L-1} p_{ij}, \qquad P_D = \sum_{i=S+1}^{L-1}\sum_{j=0}^{S} p_{ij}. \tag{26}$$

Then,

$$p_{ij}^A = \frac{p_{ij}}{P_A} = \frac{t_{ij}}{\sum\limits_{g=0}^{S}\sum\limits_{h=0}^{S} t_{gh}}, \qquad p_{ij}^B = \frac{p_{ij}}{P_B} = \frac{t_{ij}}{\sum\limits_{g=0}^{S}\sum\limits_{h=S+1}^{L-1} t_{gh}},$$

$$p_{ij}^C = \frac{p_{ij}}{P_C} = \frac{t_{ij}}{\sum\limits_{g=S+1}^{L-1}\sum\limits_{h=S+1}^{L-1} t_{gh}}, \qquad p_{ij}^D = \frac{p_{ij}}{P_D} = \frac{t_{ij}}{\sum\limits_{g=S+1}^{L-1}\sum\limits_{h=0}^{S} t_{gh}}. \tag{27}$$

The second order local entropy of the background is then defined as (assuming lower gray levels correspond to background)

$$H_A^{(2)}(S) = -\frac{1}{2}\sum_{i=0}^{S}\sum_{j=0}^{S} p_{ij}^A \, log_2 \, p_{ij}^A . \tag{28}$$

Similarly, the second order local entropy of the object region is

$$H_C^{(2)}(S) = -\frac{1}{2}\sum_{i=S+1}^{L-1}\sum_{j=S+1}^{L-1} p_{ij}^C \, log_2 \, p_{ij}^C . \tag{29}$$

Hence the total second order local entropy of object and background can be written as

$$H_W^{(2)}(S) = H_A^{(2)}(S) + H_C^{(2)}(S). \tag{30}$$

The gray level corresponding to the maximum value of $H_W^{(2)}(S)$ gives the threshold for object-background classification. Since $H_W$ considers the entropy within object and background only, it may be called within class entropy.

Similarly, based on transitional (conditional) entropy threshold for object background classification is obtained by maximizing $H_{Con}^{(2)}(S)$ where,

$$H_{Con}^{(2)}(S) = H_B^{(2)}(S) + H_D^{(2)}(S). \tag{31}$$

**Computational Steps**

*Step 1:* Compute the probabilities of occurrence of different gray values or sequences (of order $r, r = 1, 2$) of gray values for each of the assumed thresholds (say, $S$, $S = 1, 2, 3, ...., L - 2$).

*Step 2:* Using the probability values compute the entropy (e.g., global and local) of the image.

*Step 3:* Vary 'S' and select those 'S's for which the entropy values give local maxima.

Each of these local maxima corresponds to a threshold. The global optimum corresponds to object background separation.

## 3.2 Fuzzy Set Theory Based Thresholding

A fuzzy subset $A$ of the universe X is defined as a collection of ordered pairs

$$A = \{(\mu_A(x), x), \ \forall x \in X\}, \tag{32}$$

where $\mu_A(x)$, $(0 \le \mu_A(x) \le 1)$ denotes the degree of belonging of the element $x$ to the fuzzy set $A$. The support of a fuzzy set $A$ is the crisp set that contains all the elements of $X$ that have a non-zero membership value in $A$.

Since the theory of fuzzy sets is a generalization of the classical set theory, it has greater flexibility to capture faithfully the various aspects of incompleteness or imperfection in information of a situation [188]. The flexibility of fuzzy set theory is associated with the elasticity property of the concept of its membership function. The grade of membership is a measure of the compatibility of an object with the imprecise concept represented by a fuzzy set. The higher the value of membership, the lesser will be the amount (or extent) to which the concept represented by a set needs to be stretched to fit an object.

Since the regions in an image are not always crisply defined (imprecision may arise due to both grayness ambiguity and spatial ambiguity), it is natural and appropriate to avoid committing ourselves to a specific hard decision for image segmentation. Thus it is natural to consider the image segments to be fuzzy subsets of the image (first suggested by Prewitt [54]); the subsets being characterized by the possibility (degree) to which each pixel belongs to them. Moreover, (as mentioned is Section 3) for remotely sensed images the pixel intensities do not reflect the land cover types present in the corresponding area properly. Assigning unique class labels with certainty is thus a problem for remotely sensed images. Hence fuzzy set theoretic approach will be more appropriate for segmenting regions in remotely sensed images which are usually ill-defined.

**Fuzzy Entropy of an Image.** An $L$ level image $F(P \times Q)$ can be considered as an array of fuzzy singletons, each having a membership value denoting its degree of possessing some property (e.g., brightness, darkness, edginess, blurredness, texture etc.). In the notation of fuzzy sets one may therefore write [189]

$$F = \{\mu_F(p, q) : p = 1, 2, \ldots, P; \ q = 1, 2, \ldots, Q\};$$

where $\mu_F(p, q)$ denotes the grade of possessing such a property $\mu$ by the $(p, q)$th pixel. Membership can be defined based on global information, local information, positional information and a combination thereof depending on the problem.

Let us construct, say, a fuzzy subset *bright image* characterized by a membership function $\mu_F$ (Fig. 6) using the standard $S(i; a, b, c)$ function of Zadeh [53] defined as

$$
\begin{aligned}
\mu_F(i) &= 0 & \text{if } i \le a \\
&= 2\{(i-a)/(c-a)\}^2 & \text{if } a \le i \le b \\
&= 1 - 2\{(i-c)/(c-a)\}^2 & \text{if } b \le i \le c \\
&= 1 & \text{if } c \le i
\end{aligned}
\tag{33}
$$

where $\mu_F(i)$, which is a function of gray level only, represents the degree of belonging of the level $i$ to the fuzzy bright image plane F. $b = a + c/2$ is the

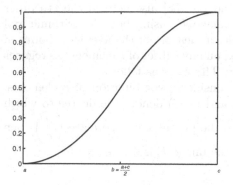

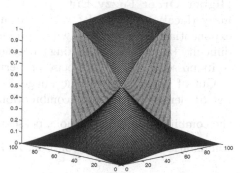

**Fig. 6.** S-type membership function (one dimensional), b is the crossover point

**Fig. 7.** Two dimensional S-type membership function, (50,50) is the crossover point

crossover point (for which the membership value is 0.5) of the membership function $\mu_F$.

Similarly one can use Z-function to construct a *dark image* plane. Where,

$$Z = (1 - S(i; a, b, c)). \tag{34}$$

Several attempts have been made in [59,190,191] to determine the appropriate membership function along with its band width and bounds for image processing problems.

Fuzzy entropy of an image (having $n = P \times Q$ pixels) using logarithmic gain function is [192]

$$H = \frac{1}{n \ln(2)} \sum_{i=1}^{n} S_n(\mu_F(p, q)). \tag{35}$$

Here,

$$S_n(\mu_F(p, q)) = -\mu_F(p, q) \ln\{\mu_F(p, q)\} - \{1 - \mu_F(p, q)\} \ln\{1 - \mu_F(p, q)\}; \tag{36}$$

and $\mu_F(p, q)$ represents the membership for the $(p, q)$th pixel.

Another definition of fuzzy entropy, given in [193,194], using exponential gain function is

$$H = \frac{1}{n(\sqrt{e} - 1)} \sum_{i=1}^{n} \{S_n(\mu_F(p, q)) - 1\} \tag{37}$$

with

$$S_n(\mu_F(p, q)) = \mu_F(p, q)e^{1-\mu_F(p,q)} + \{1 - \mu_F(p, q)\}e^{\mu_F(p,q)}. \tag{38}$$

Note that, these entropy measures, first of all, compute the fuzziness related to individual pixel of the image and then make an average over all the pixels to get a quantification of the amount of average ambiguity, the image possesses. Since their computation depends only on the histogram, they may be called global fuzzy entropy.

**Higher Order Fuzzy Entropy.** Pal and Pal [58] also defined the $r$th order fuzzy (local) entropy $H^{(r)}, r \geq 2$ of an image $F$ (using both logarithmic and exponential gain functions) which provides a measure of the average amount of difficulty (ambiguity) in making a decision on any subset of $r$ elements as regards to its possession of an imprecise property. These are as follows.

Out of the $n$ pixels of the image $F$, consider a combination of $r$ elements. Let $S_i^r$ denote the $i$th such combination and $\mu(S_i^r)$ denote the degree to which the combination $S_i^r$, as a whole, possesses the property $\mu$. There are $\binom{n}{r}$ such combinations. The entropy of order $r$ of the image $F$ is defined as [58]

$$H^{(r)} = \frac{-1}{\binom{n}{r}} \sum_{i=1}^{\binom{n}{r}} [\mu(S_i^r) \, ln\{\mu(S_i^r)\} + \{1 - \mu(S_i^r)\}ln\{1 - \mu(S_i^r)\}] \qquad (39)$$

with logarithmic gain function and

$$H^{(r)} = \frac{1}{\binom{n}{r}} \sum_{i=1}^{\binom{n}{r}} \left[\mu(S_i^r)e^{1-\mu(S_i^r)} + \{1 - \mu(S_i^r)\}e^{\mu(S_i^r)}\right] \qquad (40)$$

with exponential gain function.

$H^{(r)}$ will give a measure of the average amount of difficulty in taking a decision on any subset of size $r$ with respect to the property $\mu$. Note that Equations (35) through (38) correspond to a special case of $H^{(r)}$ for $r = 1$. $H^{(r)}, r \geq 2$ is called higher order fuzzy entropy of the image.

**Membership Function and Computation of Second Order Fuzzy Entropy.** For computing the higher order fuzzy entropy of an image, represented by a fuzzy set, one needs to choose $r$ pixels at a time and to assign a composite membership value to them. Normally these $r$ pixels are chosen as adjacent pixels. For the present investigation, we have chosen $r = 2$.

Let us consider a two dimensional S-type membership function (Fig. 7) representing fuzzy bright image plane (assuming higher gray values correspond to object region). This assigns a composite membership value to a pair of adjacent pixels as follows:

For a particular threshold $S$,

– $(S, S)$ is the most ambiguous point, i.e., the boundary between object and background. Therefore its membership value for the fuzzy bright image plane is 0.5.
– If one object pixel is followed by another object pixel (i.e., the entries of quadrant C), then its degree of belonging to object region is greater than 0.5. The membership value increases with increase in pixel intensity.

- For quadrants B and D where one object pixel is followed by one background pixel or vice versa, the membership value is less than or equal to 0.5, depending on the deviation from the boundary point $(S, S)$.
- If one background pixel is followed by another background pixel (i.e., for the entries in A), then its degree of belonging to object region is less than 0.5. The membership value decreases with decrease of pixel intensity.

Second order fuzzy entropy $(r = 2)$ computed individually over the entries in quadrants C, A, D and B of the co-occurrence matrix (Fig. 5) can be termed as object entropy $H_O^2$, background entropy $H_B^2$, transitional (object to background) $H_{O/B}^2$ entropy, and transitional (background to object) entropy $H_{B/O}^2$, respectively. Therefore, $H_O^2 + H_B^2$ gives the total within class second order local entropy and $H_{(O/B)}^2 + H_{(B/O)}^2$ gives the total second order transitional (conditional) entropy.

**Fuzzy Geometry of an Image Subset.** Entropy, as defined in Section 3.2, may be used in representing grayness ambiguity in an image, i.e., the indefiniteness in making a decision whether an individual pixel is black or white, or a collection of pixels possesses an image property or not. There is another kind of ambiguity in an image called *spatial or geometrical ambiguity* (which refers to indefiniteness in the shape and geometry of regions within the image). These can be represented by fuzzy geometrical properties. Some of them [56,185] which are used in this investigation are described below.

**Fuzzy Geometrical Properties.** Let $\mu$ represent an image fuzzy subset and $\mu$ be piecewise constant (for digital image). Then

**Area.** The area of $\mu$ is

$$a(\mu) = \sum \mu. \tag{41}$$

**Perimeter.** The perimeter of $\mu$ is

$$p(\mu) = \sum_{i=0}^{L-1} \sum_{j=0}^{L-1} |\mu(i) - \mu(j)| \times t_{ij} \tag{42}$$

where $\mu(i)$ and $\mu(j)$ are the membership values of two adjacent pixels having gray value $i$ and $j$, respectively, and $t_{ij}$ is the frequency of occurrence of the gray value $i$ followed by $j$.

**Compactness.** The compactness of a fuzzy set $\mu$ having area $a(\mu)$ and perimeter $p(\mu)$ is defined as

$$\text{comp}(\mu) = \frac{a(\mu)}{p^2(\mu)}. \tag{43}$$

Physically, compactness means the fraction of maximum area (that can be encircled by the perimeter) actually occupied by the fuzzy region (concept) represented by $\mu$. Of all possible fuzzy discs, compactness is minimum for its crisp version.

**Length.** The length of $\mu$ is

$$l(\mu) = \max_p \{\sum_q \mu(p, q)\}. \tag{44}$$

**Breadth.** Similarly, the breadth of $\mu$ is

$$b(\mu) = \max_q \{\sum_p \mu(p, q)\}. \tag{45}$$

The length (breath) of an image fuzzy subset gives its longest expansion in the $y$ direction ($x$ direction). If $\mu$ is crisp, $\mu(p, q) = 0$ or $1$; then length (breadth) is the maximum number of pixels in a column (row).

**Index of Area Coverage (IOAC).** The index of area coverage of $\mu$ is defined as

$$IOAC(\mu) = \frac{a(\mu)}{l(\mu) \times b(\mu)}. \tag{46}$$

*IOAC* of a fuzzy image subset represents the fraction (which may be improper also) of the maximum area (that can be covered by the length and breadth of the image) actually occupied by the image. The IOAC is the minimum for the non-fuzzy case of all possible fuzzy versions of a rectangle.

**Fuzzy Correlation.** Let $\Omega$ be a closed interval in $\mathbb{R}$. Let $\mu_1 : \Omega \longrightarrow [0, 1]$ and $\mu_2 : \Omega \longrightarrow [0, 1]$ be two continuous fuzzy membership functions. The correlation $C(\mu_1, \mu_2)$ between the fuzzy membership functions $\mu_1$ and $\mu_2$ (defined on the same domain) was defined by Murthy *et al.* [195], and Pal and Ghosh [186]. $C(\mu_1, \mu_2)$ basically gives a measure of relation between the natures of $\mu_1$ and $\mu_2$, i.e., with change of $x$ what happens to $\mu_1$ and $\mu_2$.

Now if the functions are discrete in nature (as applicable to digital image), the expression for correlation takes the form [186],

$$C(\mu_1, \mu_2) = \begin{cases} 1 - \frac{4}{X_1 + X_2} \sum_x \{\mu_1(x) - \mu_2(x)\}^2 & \text{if } X_1 + X_2 \neq 0 \\ 1 & \text{if } X_1 + X_2 = 0 \end{cases} \tag{47}$$

where $X_1 = \sum_x \{2\mu_1(x) - 1\}^2$ and $X_2 = \sum_x \{2\mu_2(x) - 1\}^2$.

**Correlation Between Two Fuzzy Representations (Properties) of an Image.** Fuzzy correlation between two representations of an image characterized by $\mu_1$ and $\mu_2$ is defined as [186]

$$C(\mu_1, \mu_2) = 1 - \frac{4}{X_1 + X_2} \sum_p \sum_q \{\mu_1(p, q) - \mu_2(p, q)\}^2 \tag{48}$$

with $X_1 = \sum_p \sum_q \{2\mu_1(p, q) - 1\}^2$ and $X_2 = \sum_p \sum_q \{2\mu_2(p, q) - 1\}^2$ where $\mu_1(p, q)$ and $\mu_2(p, q)$ denote the degree of possessing the property $\mu_1$ and $\mu_2$, respectively, by the $(p, q)$th pixel.

Let $\mu_2$ be the nearest two tone version of $\mu_1$ such that

$$\mu_2(x) = \begin{cases} 0 \text{ if } 0 \le \mu_1(x) \le 0.5 \\ 1 \text{ otherwise.} \end{cases} \tag{49}$$

Let $\mu_1$ (or $F_f$) denote a fuzzy *bright image* plane of $F$ having the crossover point at $S$, say, and be dependent only on gray level. Then $\mu_2$ (or $\overline{F}$) represents its closest two tone version thresholded at $S$. Then the fuzzy correlation between a fuzzy representation of an image and its nearest two tone version is expressed as:

$$C(\mu_1, \mu_2) = 1 - \frac{4}{X_1 + X_2} \left( \sum_{i=0}^{S} \{ [\mu_1(i)]^2 \, h(i) \} + \sum_{i=s+1}^{L-1} \{ [1 - \mu_1(i)]^2 \, h(i) \} \right) \tag{50}$$

with $X_1 = \sum_{i=0}^{L-1} [2\mu_1(i) - 1]^2 \, h(i)$

and $X_2 = \sum_{i=0}^{L-1} [2\mu_2(i) - 1]^2 \, h(i) = \sum_{i=0}^{L-1} h(i) = P \times Q = \text{constant.}$

Here, $h(i)$ is the frequency of the $i$th gray level.

**Correlation Measure Using Local Information.** Higher order correlation using local information (obtained from co-occurrence matrix) can also be defined [186] similarly as in higher order entropy of an image (Section 3.2).

Correlation between any two properties of $F$ computed over the entries in individual quadrants A, C, B & D of the co-occurrence matrix (Fig. 5) can be termed background correlation $[C(\mu_1, \mu_2)_B]$, object correlation $[C(\mu_1, \mu_2)_O]$, transitional (background to object) correlation $[C(\mu_1, \mu_2)_{B/O}]$, and transitional (object to background) $[C(\mu_1, \mu_2)_{O/B}]$ correlation, respectively. They may be computed by using similar expressions with different ranges of $i$ and $j$. For example,

$$C(\mu_1, \mu_2)_B = 1 - \frac{4}{X_1 + X_2} \sum_{i=0}^{S} \sum_{j=0}^{S} \left\{ [\mu_1(i,j) - \mu_2(i,j)]^2 \, t_{ij} \right\}, \tag{51}$$

with $X_1 = \sum_{i=0}^{S} \sum_{j=0}^{S} \{ [2\mu_1(i,j) - 1]^2 \, t_{ij} \}$,     and $X_2 = \sum_{i=0}^{S} \sum_{j=0}^{S} \{ [2\mu_2(i,j) - 1]^2 \, t_{ij} \}$,

$t_{ij}$ is the frequency of occurrences of the gray level $i$ followed by $j$.

Similarly for computing $C(\mu_1, \mu_2)_O$, $i$ and $j$ will range from $S+1$ to $L-1$; for $C(\mu_1, \mu_2)_{B/O}$, $i$ will range from 0 to $S$ and $j$ from $S+1$ to $L-1$; and for $C(\mu_1, \mu_2)_{O/B}$, $i$ will range from $S+1$ to $L-1$ and $j$ from 0 to $S$.

Note that $C(\mu_1, \mu_2)_O + C(\mu_1, \mu_2)_B$ gives the total within class local correlation and $C(\mu_1, \mu_2)_{O/B} + C(\mu_1, \mu_2)_{B/O}$ gives the total transitional (conditional) correlation.

From the properties of correlation we notice that if two functions $\mu_1$ and $\mu_2$ are very close then $C(\mu_1, \mu_2)$ is very high whereas, $C(\mu_1, \mu_2)$ is least when

$\mu_2 = 1 - \mu_1$. Since $\overline{F}$ is the nearest two tone version of $F_f$, $C(F_f, \overline{F})$ gives a measure of closeness of the two images $F_f$ and $\overline{F}$. The principle of maximizing fuzzy correlation for image segmentation is described in [186].

**Computational Steps of Fuzzy Thresholding.** Given an $L$ level image $F$ of dimension $P \times Q$ with minimum and maximum gray values $i_{min}$ and $i_{max}$, respectively, the algorithm for its fuzzy segmentation (through thresholding) into object and background may be described as follows:

*Step 1:* Construct the membership plane using the standard $S(i; a, b, c)$ function (Equations (33)) as

$$\mu(p, q) = \mu(i) = S(i; a, b, c) \tag{52}$$

(called bright image plane if the object regions possess higher gray values)

$$\text{or} \quad \mu(p, q) = \mu(i) = 1 - S(i; a, b, c) \tag{53}$$

(called dark image plane if the object regions possess lower gray values) with crossover point $b$, and a band width $\Delta b = c - a$.

*Step 2:* Compute the parameter $I(F)$ representing either grayness ambiguity or spatial ambiguity (as designated by $H^{(r)}$, correlation, compactness and IOAC, etc.) or both (i.e., product of grayness and spatial ambiguities).

*Step 3:* Vary $b$ between $i_{min}$ and $i_{max}$ and select those $b$ for which $I(F)$ has local minima or maxima depending on $I(F)$. (Maxima correspond for the correlation measure only.) Among the local minima/maxima, let the global one have crossover point at $S$.

The level $S$, therefore, denotes the crossover point of the fuzzy image plane $\mu_F$, which has minimum grayness and/or geometrical ambiguity. The $\mu_F$ plane then can be viewed as a fuzzy segmented version of the image $F$. For the purpose of non-fuzzy segmentation, we can take $S$ as the threshold (or boundary) for classifying or segmenting an image into object and background. In case the image has multiple regions, there will be a set of local optima corresponding to them. Faster methods of computation of the fuzzy parameters have been described by Pal and Ghosh [185].

Note that $w = \Delta b$ is the length of the window (such that $[0, w] \rightarrow [0, 1]$) which was shifted over the entire dynamic range. As $w$ decreases, the $\mu_F$ plane tends to have more intensified contrast around the crossover point, thus resulting in a decrease of ambiguity in $F$. As a result, the possibility of detecting some undesirable thresholds (spurious minima) increases because of the smaller value of $w$. On the other hand, an increase in $w$ results in a higher value of fuzziness and thus leads towards the possibility of losing some of the weak minima.

The criteria regarding the selection of membership function and the length of window (i.e., $w$) have been reported in [191], assuming continuous functions for both histogram and membership function.

## 3.3    Segmentation by Clustering

Clustering is a method of partitioning a given set of patterns into a number of homogeneous groups (clusters) depending on the similarity in features. The number of groups is normally pre-specified; but can also be varying. Initial clusters are normally chosen randomly and gradually modified to obtain the final clusters (or optimal cluster centers). A number of clustering algorithms (both non-fuzzy and fuzzy) exists in the literature [61,196]. In the present investigation we used the hard and fuzzy c-means (HCM and FCM) techniques. Here we describe them in brief.

**Hard c-Means Algorithm.** Let $X = \{x_1, x_2, \ldots, x_k, \ldots, x_n\}$,   $x_k \in I\!R^p$, be a finite data set in the $p$-dimensional space; $x_{kj}$ is the $j$th feature of the data point $x_k$.    Let $c$ $(2 \leq c < n)$ be the number of clusters and $V = (V_1, V_2, \ldots, V_i, \ldots, V_c)$, $V_i \in I\!R^p$ be the set of cluster centers (prototypes). The data set is then classified and cluster centers are updated iteratively until the classification in two successive stages remain unaltered; which can be measured by the average difference between the partitions (prototypes) computed in two successive stages. If this average difference is less than a pre-defined small positive value $\epsilon(> 0)$ then the process can be terminated and the clusters can be taken as optimal. The classification strategy is as follows:
A data point $k$ is assigned to class $i$ if,

$$d_{ik} = \min_{1 \leq s \leq c} (d_{sk}); \tag{54}$$

where $d_{ik}^2 = \sum_{j=1}^{p} (x_{kj} - v_{ij})^2$,    $v_{ij} = \frac{1}{n_i} \sum_{k=1}^{n_i} (x_{kj})$,

and  $n_i$ is the number of data points assigned to $i$th class.

**Fuzzy c-Means Algorithm.** Let $X$, $c$, and $V$ be defined as above and $U = [u_{ik}]_{c \times n}$ be a fuzzy c-partition of $X$. The membership $u_{ik}$ represents the degree of belonging of the pattern $x_k$ to the $i$th class, where $0 \leq u_{ik} \leq 1$, $\sum_{i=1}^{c} u_{ik} = 1$, $\forall k = 1, 2, \ldots, n$ and $0 < \sum_{k=1}^{n} u_{ik} < n$, $\forall i = 1, 2, \ldots, c$.
FCM finds $U$ and $V$ iteratively by minimizing

$$J_m(U, V) = \sum_{i=1}^{c} \sum_{k=1}^{n} (u_{ik})^m \|x_k - V_i\|_A^2 \tag{55}$$

where $m > 1$ and the inner product induced norm metric

$$\| \cdot \|_A^2 = (x_k - V_i)^T A (x_k - V_i) = d_{ik}^2;$$

where $A$ is any $p \times p$ positive definite matrix (in the present study we used the Euclidean norm, i.e., $A$ is an identity matrix). Based on the necessary conditions for a local minimum of $J_m$,   cluster prototypes and memberships are computed as follows:

$$\boldsymbol{V}_i = \frac{\sum\limits_{k=1}^{n}(u_{ik})^m \, \boldsymbol{x}_k}{\sum\limits_{k=1}^{n}(u_{ik})^m}, \quad 1 \le i \le c. \tag{56}$$

Let, $I_k = \{i \mid 1 \le i \le c, d_{ik} = 0\}$, and $\bar{I}_k = \{1, 2, \cdots, c\} - I_k$.
If $I_k \neq \Phi$, assign $u_{ik} = 0 \; \forall \, i \in \bar{I}_k$, and assign arbitrarily $u_{ik}$, such that
$\sum_{i \in I_k} u_{ik} = 1$.

$$\text{If } I_k = \Phi, \; u_{ik} = \left( \sum_{j=1}^{c} \left( \frac{d_{ik}}{d_{jk}} \right)^{\frac{2}{m-1}} \right)^{-1}, \quad 1 \le i \le c, \, 1 \le k \le n. \tag{57}$$

To implement the FCM algorithm we initialize either $U$ or $V$, and then FCM iterates between (56) and (57) until $U$ or $V$ or both stabilize.

**Computational Steps.** Using the above mentioned clustering techniques image pixels are labeled as one of the region types in the image; thereby providing its segmentation. We used a 2-D feature space with feature values as average and busyness of the concerned image pixel. The computational steps are described below.

*Step 1:* Choose the number of classes and the initial values for the means (for FCM an initial partition can also be chosen instead of means of the classes).
*Step 2:* Classify the samples by assigning them to the class corresponding to the closest mean (or assign membership value for each class in case of FCM).
*Step 3:* Re-compute the means of the classes (weighted means for FCM).
*Step 4:* If the change in any of the means is less than some pre-assigned small positive quantity (say, $\epsilon > 0$) then STOP else go to *Step 2*.

Here, we briefly discuss the two features, average and busyness used with FCM and HCM. Let us consider a $3 \times 3$ window centered at $(i, j)$, with gray levels as indicated in Fig. 8.

– The average gray level $\bar{a}$ over the window centered at the $(i, j)$th position (gray value $a_5$) of the image is

$$\bar{a} = \frac{1}{9} \left( \sum_{k=1}^{9} a_k \right). \tag{58}$$

| $a_1$ | $a_2$ | $a_3$ |
|-------|-------|-------|
| $a_4$ | $a_5$ | $a_6$ |
| $a_7$ | $a_8$ | $a_9$ |

**Fig. 8.** Gray values over a $3 \times 3$ window

**Fig. 9.** Enhanced (Histogram equalized) Calcutta (IRS) image

**Fig. 10.** Enhanced (Histogram equalized) Mumbai (IRS) image

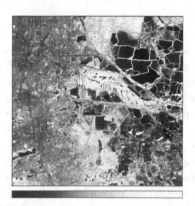

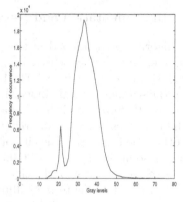

**Fig. 11.** Enhanced (Histogram equalized) Calcutta (SPOT) image

**Fig. 12.** Histogram of the original Calcutta (IRS) image

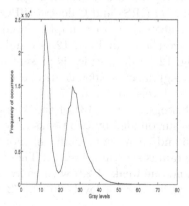

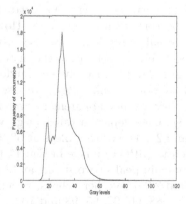

**Fig. 13.** Histogram of the original Mumbai (IRS) image

**Fig. 14.** Histogram of the original Calcutta (SPOT) image

− The busyness $B$ over the window in Fig. 8 is:

$$A_1 = |a_1 - a_2| + |a_2 - a_3| + |a_4 - a_5| + |a_5 - a_6| + |a_7 - a_8| + |a_8 - a_9|,$$

$$A_2 = |a_1 - a_4| + |a_4 - a_7| + |a_2 - a_5| + |a_5 - a_8| + |a_3 - a_6| + |a_6 - a_9|, \text{ and}$$

$$\text{busyness} = B = \frac{A_1 + A_2}{12}. \tag{59}$$

**Quantitative Measure (Index-$\beta$).** We have used several algorithms with different parameters in this investigation. Each of these gives rise to a partition (segmentation) of the image space. We evaluate the segmentation results quantitatively using the index-$\beta$, described in Section 2.3, (Equation (10))[172,173]. This measure gives the homogeneity within the regions and increases with increase in homogeneity for a fixed number of classes (say $c$).

### 3.4   Implementation and Results

The algorithms mentioned in the previous sections have been implemented on a number of remotely sensed images. We present here results on two IRS-1A (band-4) images and one SPOT (band-3) image. The IRS-1A image was taken using the scanner LISS-II (Linear Imaging Self Scanner) in the wavelength range $[0.77\mu m - 0.86\mu m]$ and it has a spatial resolution of $36.25m \times 36.25m$ [101]. One of the images is covering an area around the city of Calcutta (Fig. 9), whereas the other one is covering the city of Mumbai (Fig. 10). The SPOT images have a spatial resolution of $20.0m \times 20.0m$ and a wavelength range $[0.79\mu m - 0.89\mu m]$ [99], covering a portion of the city of Calcutta (Fig. 11). All the three images are of size $512 \times 512$. Due to poor illumination, the actual object classes present in the input image are not visible clearly. For this reason we have not included the original input images, instead an enhanced version (histogram equalized) of the input images highlighting the different object regions are shown in Figs. 9, 10 and 11, corresponding to Calcutta(IRS), Mumbai (IRS) and Calcutta (SPOT), for the convenience of readers. However, the algorithms were implemented on the actual input images whose histograms are shown in Figs. 12, 13 and 14, respectively. As seen from the histogram (Fig. 12), a deep valley is present near the gray level 23, closely surrounded by two significant peaks at gray levels 21 and 33. Some weak valleys are also present at levels 55, 62, 65, 70, 72 and 76, but they are not apparent in Fig. 12. The histogram (Fig. 13) of the Mumbai (IRS) image depicts a valley at gray level 18, surrounded by two peaks at levels 11 and 25. The other valleys at 57, 59, 64, 66 and 70 are not significant. For the Calcutta (SPOT) image in Fig. 11 the histogram is shown in Fig. 14. There are three main peaks (two strong at 18 and 29, and one weak at 22) with valleys, at 20 and 23. The other valleys, which are not visible, are at levels 66, 70, 72, 76, 78, 80, 88, 90, 92, 95, 98 and 103.

For implementing fuzzy c-means (FCM) and hard c-means (HCM) clustering algorithms we have chosen the number of clusters $c = 4, 5$ and   6. Average

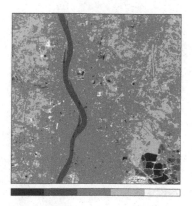

**Fig. 15.** Segmented Calcutta (IRS) image with highest $\beta(=9.949)$ value and $c=5$

**Fig. 16.** Segmented Calcutta (IRS) image with lowest $\beta(=4.357)$ value and $c=5$

**Fig. 17.** Segmented Calcutta (IRS) image using FCM when $c=5$, $\beta=5.880$

**Fig. 18.** Segmented Calcutta (IRS) image using HCM when $c=5$, $\beta=5.171$

(Equation (58)) and busyness (Equation (59)), computed over a $3 \times 3$ neighborhood incorporating local information, are used as features. For FCM, the fuzzifier $m$ (in Equation (55)) was taken as 2.

At first in Tables 2 to 4, we present the thresholds obtained by fuzzy correlation, fuzzy entropy and fuzzy geometry, respectively, for different window sizes ($w = 7, 9, 11, 13, 15, 17, 19$) for Calcutta (IRS) image (Fig. 9). The results corresponding to probabilistic entropy are shown in Table 5. The computed $\beta$ values are also shown in the tables. As expected, the number of thresholds (where * indicates global optima) decreases as $w$ increases. Note that membership functions corresponding to $w = 9$ and 11 satisfy the criteria of bounds[3] of Murthy and Pal [190].

---

[3] Bounds determine the range so that any membership function defined within this range will provide similar segmented results.

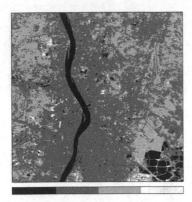

**Fig. 19.** Segmented Calcutta (IRS) image with highest $\beta(= 4.790)$ value and $c = 4$

**Fig. 20.** Segmented Calcutta (IRS) image with highest $\beta(= 2.422)$ value and $c = 2$

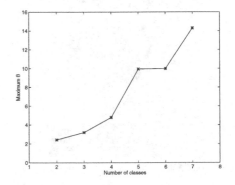

**Fig. 21.** Segmented Calcutta (IRS) image using FCM/HCM when $c = 2$, $\beta = 2.198$

**Fig. 22.** Variation of maximum value of $\beta$ with the number of classes

To demonstrate the significance of the $\beta$ value we provide segmented results corresponding to the highest and lowest $\beta$ values, obtained over all these methods, for a fixed number of regions. Figs. 15 and 16 depict such example images for $\beta = 9.949$ and $\beta = 4.357$, respectively, when the number of regions, $c$, is five. For the purpose of comparison we also show the results of FCM (Fig. 17 with $\beta = 5.880$) and HCM (Fig. 18 with $\beta = 5.171$ ) for $c = 5$. From these, the $\beta$ value is seen to reflect well the quality of segmentation. For example, the details of linear structures like roads, airport runway, water canals are seen to be more prominent in Fig. 15 (having higher $\beta$ value) as compared to others.

Let us now consider the segmented output (Fig. 19) corresponding to the best $\beta$ value $(= 4.790)$ when the number of regions, $c$, is four. Comparing the results of Fig. 15 (when $c = 5$) we see that some of the roads, canals (linear structures) which are visible in Fig. 15 are not present in Fig. 19. Also it can be seen from

**Table 2.** Thresholds based on fuzzy correlation

| Serial no. | Kind of Information | Window size $w$ | Threshold values | $\beta$ value |
|---|---|---|---|---|
| 1.1 | Global | 7 | 19*, 23, 31, 36, 39, 47 | 14.286 |
| 1.2 | | 9 | 24*, 31, 36, 47 | 9.634 |
| 1.3 | | 11 | 24*, 31, 37, 47 | 9.949 |
| 1.4 | | 13 | 24*, 32, 38, 48 | 9.612 |
| 1.5 | | 15 | 24*, 48 | 1.711 |
| 1.6 | | 17 | 24, 49* | 1.681 |
| 1.7 | | 19 | 24* | 1.466 |
| 1.8 | Local | 7 | 19, 24*, 32, 36, 47 | 9.533 |
| 1.9 | (within class) | 9 | 19, 24*, 32, 37, 48 | 9.989 |
| 1.10 | | 11 | 25*, 37, 48 | 4.486 |
| 1.11 | | 13 | 25*, 37 | 3.762 |
| 1.12 | | 15 | 25*, 36 | 4.011 |
| 1.13 | | 17 | 25*, 36 | 4.011 |
| 1.14 | | 19 | 25*, 36 | 4.011 |
| 1.15 | Local | 7 | 19, 24, 36, 40, 48* | 5.983 |
| 1.16 | (transitional) | 9 | 19, 24, 36, 48* | 4.894 |
| 1.17 | | 11 | 24, 36, 48* | 4.790 |
| 1.18 | | 13 | 24, 37, 48* | 4.358 |
| 1.19 | | 15 | 24*, 38 | 3.311 |
| 1.20 | | 17 | 24*, 38 | 3.311 |
| 1.21 | | 19 | 38* | 1.900 |

* indicates the global maximum.

Fig. 19 that the main city area consisting of dense concrete structure is merged with the sparse concrete structure class (habitation class) around the river.

Figs. 20 and 21 show a comparison of object-background partition (i.e., $c = 2$) as obtained by the thresholding with highest $\beta$ ($= 2.422$) value, and HCM/FCM ($\beta = 2.198$). Both visually and by $\beta$ values the segmented version in Fig. 20 is seen to be superior to that of Fig. 21.

Fig. 22 demonstrates the variation of maximum value of $\beta$, obtained over different methods, for $c = 2, 3, 4, 5, 6, 7$. As expected (mentioned in Section 2.3), the $\beta$ value increases with increment of $c$.

Let us now consider Fig. 20, which has highest $\beta$ value among those with $c = 2$, for object-background classification. Here the threshold is seen to be 32, distinguishing object region containing water body and city area from the background containing habitation, vegetation and open spaces. Interestingly, it is seen from the correlation based segmentation results (Table 2) that the segmented images with a threshold at or around 32 usually posses high $\beta$ value. This indicates the significance of the said threshold.

From Tables 2 and 3 it is seen that the thresholds corresponding to global correlation and global entropy measures for $\omega = 9$ and 11 are almost identical. (Note that thresholds 24, 31, 37 and 47 correspond to the boundary between water body, city area, habitation, vegetation and open spaces). Similar is the case with transitional correlation (Table 2) and transitional entropy measures (Table 3). Further, the thresholds obtained by two gain functions are seen to be similar for both global and local entropy.

In the case of IOAC and Compactness (Table 4), interestingly, only one threshold in the range 30-32 is obtained irrespective of the window size. Although the method produced only one threshold, its importance is evident from Fig. 20

**Table 3.** Thresholds based on fuzzy entropy

| Serial no. | Kind of Information | Form of gain function | Window size $w$ | Threshold values | $\beta$ value |
|---|---|---|---|---|---|
| 2.1 | Global | logarithmic | 7 | 19, 23, 31*, 36,47 | 9.793 |
| 2.2 | | | 9 | 24, 31*,37,47 | 9.949 |
| 2.3 | | | 11 | 24, 31*, 37, 47 | 9.949 |
| 2.4 | | | 13 | 24, 38*,48 | 3.788 |
| 2.5 | | | 15 | 23*, 49 | 1.639 |
| 2.6 | | | 17 | 23* | 1.432 |
| 2.7 | | | 19 | - | - |
| 2.8 | Global | exponential | 7 | 19, 23, 31*, 36,47 | 9.793 |
| 2.9 | | | 9 | 24, 31*,37,47 | 9.949 |
| 2.10 | | | 11 | 24, 31*, 37, 47 | 9.949 |
| 2.11 | | | 13 | 24, 38*, 48 | 3.788 |
| 2.12 | | | 15 | 24*, 49 | 1.681 |
| 2.13 | | | 17 | 23* | 1.432 |
| 2.14 | | | 19 | - | - |
| 2.15 | Local (within class) | logarithmic | 7 | 18, 24, 31*, 36, 49 | 9.473 |
| 2.16 | | | 9 | 25, 37*,48 | 4.486 |
| 2.17 | | | 11 | 25, 37*, 47 | 4.531 |
| 2.18 | | | 13 | 26, 36* | 4.146 |
| 2.19 | | | 15 | 26, 35* | 4.202 |
| 2.20 | | | 17 | 26, 36* | 4.146 |
| 2.21 | | | 19 | 26, 36* | 4.146 |
| 2.22 | Local (within class) | exponential | 7 | 18, 24, 31*, 36,49 | 9.473 |
| 2.23 | | | 9 | 25, 37*, 48 | 4.486 |
| 2.24 | | | 11 | 25, 37*, 47 | 4.531 |
| 2.25 | | | 13 | 25, 36* | 4.011 |
| 2.26 | | | 15 | 26, 36* | 4.146 |
| 2.27 | | | 17 | 25, 36* | 4.011 |
| 2.28 | | | 19 | 26, 36* | 4.146 |
| 2.29 | Local (transitional) | logarithmic | 7 | 19*, 24, 37, 40, 49 | 4.933 |
| 2.30 | | | 9 | 19*, 24, 37, 49 | 4.357 |
| 2.31 | | | 11 | 24, 37*, 49 | 4.274 |
| 2.32 | | | 13 | 24, 38*, 48 | 3.788 |
| 2.33 | | | 15 | 24, 31* | 2.981 |
| 2.34 | | | 17 | 23, 31* | 2.963 |
| 2.35 | | | 19 | 30* | 2.173 |
| 2.36 | Local (transitional) | exponential | 7 | 19*, 24, 37, 40, 48 | 4.945 |
| 2.37 | | | 9 | 19*, 24, 37, 49 | 4.357 |
| 2.38 | | | 11 | 24, 37*, 47 | 4.358 |
| 2.39 | | | 13 | 24, 38*, 49 | 3.764 |
| 2.40 | | | 15 | 24* | 1.466 |
| 2.41 | | | 17 | 23, 31* | 2.963 |
| 2.42 | | | 19 | 31* | 2.319 |

\* indicates the global minimum.

**Table 4.** Thresholds based on fuzzy geometry

| Serial no. | Optimizing property | Window size $w$ | Threshold values | $\beta$ value |
|---|---|---|---|---|
| 3.1 | Compactness | 7 | 32* | 2.422 |
| 3.2 | | 9 | 32* | 2.422 |
| 3.3 | | 11 | 32* | 2.422 |
| 3.4 | | 13 | 32* | 2.422 |
| 3.5 | | 15 | 32* | 2.422 |
| 3.6 | | 17 | 32* | 2.422 |
| 3.7 | | 19 | 31* | 2.319 |
| 3.8 | IOAC | 7 | 31* | 2.319 |
| 3.9 | | 9 | 30* | 2.173 |
| 3.10 | | 11 | 30* | 2.173 |
| 3.11 | | 13 | 31* | 2.319 |
| 3.12 | | 15 | 31* | 2.319 |
| 3.13 | | 17 | 31* | 2.319 |
| 3.14 | | 19 | 31* | 2.319 |

\* indicates the global minimum.

where it is seen to clearly demarcate the silhouettes of the objects present in the input image. This may be due to the incorporation of spatial ambiguities in the optimizing function. Note that this important threshold is missing in all the cases of Table 5.

**Table 5.** Thresholds based on probabilistic entropy

| Serial no. | Kind of Information | Form of gain function | Threshold values | $\beta$ value |
|------------|---------------------|-----------------------|------------------|---------------|
| 4.1 | Global | Logarithmic | 20, 26* | 1.603 |
| 4.2 | Global | Exponential | 19, 26, 34* | 4.208 |
| 4.3 | Local (within class) | Logarithmic | 20* | 1.149 |
| 4.4 | Local (transitional) | Logarithmic | 19, 24* | 1.475 |
| 4.5 | Local (within class) | Exponential | 19, 35* | 2.712 |
| 4.6 | Local (transitional) | Exponential | 19, 23*, 37 | 3.579 |

* indicates the global maximum.

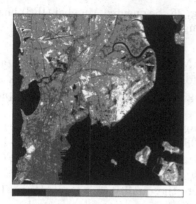    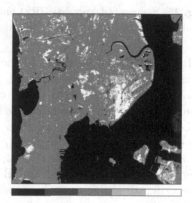

**Fig. 23.** Segmented Mumbai (IRS) image with highest $\beta(= 18.308)$ value and $c = 5$    **Fig. 24.** Segmented Mumbai (IRS) image with lowest $\beta(= 11.243)$ value and $c = 5$

From the previous discussion we therefore see that $\beta$ values provide a good quantitative index for measuring homogeneity in segmented regions. For a fixed $c$, its values increases with the quality of segmentation. Fuzzy set theoretic approaches are better than the probabilistic entropic methods. This may be due to the fact that fuzzy approaches described here exploit the ambiguities (both in grayness and in spatial domain) of the image in an effective way. Among the fuzzy techniques, fuzzy geometry based optimization (which basically optimizes the spatial ambiguities) is seen to provide a single threshold, over a wide variation of window size; which is able to segregate the basic structures in the image well. Surprisingly, the global information based fuzzy correlation and fuzzy entropy measures provided better performance (higher $\beta$ value) for extracting the structures. Among the local information based techniques, within class fuzzy correlation based algorithms, showed an upper edge. From all the results obtained by thresholding algorithms, it therefore appears that fuzzy correlation and fuzzy entropy using global information (with $w = 9$ and 11) are the best optimizing criteria from the point of possessing $\beta$ value and detecting the structural details (Fig. 15).

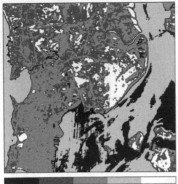
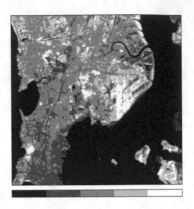
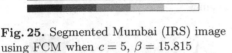

**Fig. 25.** Segmented Mumbai (IRS) image using FCM when $c = 5$, $\beta = 15.815$

**Fig. 26.** Segmented Mumbai (IRS) image using HCM when $c = 5$, $\beta = 16.056$

It may be noted here that in the case of Mumbai (IRS) image (Fig. 10) and Calcutta (SPOT) image (Fig. 11), the performances of various techniques are similar to that of Calcutta (IRS) image (Fig. 9). Therefore some of the typical results on these images are reported here.

Let us now consider the Mumbai IRS image (Fig. 10). Figs. 23 and 24 depict the segmented images for highest ($\beta = 18.308$) and lowest ($\beta = 11.243$) $\beta$ values for $c = 5$. For the purpose of comparison we also consider the results of FCM (Fig. 25 with $\beta = 15.815$) and HCM (Fig. 26 with $\beta = 16.056$) for $c = 5$. Like Calcutta IRS image, the $\beta$ value reflects the quality of segmentation. The details of linear structures (rail roads, airport runway etc.) are seen to be more prominent in Fig. 23 (with highest $\beta$ value) than others.

As in the case of Calcutta IRS image, IOAC and compactness resulted in a single threshold for all the cases irrespective of $\omega$. However, the threshold obtained by IOAC is different from that of compactness. Among the different segmented regions for $c = 2$, IOAC has highest $\beta$ value. Here the threshold at 19 is seen to discriminate the water body well from the land.

Finally, we consider the Calcutta SPOT image (Fig. 11). Figs. 27 and 28 depict the segmented versions for highest ($\beta = 9.375$) and lowest ($\beta = 6.467$) values of $\beta$ obtained over all the thresholding methods for $c = 5$. Comparing them with FCM (Fig. 29, $\beta = 6.388$) and HCM (Fig. 30, $\beta = 6.676$) we see that the details of structures are more prominent in Fig. 27. This also strengthens the fact that $\beta$ values provide a good quantitative measure of image segmentation.

# 4  Object Extraction Using Granular Computing and Rough Sets

Sections 2 and 3 dealt with image classification (supervised) and segmentation (unsupervised) methodologies based on support vector machine and fuzzy set theory, respectively. Here we explain the significance of rough sets [197] in image segmentation problem and the role of granular computing.

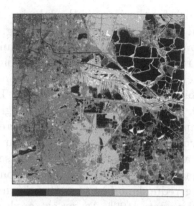 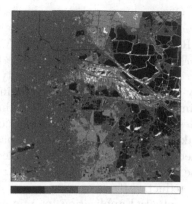

**Fig. 27.** Segmented Calcutta (SPOT) image with highest $\beta(=\ 9.375)$ value and $c=5$

**Fig. 28.** Segmented Calcutta (SPOT) image with lowest $\beta(=\ 6.467)$ value and $c=5$

 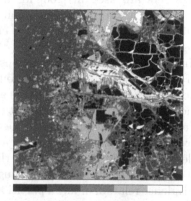

**Fig. 29.** Segmented Calcutta (SPOT) image using FCM when $c=5$, $\beta=6.388$

**Fig. 30.** Segmented Calcutta (SPOT) image using HCM when $c=5$, $\beta=6.676$

It has been argued, both from philosophical and theoretical points of views, that information granulation is essential to human problem solving, and hence has very significant impact on the design and implementation of intelligent system. Zadeh [78] identified three basic concepts that underlie the process of human cognition, namely, granulation, organization, and causation. "Granulation involves decomposition of whole into parts, organization involves integration of parts into whole, and causation involves association of causes and effects".

A granule is a clump of objects (points), in the universe of discourse, drawn together by indistinguishability, similarity, proximity, or functionality. Granulation leads to information compression/summarization. In situations involving incomplete, uncertain, or vague information, it may be difficult to differentiate different elements and instead it is convenient to consider granules, i.e., clump or group of indiscernible elements, for performing operations. Granular computing

(GrC) may be regarded as a unified framework for theories, methodologies and techniques that make use of such granules in the process of problem solving.

Recently, rough set theory [197] has become a popular mathematical framework for granular computing. The focus of rough set theory is on the ambiguity caused by limited discernibility of objects in the domain of discourse. Its key concepts are those of object 'indiscernibility' and 'set approximation'. The primary use of rough set theory has so far mainly been in generating logical rules for classification and prediction [198] using information granules; thereby making it a prospective tool for pattern recognition, image processing, feature selection, data mining and knowledge discovery process from large data sets. Use of rough set rules based on reducts has significant role for dimensionality reduction/feature selection by discarding redundant features; thereby having potential application for mining large data sets [199]. As far as rough set theoretic image processing is concerned, there is hardly any investigation reported so far. However, in related areas like pattern analysis/clustering mention may be made of the studies of Wojcik [200], and Pal and Mitra [126]. While rough sets are used in [200] for describing image features for analysis, they are used in a part in [126] for initializing EM algorithm in conjunction with minimal spanning tree (MST) for clustering with application to multi spectral images.

In the present section, we demonstrate an application of rough sets and granular computing for object extraction from gray scale image. In gray scale images boundaries between object regions are often ill-defined. This uncertainty can be handled by describing the different objects as rough sets with upper (outer) and lower (inner) approximations. The set approximation capability of rough sets is exploited in the present investigation [174] to formulate an entropy measure, called rough entropy, quantifying the uncertainty in an object-background image. This has been done by defining an image as a collection of pixels and the equivalence relation induced partition as pixels lying within each non-overlapping window over the image. With this definition the roughness of various transforms (or partitions) of the image can be computed using image granules, i.e., windows, of different sizes.

Maximization of the said rough entropy measure minimizes the uncertainty arising from vagueness of the boundary region of the object. Therefore, for a given granule size, the threshold for object-background classification can be obtained through its maximization with respect to different image partitions. A guideline for selecting the appropriate granule size from gray level distribution is given, as well as a way of computing the rough entropy efficiently only in one pass (or scan) of the image. Effectiveness of the method is demonstrated on different kinds of remotely sensed images.

The section is organized as follows: Basic definitions of rough sets, description of an image as a rough set, and a definition of rough entropy are provided in Section 4.1. The problem of object extraction minimizing roughness is addressed in Section 4.2. Results of the experiments are provided in Section 4.5.

## 4.1  Rough Entropy Measure of an Image

**Rough Sets.** Let $\mathcal{A} = < U, A >$ be an information system, and let $B \subseteq A$ and $X \subseteq U$. We can approximate the set $X$ using only the information contained in $B$ by constructing the lower and upper approximations of $X$. If $X \subseteq U$, the sets $\{\mathbf{x} \in U : [\mathbf{x}]_B \subseteq X\}$ and $\{\mathbf{x} \in U : [\mathbf{x}]_B \cap X \neq \emptyset\}$, where $[\mathbf{x}]_B$ denotes the equivalence class of the object $\mathbf{x} \in U$ relative to $I_B$ (the equivalence relation), are called the $B$-*lower* and $B$-*upper approximations* of $X$ in $U$. They are denoted by $\underline{B}X$ and $\overline{B}X$, respectively [197]. The objects in $\underline{B}X$ can be certainly classified as members of $X$ on the basis of knowledge in $B$, while objects in $\overline{B}X$ can only be classified as possible members of $X$ on the basis of $B$. These are illustrated in Fig. 31 where the sets of dark-gray granules represent lower approximation, while those of both dark-gray and light-gray granules together denote upper approximation. Therefore, a rough set is nothing but a crisp set with rough representation.

The roughness of a set $X$ with respect to $B$ can be characterized numerically [197] as $R_\alpha = 1 - \frac{|\underline{B}X|}{|\overline{B}X|}$. This means if roughness of the set $X$ is 0 then $X$ is crisp with respect to $B$, and if $R_\alpha > 0$ then $X$ is rough (i.e., X is vague with respect to $B$). For details one may refer to Pawlak [197], Skowron and Rauszer [198] and Komorouski *et al.* [199].

**Image as a Rough Set.** In gray scale images boundaries between object regions are often ill defined because of grayness and/or spatial ambiguities [201]. This uncertainty can be handled by describing the different objects as rough sets with upper (or outer) and lower (or inner) approximations. Here the concepts of upper and lower approximation can be viewed, respectively, as outer and inner approximations of an image region in terms of granules.

Let the universe $U$ be an image consisting of a collection of pixels. Then if we partition $U$ into a collection of non-overlapping windows (of size $m \times n$, say), each window can be considered as a granule $G$. In other words, the induced equivalence classes $I_{m \times n}$ have $m \times n$ pixels in each non-overlapping window.

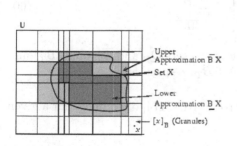

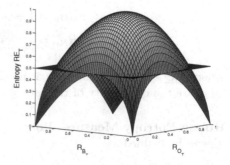

**Fig. 31.** Rough representation of a set with upper and lower approximations

**Fig. 32.** Plot of rough entropy for various values of roughness of the object and background

Given this granulation, object regions in the image can be approximated by rough sets [174].

Let us consider an object background separation (a two class) problem of an $M \times N$, L level image. Let prop(B) and prop(O) represent two properties (say, gray level intervals $0, 1, \cdots, T$ and $T+1, T+2, \cdots, L-1$) that characterize background and object regions, respectively. Given this framework, object and background can be viewed as two sets with their rough representation as follows:

The inner approximation of the object ($\underline{O}_T$):

$$\underline{O}_T = \left\{ \bigcup_i G_i \mid P_j > T, \forall j = 1, \cdots mn, \text{ and } P_j \text{ is a pixel belonging to } G_i \right\}$$

Outer approximation of the object ($\overline{O}_T$):

$$\overline{O}_T = \left\{ \bigcup_i G_i, \exists j, j = 1 \cdots mn \ s.t. \ P_j > T, \text{ where } P_j \text{ is a pixel in } G_i \right\}$$

Inner approximation of the background ($\underline{B}_T$):

$$\underline{B}_T = \left\{ \bigcup_i G_i \mid P_j \leq T, \forall j = 1, \cdots mn, \text{ and } P_j \text{ is a pixel belonging to } G_i \right\}$$

Outer approximation of the background ($\overline{B}_T$):

$$\overline{B}_T = \left\{ \bigcup_i G_i, \exists j, j = 1 \cdots mn \ s.t. \ P_j \leq T, \text{ where } P_j \text{ is a pixel in } G_i \right\}$$

Therefore, the rough set representation of the image (i.e, object $O_T$ and background $B_T$) for a given $I_{m \times n}$ depends on the value of T.

Let the roughness of object $O_T$ and background $B_T$ be defined as

$$\begin{aligned} R_{O_T} &= 1 - \frac{|\underline{O}_T|}{|\overline{O}_T|} = \frac{|\overline{O}_T| - |\underline{O}_T|}{|\overline{O}_T|} \\ R_{B_T} &= 1 - \frac{|\underline{B}_T|}{|\overline{B}_T|} = \frac{|\overline{B}_T| - |\underline{B}_T|}{|\overline{B}_T|} \end{aligned} \tag{60}$$

where $|\underline{O}_T|$ and $|\overline{O}_T|$ are the cardinality of the sets $\underline{O}_T$ and $\overline{O}_T$, and $|\underline{B}_T|$ and $|\overline{B}_T|$ are the cardinality of the sets $\underline{B}_T$ and $\overline{B}_T$, respectively.

**Rough Entropy Measure.** Rough Entropy (RE) [174] of an image for a given $T$, can be defined as

$$RE_T = -\frac{e}{2}[R_{O_T} \ log_e(R_{O_T}) + R_{B_T} log_e(R_{B_T})]. \tag{61}$$

Its plot for various values of $R_{O_T}$ and $R_{B_T}$ is shown in Fig. 32.

(1) The value of $RE_T$ lies between 0 and 1.

(2) $RE_T$ has a maximum value of unity when $R_{O_T} = R_{B_T} = 1/e$, and minimum value of zero when $R_{O_T}, R_{B_T} \in \{0,1\}$.

(3) (a) Since the boundary pixels are common for both object and background, we have $\overline{O}_T - \underline{O}_T = \overline{B}_T - \underline{B}_T = Q_T$, say. Therefore,

$$R_{O_T} = R_{B_T}, \; \textit{iff} \; |\underline{O}_T| = |\underline{B}_T|.$$

Under this condition, the distribution of $RE_T$ on the diagonal (joining (0,0) and (1,1)) is shown in Fig. 33, where $RE_T$ attains a maximum value of unity at $R_{O_T} = R_{B_T} = 1/e$.

(b) When $|\underline{O}_T| < |\underline{B}_T|$, then $R_{O_T} > R_{B_T}$ and $|\underline{O}_T| > |\underline{B}_T|$, then $R_{O_T} < R_{B_T}$.

In either case, $RE_T$ will decrease from its maximum value of unity and will reach a value of zero at $(0,0), (0,1), (1,0)$ and $(1,1)$ in the $(R_{O_T}, R_{B_T})$ plane (Fig. 32).

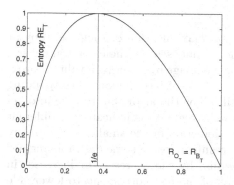

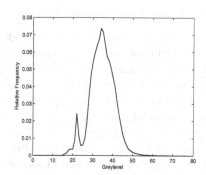

**Fig. 33.** Plot of rough entropy for the values (0,0) to (1,1) on the diagonal of Fig. 32 (i.e., when $R_{O_T} = R_{B_T}$)

**Fig. 34.** Histogram of the Calcutta (IRS, NIR) image, minimum estimated base width is 8 between 16 and 24 graylevel

## 4.2  Object Extraction by Minimizing Roughness

Let us describe a method of object enhancement/extraction based on the principle of minimizing the roughness of both object and background regions, i.e., maximizing $RE_T$. As explained in Section 4.1, one can compute for every $T$ the $RE_T$ of the image, representing the background and object regions $(0, \cdots, T)$ and $(T + 1, \cdots L - 1)$, respectively, and select the one for which $RE_T$ is maximum. In other words, select

$$T^* = \arg \max_T RE_T, \tag{62}$$

as the optimum threshold to provide the object background segmentation. Note that maximizing the rough entropy to get the required threshold basically implies minimizing both the object *roughness* and background *roughness*.

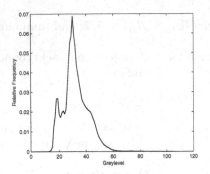

**Fig. 35.** Histogram of the Calcutta (SPOT, NIR) image, minimum estimated base width is 6 between 14 and 20 graylevel

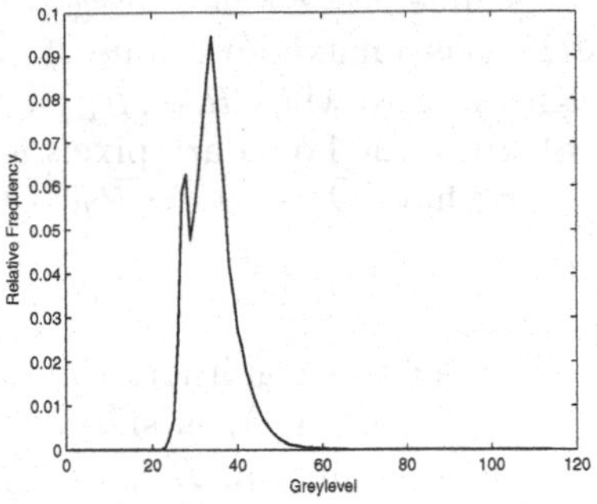

**Fig. 36.** Histogram of the Calcutta (SPOT, PAN) image, minimum estimated base width is 5 between 24 and 29 graylevel

## 4.3    Choice of Granule Size

As can be seen, the determination of $T^*$ by maximization of rough entropy or minimization of roughness depends on the granule size. A choice of granule size can be made from gray level distribution of the image by selecting a value approximately equal to the minimum of half the width of base regions corresponding to all the peaks in the histogram. This will allow the algorithm to take into account the local information (details) of all the regions, as indicated by different peaks in the histogram, and facilitate the detection of the smallest region. Any granule larger (or smaller) than this may result in losing some desirable regions (or detection of spurious undesirable regions) by the decrease (or increase) in the value of $T^*$, assuming that the regions of interest correspond to lower side of the histogram. The details of the selection procedure are shown in Section 4.5 with histograms of different images.

## 4.4    Algorithm for Threshold Selection

Following is the algorithm for efficient implementation of the aforesaid methodology for selecting $T^*$:

Let $max\_gray$ and $min\_gray$ be the maximum and minimum gray level values of the image, respectively. Let $granule_i$ represent a window of $m \times n$ pixels. Let total number of granules be $total\_no\_granule$.

Initialize: Four integer arrays namely $object\_lower$, $object\_upper$, $background\_lower$, $background\_upper$ each of size $(max\_gray - min\_gray + 1)$ to zero.

Step 1: for $i = 1$ to $total\_no\_granule$
$max\_granule_i$ = maximum gray value of pixels in $granule_i$
$min\_granule_i$ = minimum gray value of pixels in $granule_i$
(a)  for $max\_granule_i \leq j \leq max\_gray$
$object\_lower(j) = object\_lower(j) + 1$

(b) for $min\_granule_i \leq j \leq max\_gray$
   $object\_upper(j) = object\_upper(j)+1$
(c) for $min\_gray \leq j \leq min\_granule_i$
   $background\_lower(j)= background\_lower(j)+1$
(d) for $min\_gray \leq j \leq max\_granule_i$
   $background\_upper(j)= background\_upper(j)+1$

Step 2: for $l = min\_gray$ to $max\_gray$
   $object\_roughness(l) = 1 - [ object\_lower(l)/object\_upper(l)]$
   $background\_roughness(l) = 1-[background\_lower(l)/background\_upper(l)]$
   $Rough\_entropy(l) = - \lceil\frac{e}{2}\rceil \times [object\_roughness(l) \log_e (object\_roughness(l))$
   $\qquad\qquad + background\_roughness(l) \log_e (background\_roughness(l))]$

Step 3: $Threshold$(optimal) $= \arg \max_{l} [rough\_entropy(l)]$.    ◇

Remark: Given the max_gray and min_gray values, the computation of rough entropy (and hence the algorithm) requires only a single scan of pixels in the image, since $max\_granule_i$ and $min\_granule_i$ are computed exactly once for each $i$. Therefore the computational complexity of the algorithm is same as that of histogram computation (i.e., algorithm has a linear complexity in term of size of an image).

## 4.5  Experimental Results

The effectiveness of the methodology for object extraction based on rough entropy (Equation (61)) is demonstrated on three different types of satellite images taken over some parts of Calcutta City. The first one is from Indian Remote Sensing (IRS) Satellite, whereas the second and third ones are from Satellite Pour d'Observation de la Terre (SPOT). The first and second correspond to a single band of multispectral (MSS) images in the Near-Infrared (NIR) region of spectrum and the third one is a Panchromatic (PAN) image, covering a large range of the visible region of spectrum. Also the spatial resolution and dynamic range (another representation of maximum gray level) for the images are quite different, the details are given in Table 6.

The Calcutta (IRS, NIR) image is covering an area around the city of Calcutta (Fig. 37(a)). The SPOT images using HRV multispectral, and HRV Panchromatic (High Resolution Visible) cover some portions of the extended Calcutta City. These areas mainly represent water bodies and vegetation. Each of these images is of size 512 × 512.

**Table 6.** Details of the satellite images used with rough entropy[99,101]

| Image Name | Satellite | Imaging Instrument | Wavelength ($\mu m$) | Spatial Resolution (metres) | Dynamic Range (bits) |
|---|---|---|---|---|---|
| Calcutta (IRS, NIR) | IRS | LISS-II$^m$ | $0.77 - 0.86$ | $36.25 \times 36.25$ | 7 |
| Calcutta (SPOT, NIR) | SPOT | HRV$^m$ | $0.79 - 0.89$ | $20 \times 20$ | 8 |
| Calcutta (SPOT, PAN) | SPOT | HRV$^p$ | $0.51 - 0.73$ | $10 \times 10$ | 8 |

m = multispectral mode
p = panchromatic mode

It may be noted here that all these images are having very poor illumination, the actual object classes present in the input images are not visible clearly. This can be seen in the histograms provided in Figs. 34, 35 and 36. Therefore an enhanced version of each of the input images highlighting the different object regions is shown in Figs. 37(a), 38 (a) and 39(a), for convenience.

At first we discuss about the selection of the granule size (window size) for the computation of rough entropy on these images. The histograms (Figs. 34, 35 and 36 ) of all these images are almost bimodal. For the Calcutta (IRS, NIR) image (Fig. 34), the base width of the two regions are approximately 8 and 30, and therefore according to the criterion in Section 4.3, $4 \times 4$ (half of the smaller base width) is the choice of the granule size. Similarly, for the Calcutta SPOT (MSS) image it is $3 \times 3$, because the base widths are 6 and 22. The SPOT Panchromatic image has base widths 5 and 12, therefore the choice again is $3 \times 3$.

Note that all these images (Figs. 37(a), 38(a) and 39(a)) have many more object regions. However, considering it as a two class (object and background) problem, different segmented results are shown in Figs. 37, 38 and 39. For

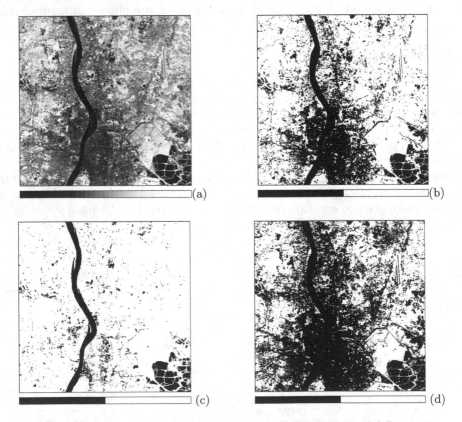

**Fig. 37.** Thresholds by maximizing rough entropy using granule of different sizes on Calcutta (IRS, NIR) image: (a) original, (b) threshold = 30, granule size = $4 \times 4$, (c) threshold = 26, granule size = $6 \times 6$, and (d) threshold = 33, granule size = $2 \times 2$

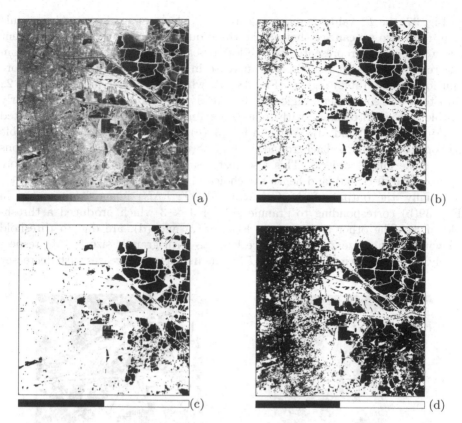

**Fig. 38.** Thresholds by maximizing rough entropy using granule of different sizes on Calcutta (SPOT, NIR) image: (a) original, (b) threshold = 25, granule size = 3 × 3, (c) threshold = 22, granule size = 6 × 6, and (d) threshold = 30, granule size = 2 × 2

Calcutta(IRS, NIR) image the output shown in Fig. 37(b) corresponds to granule size of 4 × 4 which produced a threshold of 30. The other two outputs, Figs. 37(c) & (d), are due to threshold 33 with granule size 2 × 2 and threshold 26 with granule size 6 × 6, respectively. It may be noted here that the value of $T^*$ increases/ decreases with decrease/ increase in granule size. Here we can see that while all the three output images are able to segment the water bodies (represented by the lower peak region in the histogram) from the rest of the objects, increase in $T^*$ value to 33 introduces more spurious (undesirable) regions (Fig. 37(d)), whereas decrease in $T^*$ value to 26 fails to detect some useful regions (e.g., airport runways, roads, canals) as object (Fig. 37(c)). This justifies the selection of 30 as the more appropriate threshold, and hence the choice of granule size 4 × 4.

Similarly, for the Calcutta (SPOT, NIR) image, the output shown in Fig. 38(b) corresponds to granule size of 3 × 3 which produced a threshold of 25. The other two outputs, Figs. 38(c) & (d), are due to threshold 30 with granule size 2 × 2 and threshold 22 with granule size 6 × 6, respectively. Like the previous image in Fig. 37, the $T^*$ value increases/ decreases with decrease/ increase in granule size. Here also we can see that all the three output images are able to segment the water bodies (represented by the lower peak region in the histogram) from the rest of the object regions. The increase in $T^*$ value to 30 introduces more spurious (undesirable) regions (Fig. 38(d)), whereas decrease in $T^*$ value to 22 fails to detect some useful regions. Particularly, the water canal which is clearly visible in Fig. 38(b) along with some other linear structures, is not segmented well in Figs. 38(c) and (d). While, in Fig. 38(c) some of these objects are unable to come out from the background, in Fig. 38(d) too many undesirable regions have come out as objects. This also justifies the selection of 25 as the more appropriate threshold, and hence the choice of granule size $3 \times 3$.

Finally, the output of the Calcutta (SPOT, PAN) image is presented in Fig. 39(b) corresponding to granule size of $3 \times 3$ which produced a threshold of 31. The other two outputs, Figs. 39(c) and (d), are due to threshold 32 with granule size $2 \times 2$ and threshold 28 with granule size $5 \times 5$, respectively. Like the previous images, the $T^*$ value increases/ decreases with decrease/

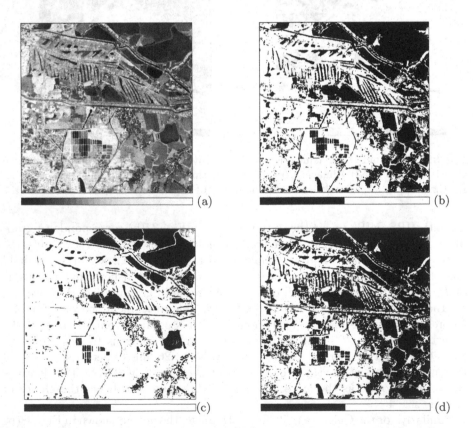

**Fig. 39.** Thresholds by maximizing rough entropy using granule of different sizes on Calcutta (SPOT, PAN) image: (a) original, (b) threshold = 31, granule size = $3 \times 3$, (c) threshold = 28, granule size = $5 \times 5$, and (d) threshold = 32, granule size = $2 \times 2$

increase in granule size. Here also we can see that all the three output images are able to segment the water bodies (represented by the lower peak region in the histogram) from the rest of the objects. Increase in $T^*$ value to 31 introduces more spurious (undesirable) regions (Fig. 39(d)), whereas decrease in $T^*$ value to 28 fails to detect some useful regions, particularly some structures. These are evident from the point of visibility of various curvilinear objects and regions like, water lakes, which came out nicely in Fig. 39(b). Thus the results here also justify the selection of 31 as the more appropriate threshold, and hence the choice of granule size $3 \times 3$.

## 5   Gray Level Based Hough Transform for Region Extraction

While Sections 3 and 4 concern with the problem of segmenting an image using gray level thresholding into homogeneous regions, the present one deals with the said problem in terms of homogeneous line segments.

Hough transform (HT) [5,202] is used for finding straight lines or analytic curves in a binary image. There are several ways in which Hough Transform (HT) for straight lines can be formulated and implemented. Risse [203] has listed some of these forms and analyzed them alongwith their complexities. The most popular one is $(\rho, \theta)$ form, which is given by Duda and Hart [204]. This parametric form specifies a straight line in terms of the angle $\theta$ (with the abscissa) of its normal and its algebraic distance $\rho$ from the origin. The equation of such a line in x-y plane is,

$$x \cos \theta + y \sin \theta = \rho \tag{63}$$

where $\theta$ is restricted to the interval $[0, \pi]$. Some of the advantages of this parametric form are its simplicity and ease in implementation.

Hough transforms are applied usually on binary images. Hence, one needs to convert, initially, the gray level image to a binary one (through thresholding, edge detection, thinning etc.) to apply Hough Transform (HT). Note that, in the process of binarization, some information regarding line segments in the image may get lost. Thus, it becomes appropriate and necessary to find a way of making Hough Transform (HT) applicable directly on gray level images.

In this section, we present a technique [175] for extracting homogeneous regions of arbitrary shape and size in a gray level image based on Hough transform. The regions are defined in terms of homogeneous line segments. The technique includes some operations which are performed in a window to obtain homogeneous line segments. For every quantized $(\rho, \theta)$ cell in the Hough space ($\rho$ represents perpendicular distance of the straight line from the origin, $\theta$ represents angle made by normal to the straight line with the x-axis), the variance of the pixel intensities contributing to the $(\rho, \theta)$ cell is computed. The cells whose variances are less than a pre-specified threshold, are found. Each such cell would represent a homogeneous line segment in the image. The window is then moved over the entire image, so as to result in an output consisting of only the homogeneous line segments; thereby constituting different homogeneous regions. The performance of the method has been demonstrated on Indian Remote Sensing (IRS) Satellite images for different parameter values.

In this connection we mention the methods of gray scale Hough transform (GSHT) of Lo and Tsai [205], fuzzy Hough transform (FHT) of Basak and Pal [206], and generalized Hough transform (GHT) of Ballard [207]. GSHT enables one to find thick lines (called bands) from gray scale images. Therefore, it can be used for detecting road like structures only in remote-sensing images. FHT can handle the impreciseness/ ill-definedness in shape description. GHT is able to extract arbitrary shapes from the edge map of a gray level image using prototype information of the objects to be extracted. Note that our method does not need this information and is thus able to extract objects of irregular shapes and arbitrary sizes as found in remote sensing images.

The rest of this section is organized as follows: Section 5.1 provides the definition and formulation of the region. A Strategies of region extraction in gray level image using Hough transform is described in Section 5.2. Algorithm and implementation are described in Section 5.3. The Section 5.4 presents results on the remote sensing images.

## 5.1   Definition and Formulation

A region in a gray level image can be viewed as a union of several line segments, so that it consists of a connected set of pixels having low gray level variation. Therefore, to extract a region, we need to define a line segment in gray level image. A line segment in a gray level image is defined using two threshold parameters, minimum length of the line ($l$) and maximum variation of the line ($v$). The mathematical formulation of the region in terms of line segments is stated below.

**Def. 1:** A pixel $P = (i, j)$ is said to fall on a line segment joining the pixels $P_1 = (i_1, j_1)$ and $P_2 = (i_2, j_2)$   if $\exists \ \lambda_0, \ \ 0 \leq \lambda_0 \leq 1$ such that

$$\lambda_0 \, P_1 + (1 - \lambda_0) \, P_2 = P.$$

**Def. 2:** A collection of pixels $L(l, v)$ is said to be a line segment in a gray level image if,

> there exist $P_1 = (i_1, j_1)$ and $P_2 = (i_2, j_2)$, $P_1 \neq P_2$;   such that
> $L(l, v) = \{ \ P : P$ is a pixel falling on the line segment joining $P_1$ and $P_2 \ \}$,
> - The number of pixels in $L(l, v)$ is $\geq l$, and
> - The variance of the gray values of pixels in $L(l, v) \leq v$.

Let $\mathcal{A}_{l,v} = \{L(l, v) : L(l, v)$ is a line segment in the image$\}$. That is $\mathcal{A}_{l,v}$ represents the collection of all line segments in the image.

**Def. 3.1:** A pixel $P$ is said to be in the homogeneous region if $P \in L(l, v)$, at least for one $L(l, v) \in \mathcal{A}_{l,v}$.

**Def. 3.2:** A pixel $P$ is said to be in the non-homogeneous region if $P \notin L(l, v)$, $\forall \ L(l, v) \in \mathcal{A}_{l,v}$.

Let $N_H = \{P : P$ is a pixel in the non-homogeneous region $\}$.

The region $R$ is defined as follows.

**Def. 4.1:** Let $L_1(l,v)$ and $L_2(l,v) \in \mathcal{A}_{l,v}$. Then $L_1(l,v)$ and $L_2(l,v)$ are said to be directly connected if either $L_1(l,v) \cap L_2(l,v) \neq \phi$ or $\exists$ pixels $P_1 \in L_1(l,v)$ and $P_2 \in L_2(l,v)$ such that $P_1$ is one of the eight neighbors of $P_2$. Note that a line segment $L(l,v)$ is directly connected to itself.

**Def. 4.2:** Two line segments $L_\alpha(l,v)$ and $L_\beta(l,v)$ belonging to $\mathcal{A}_{l,v}$ are said to be connected if they are directly connected or there exist $L_i(l,v) \in \mathcal{A}_{l,v}$ ; $i = 1,2,...,k$ where $k \geq 3$ such that $L_i(l,v)$ and $L_{i+1}(l,v)$ are directly connected $\forall i = 1,2,...,(k-1)$
where $L_1(l,v) = L_\alpha(l,v)$ and $L_k(l,v) = L_\beta(l,v)$.

**Def. 4.3:** Let $B_{L_\alpha}(l,v) = \{L(l,v) : L(l,v) \in \mathcal{A}_{l,v}, \text{ and } L(l,v) \text{ and } L_\alpha(l,v)$ are connected $\}$, $L_\alpha(l,v) \in \mathcal{A}_{l,v}$.
Note that for $L_\alpha(l,v), L_\beta(l,v) \in \mathcal{A}_{l,v}$ either
$B_{L_\alpha}(l,v) = B_{L_\beta}(l,v)$ or $B_{L_\alpha}(l,v) \cap B_{L_\beta}(l,v) = \phi$
Note also that $\bigcup\limits_{L_\alpha(l,v) \in \mathcal{A}_{l,v}} B_{L_\alpha}(l,v) = \mathcal{A}_{l,v}$.

That is $\mathcal{A}_{l,v}$ is partitioned into finitely many sets using $B_{L_\alpha}(l,v)$'s.

**Def. 4.4:** Let $R_{L_\alpha}(l,v) = \cup_{L(l,v) \in B_{L_\alpha}(l,v)} L(l,v)$. Then $R_{L_\alpha}(l,v)$ is said to be a region generated by $L_\alpha(l,v)$. Note that $R_{L_\alpha}(l,v)$ is a set consisting of pixels in the given image. Observe also that the same region can be generated by different line segments (follows from Def. 4.3)

## Observations on the Above Definitions

(a) A line segment may be termed as a *region* according to the above definitions.

(b) The variance of $n$ points $x_1, x_2, ..., x_n$, is given by $\frac{1}{n}\sum(x_i - \bar{x})^2$ where $\bar{x} = \frac{1}{n}\sum_{i=1}^{n} x_i$. Now, variance $< v \implies \sum(x_i - \bar{x})^2 < nv$.

Observe that there may be a point (say, $x'$) among $x_1, x_2, ..., x_n$, such that $(x' - \bar{x})^2$ may be high (say, $(x' - \bar{x})^2 > kv$, where $k > 1$). Even then $\sum(x_i - \bar{x})^2$ can still be less than $nv$. This observation indicates the removal of noise upto some extent by the proposed variance based definition of line segment.

(c) The values for $l$ and $v$ are to be chosen "appropriately" to obtain the actual regions in an image. Some portions of the actual regions may not be obtained if the value of $l$ is high. If the value of $v$ is high, the number of pixels detected to constitute various regions increases, and therefore the possibility of some spurious collection of pixels being termed as region increases. Reducing the value of $v$, on the other hand, decreases the number of detected pixels; thereby increasing the possibility of losing some actual regions. Similarly if the value of $l$ is low then some unwanted regions may arise.

(d) Observe that a pixel $P$ does not fall into any region $\implies$ There exists no line segment passing through $P$ with low gray level variation (i.e. variance of the gray level values of the pixels on any line segment passing through $P$ is greater than $v$). Hence the above stated definitions intend to suppress the pixels in the non-homogeneous region of the image.

(e) Note that the two adjacent collection of connected pixel, each collection having different average gray value, may fall into the same region; thereby losing their identity according to the definition. (However, in such cases, a further processing may be necessary to separate (partition) the said collection.) ♠

The definitions regarding *regions* have been stated above. There may exist several other ways of obtaining the said regions from the image. Note that regions have been defined as a union of line segments, and Hough transform is a standard method of obtaining line segments. But the Hough transform finds line segments in a binary image. In order to make Hough Transform (HT) applicable directly on a gray level image, we formulate a method in the next section which is able to find line segments (and hence the regions) in a gray level image.

## 5.2    Strategies of Region Extraction in Gray Level Image Using Hough Transform

*Extraction of line segments [175]:*

(i) Consider the equation for a straight line to be $x \cos \theta + y \sin \theta = \rho$. Apply suitable sampling on $\theta$ and $\rho$, and construct the Hough accumulator. Transform each point of the image (pixel) using different values of $\theta$ (and its corresponding $\rho$ values). Note that a point in the image space is mapped to more than one cell in the Hough space and each of these cells represents a line in the image space.

(ii) Compute, for each cell, the length of the corresponding line ($\ell$) as the total number of image points (pixels) mapped into that cell, i.e., the cell count. Variance of the said pixel values may be termed as the variance ($V$) of the corresponding line.

(iii) For a cell in the Hough space if the length of the line is less than $l$ or variance of the line is greater than $v$ then suppress the cell in the Hough space.

(iv) Remap this Hough space (containing unsuppressed cells) to image space. This process of remapping preserves all those pixels which are not suppressed in at least one of the cells of the Hough space. Since this transformation preserves only the location of pixels, not their gray values, they may be restored from the original image.

Let us consider an image of size $M \times M$. If $M$ is too large compared to the threshold parameter $l$, then $\ell$ values (cell counts) will be larger, and as a result, variance of the gray levels on a line may exceed the threshold value $v$. Many genuine line segments, therefore, may not be detected in such a case. To avoid this, the search process for obtaining the line segments is to be conducted locally. That is, a window of size $\omega \times \omega$ needs to be moved over the entire image to search for line segments. Here $\omega$ may be taken as, $2l > \omega \geq l$, because $\omega \geq 2l$ may still lead to the suppression of actual lines in the image.

*Extraction of Regions [175]:*

One can clearly see that the above mentioned process extracts line segments which are connected according to the *Definition 4*. Therefore the collection of these line segments will result in regions of different sizes and shapes.

*Note:*

- There is no restriction on the shape of the "region" thus obtained. The only restriction, we used on the size of the region (i.e., length of the line $\geq l$), is a weak one.
- The method does not need any prior representation of the shape of region to be detected. Therefore, it can extract regions of arbitrary shape and size.

## 5.3  Algorithm and Implementation

It has been mentioned in the earlier section that values for $l$ (length of the line), $v$ (variance threshold) and $\omega$ (window size) are to be selected and the obtained lines are to be remapped to image domain to procure regions. This process of remapping the cells in the Hough space is to be carried out on every window. The steps of the entire algorithm are stated below.

Step 1:  For a window ($\omega \times \omega$) of the image obtain the Hough accumulator values for different $\rho$ and $\theta$. $\rho$ values and $\theta$ values are sampled suitably in their respective domains. For each cell in Hough space, mean and variance of corresponding pixel values in the image domain are computed using two more accumulators (one for sum of gray values $\sum x$ and one for sum of squares of gray values $\sum x^2$). These sum and sum of squares along with the count of cells ($\ell$) will be used for computing the variance $V = \frac{\sum x^2}{\ell} - \left(\frac{\sum x}{\ell}\right)^2$ of the cell. If the cell count is $< l$ then replace the cell count by zero (i.e., the cell is suppressed). If $V > v$ then also replace the cell count by zero. The cells with count non-zero are remapped to image domain preserving the position of window.

Step 2:  Repeat Step 1 for all possible windows of size $\omega \times \omega$ in the image.

Step 3:  Restore the gray values of remapped pixels from the original image.

The number of computations in the aforesaid algorithm can be reduced drastically in the following way.

(i) Note that for a given window size $\omega$, Hough transformation of the pixel does not change with its location, because the reference frame for computing $\theta$ (and hence $\rho$) remains the same. Thus, the Hough accumulator values can be calculated only once and be used for every position of the window.

(ii) Again, all possible windows of size $\omega \times \omega$ need not be considered. The window can be moved by half of its size, both horizontally and vertically. This process, though marginally reduces the accuracy of the regions obtained, decreases computations drastically.

**(iii)** Keeping the point (iv) of Section 5.2 in mind, the process of restoring gray values in the aforesaid Step 3 can be combined with Step 1 by preserving only those pixels in the image domain which are not getting suppressed in at least one cell of the Hough space.

## 5.4    Experimental Results

We have applied the proposed method on IRS images to demonstrate its usefulness. The IRS images considered here have spatial resolution of 36.25m × 36.25m, wavelength range $0.77\mu m$ - $0.86\mu m$ and gray level values in the range 0-127. Size of the images is $512 \times 512$. An enhanced (linearly stretched) image is provided in Fig. 40(a) (city of Calcutta) and Fig. 41(a) (city of Mumbai) for the convenience of readers, since the original images are poorly illuminated. However the method has been implemented on the original images.

In the present investigation, we have used $w = 16$ and $l = 14$ for various values of $v$. The output corresponding to Fig. 40(a) and 41(a) for $v = 0.2$, 0.4 and 0.6 are shown in Figs. 40(b), 40(c) and 40(d), and Figs. 41(b), 41(c)

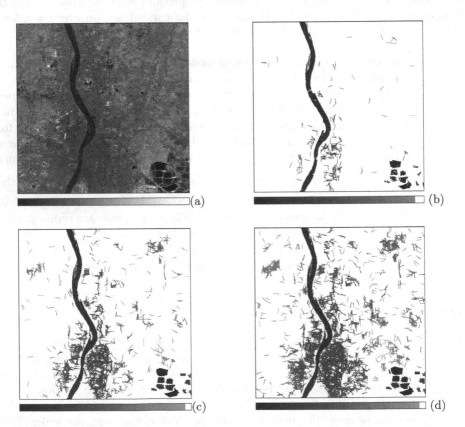

**Fig. 40.** Input and output Calcutta IRS images: (a) Input image, (b) Output with $v$ = 0.2, (c) Output with $v = 0.4$, and (d) Output with $v = 0.6$

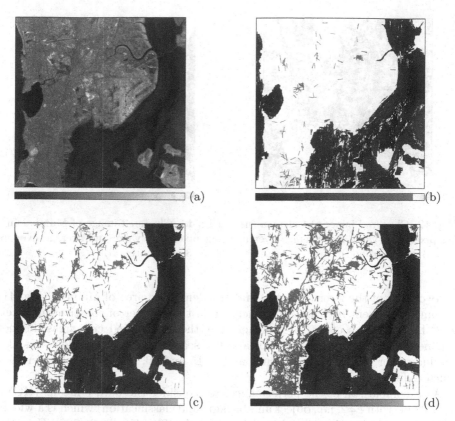

**Fig. 41.** Input and output Mumbai IRS images: (a) Input image, (b) Output with $v$ = 0.2, (c) Output with $v = 0.4$, and (d) Output with $v = 0.6$

and 41(d), respectively. As stated in Step 3 of the Algorithm *(Section 5.3)* the extracted regions have their original gray values restored. Justification of the result (output image) is evident from the observations laid down in Section 5.1 and under the note in Section 5.2. For example, as $v$ increases, the number and size of the detected regions are seen to increase. (Note that the set of pixels in Fig. 40(b)(41(b)) is a subset of that of Fig. 40(c)(41(c)), and that in Fig. 40(c)(41(c)) is a subset of Fig. 40(d)(41(d)). Similarly, decreasing $v$ enables one to detect tiny homogeneous regions as separate classes, even when they are embedded in a different wide homogeneous region. That is why the two bridges over the river Ganges in Calcutta image (Fig. 40(b)) and a bridge on the Arabian sea (Thane creek) in Mumbai image (Fig. 41(b)) became prominent as separate regions for $v = 0.2$. For $v = 0.4$ and 0.6, they disappeared in Fig. 40 and became faint in Fig. 41. (This makes "the evidence", for the existence of a bridge, available for a further stage of the vision process to recognize or deal with.) In other words, as $v$ increases the Ganges in Calcutta image comes out as a single region and the sea in Mumbai image becomes smoother. Similarly, note from the lower parts of Figs. 40 and 41 that the two dense city areas (namely, Howrah and Calcutta)

**Fig. 42.** Clustered image of Calcutta (using pixel gray value and average over its 3 × 3 neighborhood)

**Fig. 43.** Clustered image of Mumbai (using pixel gray value and average over its 3 × 3 neighborhood)

on two sides of the river (Fig. 40), and the dense city area of Mumbai (Fig. 41) become prominent because of the increase in the number of constituting pixels with higher values of $v$. Note further that these extracted city areas did not get merged with the river (Fig. 40) and the sea (Fig. 41). Experiment was also conducted for other values of $l$ such as $l = 15$ and 16, but the results were not much different.

As a comparison of the performance, we consider the hard c-means (HCM) algorithm (with $c=2$, i.e., object and background classification) which is a widely used segmentation algorithm based on pixel classification [89,99,208]. Here the input features are considered to be the gray level value of the pixel and the average value over its 3 × 3 neighborhood. (This method with $c = 2$ is chosen for comparison because this also provides object background classification like ours.) From the segmented output (Figs. 42 and 43) one may note that the algorithm failed to isolate the respective city areas from the river and sea. It also could not, unlike Fig. 40(b) and Fig. 41(b), enhance the bridge regions.

To restrict the size of the article, we have presented the results corresponding to $v = 0.2$, 0.4 and 0.6 for $\omega = 16$ and $l = 14$ only, although the experiment was also conducted for other values of $l$ (e.g., 15 and 16) and $\omega$ (e.g., 8, 24 and 32). Variance of pixel values is used here as a measure of homogeneity of line segment. One may use any other homogeneity measure for this purpose.

## 6   Conclusions and Scope of Future Research

In Section 2, a new method known as active support vector learning algorithm for supervised pixel classification in remote sensing images is presented. The goal is to minimize the number of labeled points required to design the classifier. The algorithm uses an initial set of small number of labeled pixels to design a crude classifier, which is subsequently refined by using more number of points obtained by querying from a pool of unlabeled pixels. It is seen that the number of labeled points required by the active learning algorithm is far less compared to the conventional support vector machine. It also provides better accuracy compared to completely unsupervised segmentation algorithms or a supervised algorithm having access to only partially accurate class labels of a large number of pixels. An index-$\beta$ to evaluate the classification/ segmentation in terms of homogeneity of regions is introduced. For a given $k$ (number of classes), the higher the value of $\beta$, the better is the homogeneity within the classified/ segmented regions.

Fuzzy thresholding provides a useful segmentation technique for remote sensing images. This is the objective of Section 3. Among the various thresholding techniques fuzzy correlation provided the best performance, followed by fuzzy entropy as far as $\beta$ value and the detection of various land cover types are concerned. FCM provided superior performance to HCM. The time requirement of a thresholding techniques considered here, on an average, is seen to be of the order of $1/100^{th}$ of HCM and $1/1000^{th}$ of FCM. Significance of the index-$\beta$ in evaluating the segmentation is also demonstrated here successfully.

Rough entropy of an image is defined using the concept of image granules in Section 4. Based on this measure, a method of extracting object regions from an image is described by minimizing both object and background roughness. Here granules carry local information and reflect the inherent spatial relation of the image by treating pixels of a window as indiscernible or homogeneous. Maximization of homogeneity in both object and background regions during their partitioning is achieved through maximization of rough entropy; thereby providing optimum results for object background classification. The guideline described for the choice of granule size from gray level distribution is seen to be appropriate. Note that as far as rough set theoretic image processing is concerned, there is hardly any investigation reported so far. In this context the aforesaid investigation is unique.

A method of extracting regions in a gray level image using the principle of Hough transform has been described in Section 5. A definition of "region" in terms of homogeneous line segments (instead of clusters) is provided. Since the methodology does not involve any parametric form or representation in terms of template of the shape of regions, it has the ability to detect regions of any arbitrary shape and size.

Although the methods described here performed well for the test images, they need further investigation for wide utility.

The active learning strategy of Section 2 adopted the method of queries for the most interesting/ambiguous unlabeled point as measured by its distance from the current separating hyper-surface. Other query strategies based on version space splitting, logistic regression may be used in the future study. Also, besides active learning, other semi-supervised learning techniques, and co-training may help in circumventing the problem arising from the scarcity of labeled data in remote sensing image analysis.

Fuzzy thresholding described in Section 3 provided a useful segmentation technique for single band remotely sensed images. However, there is no way to apply these thresholding techniques to a multi-feature images, like color and multispectral images. On the other hand, one can segment the images in different bands and combine them to take a decision based on the multifeatured images. Therefore, extending the fuzzy thresholding methods to multispectral data sets, can be an interesting area to explore.

The method of segmentation by rough entropy as described in Section 4 divides the image only into two regions (i.e., object and background). Extension of the algorithm to multi-class segmentation problem is therefore natural for their wide applications and constitute a part of further investigation. Moreover, the sensitivity of the method to noise needs to be investigated. The use of granular computing for image segmentation using other rough set methods (e.g., granular clustering) can also be attempted. It may be mentioned here that there exist some definitions of rough entropy useful for other applications. For example, the one defined by Beaubouef *et al.* [209] is applicable to relational database and the one of Düntsch and Gediga [210] for optimal granulation and feature selection.

The investigation using Hough transform (Section 5) needs two parameters, one is length and the other is the variation $v$. An investigation for detecting automatically $v$ will constitute a part of further research. A strategy for using this method for multispectral image needs to be formulated to make it more useful for real life applications. A comparatively fast method may be developed for finding the homogeneous line segments; thereby making the overall algorithm computationally inexpensive.

## Acknowledgements

The article is based on the Doctoral dissertation of the author which was written under the supervision of Prof. Sankar K. Pal, Director, Indian Statistical Institute. The author would like to sincerely thank *Prof. C. A. Murthy, Dr. Ashish Ghosh*, and *Dr. Pabitra Mitra*, co-authors of different articles contributing the thesis work. This work was partially supported by *Department of Science and Technology, Government of India* by sponsoring a project titled "Advanced Techniques for Remote Sensing Image Processing".

## References

1. Osteaux, M., Meirleir, K.D., Shahabpour, M., eds.: Magnetic Resonance Imaging and Spectroscopy in sports medicine. Springer, Berlin, Germany (1991)
2. Gonzalez, R.C., Woods, R.E.: Digital Image Processing. 2nd edn. Pearson Education, Inc., Singapore (2002)
3. Hall, E.L.: Computer Image Processing and Recognition. Academic Press, New York, USA (1979)
4. Marr, D.: Vision. Freeman, San Fransicsco, USA (1982)
5. Rosenfeld, A., Kak, A.C.: Digital Picture Processing. volumes I & II. Academic Press, New York (1982)
6. Horowitz, S.L., Pavlidis, T.: Picture segmentation by directed split and merge procedure. In: Proc. 2nd Int. Joint Conf. Pattern Recognition. (1974) 424–433

7. Perez, A., Gonzalez, R.C.: An iterative thresholding algorithm for image segmentation. IEEE Trans. Pattern Analysis and Machine Intelligence **9** (1987) 742–751

8. Pal, N.R., Pal, S.K.: Image model, poisson distribution and object extraction. Int. J. Pattern Recognition and Artificial Intelligence **5** (1991) 459–483

9. Besl, P.J., Jain, R.C.: Segmentation through variable order surface fitting. IEEE Trans. Pattern Analysis and Machine Intelligence **10** (1988) 167–192

10. Taxt, T., Flynn, P.J., Jain, A.K.: Segmentation of document images. IEEE Trans. Pattern Analysis and Machine Intelligence **11** (1989) 1322–1329

11. Chow, C., Kaneko, T.: Automatic boundary detection of the left ventricle from cineangiograms. Computers and Biomedical Research **5** (1972) 388–341

12. Nakagawa, Y., A.Rosenfeld: Some experiments on variable thresholding. Pattern Recognition **11** (1979) 191–204

13. Yanowitz, S.D., Bruckstein, A.M.: A new method for image segmentation. Computer Vision, Graphics and Image Processing **46** (1989) 82–95

14. Mardia, K.V., Hainsworth, T.J.: A spatial thresholding method for image segmentation. IEEE Trans. Pattern Analysis and Machine Intelligence **10** (1988) 919–927

15. Ridler, T., Calvard, S.: Picture thresholding using an iterative selection method. IEEE Trans. System, Man and Cybernetics **8** (1978) 630–632

16. Lloyd, D.E.: Automatic target classification using moment invariants of image shapes. Report RAE IDN AW126, Farnborough, UK (1985)

17. Otsu, N.: A threshold selection method from grey-level histograms. IEEE Trans. Systems, Man and Cybernetics **9** (1979) 62–66

18. Kittler, J., Illingworth, J.: Minimum error thresholding. Pattern Recognition **19** (1986) 41–47

19. Pal, N.R., Bhandari, D.: On object-background classification. Int. J. Systems Science **23** (1992) 1903–1920

20. Pun, T.: A new method for gray level picture thresholding using the entropy of the histogram. Signal Processing **2** (1980) 223–237

21. Kapur, J.N., Sahoo, P.K., Wong, A.K.C.: A new method for gray level picture thresholding using the entropy of histogram. Computer Vision, Graphics and Image Processing **29** (1985) 273–285

22. Wong, A.K.C., Sahoo, P.K.: A gray level threshold selection method based on maximum entropy principle. IEEE Trans. Systems, Man and Cybernetics **19** (1989) 866–871

23. Levine, M.D., Nazif, A.M.: Dynamic measurement of computer generated image segmentation. IEEE Trans. Pattern Analysis and Machine Intelligence **7** (1985) 155–164

24. Weszka, J.S., Rosenfeld, A.: Threshold evaluation techniques. IEEE Trans. Systems, Man and Cybernetics **8** (1978) 622–629

25. Deravi, F., Pal, S.K.: Gray level thresholding using second order statistics. Pattern Recognition Letters **1** (1983) 417–422

26. Chanda, B., Chaudhuri, B.B., Majumder, D.D.: On image enhancement and threshold selection using the gray level co-occurrence matrix. Pattern Recognition Letters **3** (1985) 243–251

27. Pal, S.K., Pal, N.R.: Segmentation based on measures of contrast, homogeneity, and region size. IEEE Trans. Systems, Man and Cybernetics **17** (1987) 857–868

28. Pal, N.R., Pal, S.K.: Entropic thresholding. Signal Processing **106** (1989) 97–108

29. Abutaleb, A.S.: Automatic thresholding of gray level pictures using two-dimensional entropy. Computer Vision, Graphics and Image Processing **47** (1989) 22–32

30. Peleg, S.: A new probabilistic relaxation scheme. IEEE Trans. Pattern Analysis and Machine Intelligence **2** (1980) 362–369
31. Rosenfeld, A., Hummel, R.A., Zucker, S.W.: Scene labeling by relaxation operations. IEEE Trans. Systems, Man and Cybernetics **6** (1976) 420–433
32. Asker, M., Derin, H.: A recursive algorithm for the Bayes solution of the smoothing problem. IEEE Trans. Automatic Control **26** (1981) 558–561
33. Derin, H., Elliot, H., Cristi, R., Geman, D.: Bayes smoothing algorithms for segmentation of binary images modeled by markov random fields. IEEE Trans. Pattern Analysis and Machine Intelligence **6** (1984) 707–720
34. Geman, S., Geman, D.: Stochastic relaxation, Gibbs distribution, and the Bayesian restoration of images. IEEE Trans. Pattern Analysis and Machine Intelligence **6** (1984) 707–720
35. Kohonen, T.: Self-organization and Associative Memory. Springer, New York (1989)
36. Pao, Y.H.: Adaptive Pattern Recognition and Neural Networks. Addison-Wesley, New York (1989)
37. Babaguchi, N., Yamada, K., Kise, K., Tezuka, T.: Connectionist model binarization. In: Proc. 10th ICPR. (1990) 51–56
38. Blanz, W.E., Gish, S.L.: A connectionist classifier architecture applied to image segmentation. In: Proc. 10th ICPR. (1990) 272–277
39. Chen, C.T., Tsao, E.C., Lin, W.C.: Medical image segmentation by a constraint satisfaction neural network. IEEE Trans. Nuclear Science **38** (1991) 678–686
40. Ghosh, A., Pal, N.R., Pal, S.K.: Image segmentation using neural networks. Biological Cybernetics **66** (1991) 151–158
41. Ghosh, A., Pal, N.R., Pal, S.K.: Neural network, Gibbs distribution and object extraction. In Vidyasagar, M., Trivedi, M., eds.: Intelligent Robotics, New Delhi (1991) 95–106
42. Ghosh, A., Pal, N.R., Pal, S.K.: Object background classification using hopfield type neural network. Int. J. Pattern Recognition and Artificial Intelligence **6** (1992) 989–1008
43. Kuntimad, G., Ranganath, H.S.: Perfect image segmentation using pulse coupled neural networks. IEEE Trans. Neural Networks **10** (1999) 591–598
44. Manjunath, B.S., Simchony, T., Chellappa, R.: Stochastic and deterministic network for texture segmentation. IEEE Trans. Acoustics Speech Signal Processing **38** (1990) 1039–1049
45. Eckhorn, R., Reitboeck, H.J., Arndt, M., Dicke, P.: Feature linking via synchronization among distributed assemblies: Simulation of results from cat cortex. Neural Computation **2** (1990) 293–307
46. Ghosh, S., Ghosh, A.: A GA-FUZZY approach to evolve hopfield type optimum networks for object extraction. In Pal, N.R., Sugeno, M., eds.: Advances in Soft Computing - AFSS 2002: Int. Conf. on Fuzzy Systems. Volume LNCS-2275., Springer-Verlag, Germany (2003) 444–449
47. Pal, S.K., De, S., Ghosh, A.: Designing hopfield type networks using genetic algorithms and its comparison with simulated annealing. Int. J. Pattern Recognition and Artificial Intelligence **11** (1997) 447–461
48. Jiang, Y., Zhou, Z.: SOM ensemble-based image segmentation. Neural Processing Letters **20** (2004) 171–178
49. Gonzalez, R.C., Wintz, P.: Digital Image Processing. Addison-Wesley, Reading, Massachusetts (1987)
50. Davis, L.S.: A survey of edge detection techniques. Computer Graphics and Image Processing **4** (1975) 248–270

51. Kundu, M.K., Pal, S.K.: Thresholding for edge detection using human psychovisual phenomena. Pattern Recognition Letters **4** (1986) 433–441
52. Haddon, J.F.: Generalized threshold selection for edge detection. Pattern Recognition **21** (1988) 195–203
53. Zadeh, L.A.: Fuzzy sets. Information and Control **8** (1965) 338–353
54. Prewitt, J.M.S.: Object enhancement and extraction. In Lipkin, B.S., Rosenfeld, A., eds.: Picture Processing and Psycho-Pictorics, Academic Press, New York (1970) 75–149
55. Pal, S.K., King, R.A.: Image enhancement using fuzzy sets. Electronic Letters **16** (1980) 376–378
56. Pal, S.K., Rosenfeld, A.: Image enhancement and thresholding by optimization of fuzzy compactness. Pattern Recognition Letters **7** (1988) 77–86
57. Murthy, C.A., Pal, S.K.: Fuzzy thresholding: Mathematical framework, bound functions and weighted moving average technique. Pattern Recognition Letters **11** (1990) 197–206
58. Pal, S.K., Dasgupta, A.: Spectral fuzzy sets and soft thresholding. Information Sciences **65** (1992) 65–97
59. Xie, W.X., Bedrosian, S.D.: Experimentally driven fuzzy membership function for gray level images. J. Franklin Institute **325** (1988) 154–164
60. Tobias, O.J., Seara, R.: Image segmentation by histogram thresholding using fuzzy sets. IEEE Trans. Image Processing **12** (2002) 1457–1465
61. Bezdek, J.C.: Pattern Recognition with Fuzzy Objective Function Algorithms. Plenum Press, New York (1981)
62. Hall, L.O., Bensaid, A.M., Clarke, L.P., Velthuizen, R.P., Silbiger, M., Bezdek, J.C.: A comparison of neural network and fuzzy clustering techniques in segmenting magnetic resonance images of the brain. IEEE Trans. Neural Networks **3** (1992) 672–681
63. Huntsberger, T.L., Jacobs, C.L., Cannon, R.L.: Iterative fuzzy image segmentation. Pattern Recognition **18** (1985) 131–138
64. Trivedi, M.M., Bezdek, J.C.: Low-level segmentation of aerial images with fuzzy clustering. IEEE Trans. Systems, Man and Cybernetics **16** (1986) 589–598
65. Cannon, R.L., Dave, J.V., Bezdek, J.C.: Efficient implementation of fuzzy c-means clustering algorithms. IEEE Trans. Pattern Analysis and Machine Intelligence **8** (1986) 248–255
66. Keller, J.M., Carpenter, C.L.: Image segmentation in presence of uncertainty. Int. J. Intelligent Systems **5** (1990) 193–208
67. Couprie, M., Najman, L., Bertrand, G.: Quasi-linear algorithms for the topological watershed. J. Mathematical Imaging and Vision **22** (2005) 231–249
68. Meyer, F.: Topographic distance and watershed lines. Signal Processing **38** (1994) 113–125
69. Patras, I., Lagendijk, R.L., Hendriks, E.A.: Video segmentation by MAP labeling of watershed segments. IEEE Trans. Pattern Analysis and Machine Intelligence **23** (2001) 326–332
70. Bhanu, B., Fonder, S.: Functional template-based SAR image segmentation. Pattern Recognition **37** (2004) 61–77
71. Bhanu, B., Lee, S.: Genetic Learning for Adaptive Image Segmentation. Kluwer Academic Publishers, Norwell, MA, USA (1994)
72. Acharyya, M., De, R.K., Kundu, M.K.: Segmentation of remotely sensed images using wavelets features and their evaluation in soft computing framework. IEEE Trans. Geoscience and Remote Sensing **41** (2003) 2900–2905

73. Heiler, M., Schnörr, C.: Natural image statistics for natural image segmentation. Int. J. Computer Vision **63** (2005) 5–19

74. Ho, S., Bullitt, E., Gerig, G.: Level-set evolution with region competition: Automatic 3-D segmentation of brain tumors. In: ICPR '02: Proc. 16th Int. Conf. on Pattern Recognition. Volume 1., Washington, DC, USA (2002) 532–535

75. Kervrann, C., Trubuil, A.: Optimal level curves and global minimizers of cost functionals in image segmentation. J. Mathematical Imaging **17** (2002) 153–174

76. Lin, P., Zheng, C., Yang, Y., Gu, J.: Statistical model based on level set method for image segmentation. In: Proc. 4th Int. Conf. on Computer and Information Technology (CIT'04), Washington, DC, USA (2004) 143–148

77. Pal, S.K., Ghosh, A., Kundu, M.K., eds.: Soft Computing for Image Processing. Physica-Verlag, Heidelberg, Germany (2000)

78. Zadeh, L.A.: Toward a theory of fuzzy information granulation and its centrality in human reasoning and fuzzy logic. Fuzzy Sets and Systems **90** (1997) 111–127

79. Ghosh, A., Pal, N.R., Pal, S.K.: Self-organization for object extraction using multilayer neural network and fuzziness measures. IEEE Trans. Fuzzy Systems **1** (1993) 54–68

80. Ohlander, R.B.: Analysis of natural scenes. PhD thesis, Department of Computer Science, Carnegie Mellon University, Pittsburgh, Pennsylvania (1975)

81. Overheim, R.D., Wagner, D.L.: Light and Color. Wiley, New York (1982)

82. Cheng, H.D., Jiang, X.H., Sun, Y., Wang, J.: Color image segmentation: Advances and prospects. Pattern Recognition **34** (2001) 2259–2281

83. Naik, S.K., Murthy, C.A.: Standardization of edge magnitude in color images. IEEE Trans. Image Processing **Communicated** (2005)

84. Naik, S.K., Murthy, C.A.: Distinct multi-colored region descriptors for object recognition. IEEE Trans. Pattern Analysis and Machine Intelligence **Communicated** (2005)

85. Levine, M.D., Nazif, A.M.: An experimental rule based system for testing low level segmentation strategy. In Uhr, L., Preston, K., eds.: Multi-Computer Architectures and Image Processing: Algorithms and Programs, Academic Press, New York, Also available as Report No. 81-6, Department of Electrical Engineering, McGill University, June 1981 (1982)

86. Lim, Y.W., Lee, S.U.: On the color image segmentation algorithm based on the thresholding and fuzzy c-means techniques. Pattern Recognition **23** (1990) 935–952

87. Pal, N.R., Bhandari, D.: Object background classification: Some new techniques. Signal Processing **33** (1993) 139–158

88. Brink, A.B.: Gray level thresholding of images using a correlation criterion. Pattern Recognition Letters **9** (1989) 335–341

89. Pal, N.R., Pal, S.K.: A review on image segmentation techniques. Pattern Recognition **26** (1993) 1277–1294

90. Sezgin, M., Sankur, B.: Survey over image thresholding techniques and quantitative performance evaluation. J. Electronic Imaging **13** (2004) 146–165

91. Egmont-Petersen, M., de Ridder, D., Handels, H.: Image processing with neural networks - a review. Pattern Recognition **35** (2002) 2279–2301

92. Freixenet, J., Munoz, X., Raba, D., Marti, J., Cufi, X.: Yet another survey on image segmentation: Region and boundary information integration. In: ECCV '02: Proc. 7th European Conf. on Computer Vision-Part III, London, UK (2002) 408–422

93. Fu, K.S., Mui, J.K.: A survey on image segmentation. Pattern Recognition **13** (1981) 3–16

94. Haralick, R.M., Shapiro, L.G.: Survey, image segmentation techniques. Computer Vision, Graphics and Image Processing **29** (1985) 100–132
95. Karmakar, G.C., Dooley, L., Syed, M.R.: Review of fuzzy image segmentation techniques. In: Design and management of multimedia information systems: Opportunities and challenges, Hershey, PA, USA (2001) 282–314
96. Sahoo, P.K., Soltani, S., Wong, A.K.C., Chen, Y.C.: A survey of thresholding techniques. Computer Vision, Graphics and Image Processing **41** (1988) 233–260
97. Zhang, Y.J.: A survey on evaluation methods for image segmentation. Pattern Recognition **29** (1996) 1335–1346
98. Mather, P.M.: Computer Processing of Remotely-Sensed Images: An Introduction. John Wily & Sons Ltd, England (1999)
99. Richards, J.A., Jia, X.: Remote Sensing Digital Image Analysis: An Introduction. 3rd edn. Springer, Germany (1999)
100. Schowengerdt, R.A.: Remote Sensing: Models and Methods for Image Processing. 2nd edn. Academic Press, San Diego, CA, USA (1997)
101. Thiruvengadachari, S., Kalpana, A.R., by: S. Adiga, R., Sreenivasi, M.: IRS Data Users Handbook (Revision 1), NRSA Data Centre, Dept. of Space, Govt. of India (1989)
102. Swain, P.H., Davis, S.M., eds.: Remote Sensing: The Quantitative Approach. McGraw Hill Inc., New York, USA (1978)
103. Bauer, M.E., Cipra, J.: Identification of agricultural crops by computer processing of ERTS MSS data. In: Symposium on Significant Results Obtained from ERTS-1,, NASA Document no. SP-327, Washington, DC (1973) 205–212
104. Kettig, R.L., Landgrebe, D.A.: Computer classification of remotely sensed multispectral image data by extraction and classification of homogeneous objects. IEEE Trans. Geoscience Electronics **GE-14** (1976) 19–26
105. Lee, C., Landgrebe, D.A.: Fast multistage likelihood classification. IEEE Trans. Geoscience and Remote Sensing **29** (1991) 509–517
106. Sun, W., Heidt, V., Gong, P., Xu, G.: Information fusion for rural land-use classification with high-resolution satellite imagery. IEEE Trans. Geoscience and Remote Sensing **41** (2003) 883–890
107. Bandyopadhyay, S., Pal, S.K.: Pixel classification using variable string genetic algorithms with chromosome differentiation. IEEE Trans. Geoscience and Remote Sensing **29** (2001) 303–308
108. Bischof, H., Schneider, W., Pinz, A.J.: Multispectral classification of landsat-images using neural networks. IEEE Trans. Geoscience and Remote Sensing **30** (1992) 482–490
109. Erbek, S.F., Ozkan, C., Taberner, M.: Comparison of maximum likelihood classification method with supervised artificial neural network algorithms for land use activities. Int. J. Remote Sensing **25** (2004) 1733–1748
110. Pal, M., Mather, P.M.: Assessment of the effectiveness of support vector machines for hyperspectral data. Future Generation Computer Systems **20** (2004) 1215–1225
111. Murthy, C.A., Chatterjee, N., Shankar, B.U., Majumder, D.D.: IRS image segmentation: Minimum distance classifier approach. In: Proc. 11th ICPR, The Hague, The Netherlands (1992) 781–784
112. Wacker, A.G., Landgrebe, D.A.: Minimum distance classification in remote sensing. In: First Canadian Symposium on Remote Sensing, Ottawa, Canada, Also available as LARS Technical Note 030772 (25 pages), Purdue University, Lafayette Indiana (1972)

113. Khazenie, N., Crawford, M.M.: Spatial-temporal autocorrelated model for contextual classification. IEEE Trans. Geoscience and Remote Sensing **28** (1990) 529–539

114. Li, F., Peng, J.: Double random field models for remote sensing image segmentation. Pattern Recognition Letters **25** (2004) 129–139

115. Jhung, Y., Swain, P.H.: Bayesian contextual classification based on modified M-estimates and markov random fields. IEEE Trans. Geoscience and Remote Sensing **34** (1996) 67–75

116. Gong, P., Howarth, P.J.: Performance analysis of probabilistic relaxation methods for land-cover classification. Remote Sensing of Environment **30** (1989) 33–42

117. Richards, J.A., Landgrebe, D.A., Swain, P.: Pixel labeling by supervised probabilistic relaxation. IEEE Trans. Pattern Analysis and Machine Intelligence **3** (1981) 188–191

118. Solberg, A.H.S., Taxt, T., Jain, A.K.: A markov random field model for classification of multisource satellite imagery. IEEE Trans. Geoscience and Remote Sensing **34** (1996) 100–113

119. Swain, P.H., Hauska, H.: The decision tree classifier: Design and potential. IEEE Trans. Geoscience Electronics **15** (1977) 142–147

120. Pal, M., Mather, P.M.: An assessment of the effectiveness of decision tree methods for land cover classification. Remote Sensing of Environment **86** (2003) 554–565

121. Parui, S.K., Shanka, B.U., Dutta, A., Majumder, D.D.: Unsupervised classification of Indian remote sensing satellite imagery. In: Proc. ICAPRDT, Indian Statistical Institute, Calcutta (1993) 68–74

122. Ho, S., Lee, K.: Design and analysis of an efficient evolutionary image segmentation algorithm. J. VLSI Signal Processing Systems **35** (2003) 29–42

123. Sahasrabudhe, S.C., Dasgupta, K.S.: A valley-seeking threshold selection technique. In Shapiro, L., Rosenfeld, A., eds.: Computer Vision and Image Processing: CVIP92, Academic Press, Boston (1992) 55–65

124. Laprade, R.H.: Split-and-merge segmentation of aerial photographs. Computer Vision, Graphics and Image Processing **44** (1988) 77–86

125. Baraldi, A., Parmiggiani, F.: Single linkage region growing algorithms based on the vector degree of match. IEEE Trans. Geoscience and Remote Sensing **34** (1996) 137–148

126. Pal, S.K., Mitra, P.: Multispectral image segmentation using the rough-set-initialized EM algorithm. IEEE Trans. Geoscience and Remote Sensing **40** (2002) 2495–2501

127. Shah, C.A., Arora, M.K., Robila, S.A., Varshney, P.K.: ICA mixture model based unsupervised classification of hyperspectral imagery. In: Proc. 31st Applied Image Pattern Recognition Workshop (AIPR 2002), From Color to Hyperspectral: Advancements in Spectral Imagery Exploitation, Washington, DC, USA (2002) 29–35

128. Myers, V.I.: Remote sensing applications in agriculture. In Colwell, J.E., Colwell, R.N., eds.: Manual of Remote Sensing, American Society of Photogrammetry, Falls Church, VA. (1983) 2111–2228

129. Vaiopoulos, D., Skianis, G.A., Nikolakopoulos, K.: The contribution of probability theory in assessing the efficiency of two frequently used vegetation indices. Int. J. Remote Sensing **25** (2004) 4219–4236

130. McFeeters, S.K.: The use of the normalized difference water index in the delineation of open water features. Int. J. Remote Sensing **17** (1996) 1425–1432

131. Wang, F.: Fuzzy supervised classification of remote sensing images. IEEE Trans. Geoscience and Remote Sensing **28** (1990) 194–201

132. Melgani, F., Hashemy, B.A.R.A., Taha, S.M.R.: An explicit fuzzy supervised classification method for multispectral remote sensing images. IEEE Trans. Geoscience and Remote Sensing **38** (2000) 287–295
133. Mandal, D.P., Murthy, C.A., Pal, S.K.: Utility of multiple choices is detecting ill-defined roadlike structures. Fuzzy Sets and Systems **64** (1994) 213–228
134. Pal, S.K., Murthy, C.A., Shankar, B.U.: Pixel classification in remotely sensed images using shape estimation with fuzzy sets. In Mardia, K.V., Gill, C.A., Dryden, I.L., eds.: Proc. Image Fusion and Shape Variability Techniques, Leeds, U. K. (1996) 141–145
135. Cannon, R.L., Dave, J.V., Bezdek, J.C., Trivedi, M.: Segmentation of a thematic mapper image using the fuzzy c-means clustering algorithm. IEEE Trans. Geoscience and Remote Sensing **24** (1986) 400–408
136. Shankar, B.U., Pal, N.R.: FFCM: An effective approach for large data sets. In: Proc. 3rd Int. Conf. on Fuzzy Logic, Neural nets and Soft Computing, Iizuka, Japan (1994) 331–332
137. Maulik, U., Bandyopadhyay, S.: Fuzzy partitioning using real coded variable length genetic algorithm for pixel classification. IEEE Trans. Geosciences and Remote Sensing **41** (2003) 1075–1081
138. Mecocci, A., Gamba, P., Marazzi, A., Barni, M.: Texture segmentation in remote sensing images by means of packet wavelets and fuzzy clustering. In Franceschetti, G., Oliver, C.J., Shiue, J.C., Tajbakhsh, S., eds.: Proc. SPIE: Synthetic Aperture Radar and Passive Microwave Sensing. Volume 2584. (1995) 142–151
139. Lorette, A., Descombes, X., Zerubia, J.: Texture analysis through a markovian modelling and fuzzy classification: Application to urban area extraction from satellite images. Int. J. Computer Vision **36** (2000) 221–236
140. Baraldi, A., Parmiggiani., F.: Neural network for unsupervised categorization of multivalued input patterns: An application to satellite image clustering. IEEE Trans. Geoscience and Remote Sensing **33** (1990) 305–316
141. Benediktsson, J.A., Swain, P.H., Ersoy, O.K.: Neural network approach versus statistical methods in classification of multisource remote sensing data. IEEE Trans. Geoscience and Remote Sensing **28** (1990) 540–552
142. Benediktsson, J.A., Swain, P.H., Ersoy, O.K.: Conjugate - gradient neural networks in classification of multisource and very-high-dimension remote sensing data. Int. J. Remote Sensing **14** (1993) 2883–2903
143. Decatur, S.E.: Application of neural network to terrain classification. In: Proc. IJCNN'89, vol. I, Washington DC, USA (1989) 283–288
144. Lee, J., Weger, R.C., Sengupta, S.K., Welch, R.M.: A neural network approach to cloud classification. IEEE Trans. Geoscience and Remote Sensing **28** (1990) 846–855
145. Liu, Z., Liu, A., Wang, C., Niu, Z.: Evolving neural network using real coded genetic algorithm (GA) for multispectral image classification. Future Generation Computer Systems **20** (2004) 1119–1129
146. Villmann, T., Merényi, E.: Extensions and modifications of the Kohenen-SOM and applications in remote sensing image analysis. In: Self-Organizing neural networks: Recent advances and applications, Springer-Verlag, New York, USA (2002) 121–144
147. Villmann, T., Merényi, E., Hammer, B.: Neural maps in remote sensing image analysis. Neural Networks **16** (2003) 389–403
148. Xue, X., Zhang, Y., Zhao, R., Duan, F., Chen, Y.: A new method of SAR image segmentation based on neural network. In: ICCIMA '03: Proc. 5th Int. Conf. on

Computational Intelligence and Multimedia Applications, Washington, DC, USA (2003) 149–153

149. Mandal, D.P., Murthy, C.A., Pal, S.K.: Analysis of IRS imagery for detecting man-made objects with a multivalued recognition system. IEEE Trans. Systems, Man and Cybernetics, Part A **26** (1996) 241–247

150. Ton, J.: A Knowledge Based Approach for Landsat Image Interpretation. PhD thesis, Michigan State University, Michigan, USA (1988)

151. Ton, J., Sticklen, J., Jain, A.K.: Knowledge-based segmentation of landsat images. IEEE Trans. Geoscience and Remote Sensing **29** (1991) 222–232

152. Pal, S.K., Bandyopadhyay, S., Murthy, C.A.: Genetic classifiers and remotely sensed images: Comparison with standard method. Int. J. Remote Sensing **22** (2001) 2445–2569

153. Bandyopadhyay, S., Maulik, U.: Genetic clustering for automatic evolution of clusters and application to image classification. Pattern Recognition **35** (2002) 1197–1208

154. Xie, X.L., Beni, G.: A validity measure for fuzzy clustering. IEEE Trans. Pattern Analysis and Machine Intelligence **13** (1991) 841–847

155. Brown, M., Lewis, H.G., Gunn, S.R.: Linear spectral mixture models and support vector machine for remote sensing. IEEE Trans. Geoscience and Remote Sensing **38** (2000) 2346–2360

156. Huang, C., Davis, L.S., Townshend, J.R.G.: An assessment of support vector machine for land cover classification. Int. J. Remote Sensing **23** (2002) 725–749

157. Varshney, P.K., Arora, M.K.: Advanced Image Processing Techniques for Remote Sensed Hyperspectral Data. Springer, Germany (2004)

158. Kundu, M.K., Acharyya, M.: M-band wavelets: Application to texture segmentation for real life images analysis. Int. J. Wavelets, Multiresolution and information Processing **1** (2003) 115–149

159. Lindsay, R.W., Percival, D.B., Rothrock, D.A.: The discrete wavelet transform and the scale analysis of the surface properties of sea ice. IEEE Trans. Geoscience and Remote Sensing **34** (1996) 771–787

160. Niedermeier, A., Romaneessen, E., Lehner, S.: Detection of coastlines in SAR images using wavelet methods. IEEE Trans. Geoscience and Remote Sensing **36** (2000) 2270–2281

161. Parui, S.K., Shankar, B.U., Mukherjee, A., Majumder, D.D.: A parallel algorithm for detection of linear structures in satellite images. Pattern Recognition Letters **12** (1991) 765–770

162. Hu, J., Sakoda, B., Pavlidis, T.: Interactive road finding for aerial images. In: Proc. IEEE Workshop on Applications of Computer Vision. (1992) 56–63

163. Zlotnick, A., Carnine, Jr., P.D.: Finding road seeds in aerial images. CVGIP: Image Understanding **57** (1993) 243–260

164. Barzohar, M., Cooper, D.B.: Automatic finding of main roads in aerial images by using geometric-stochastic models and estimation. In: IEEE Computer Society Conf. on Computer Vision and Pattern Recognition. (1993) 459–464

165. Geman, D., Jedynak, B.: An active testing model for tracking roads in satellite images. IEEE Trans. Pattern Analysis and Machine Intelligence **18** (1996) 1–14

166. Gruen, A., Li, H.: Semi-automatic linear feature extraction by dynamic programming and LSB-Snakes. Photogrammetric Engineering and Remote Sensing **63** (1997) 985–995

167. Park, S.R., Kim, T.: Semi-automatic road extraction algorithm from IKONOS images using template matching. In: Proc. 22nd Asian Conf. on remote Sensing. (2001) 1209–1213

168. Stoica, R., Descombes, X., Zerubia, J.: A Gibbs point process for road extraction from remotely sensed images. Int. J. Computer Vision **57** (2004) 121–136
169. Mena, J.B.: State of the art on automatic road extraction for GIS update: A novel classification. Pattern Recognition Letters **24** (2003) 3037–3058
170. Mitra, P., Shankar, B.U., Pal, S.K.: Active support vector machines for pixel classification in remote sensing images. In: Proc. 1st Indian International Conference on Artificial Intelligence, IICAI-03. (2003) 543–553
171. Mitra, P., Shankar, B.U., Pal, S.K.: Segmentation of multispectral remote sensing images using active support vector machines. Pattern Recognition Letters **25** (2004) 1067–1074
172. Pal, S.K., Ghosh, A., Shankar, B.U.: Segmentation of remotely sensed images with fuzzy thresholding, and quantitative evaluation. Int. J. Remote Sensing **21** (2000) 2269– 2300
173. Shankar, B.U., Ghosh, A., Pal, S.K.: On fuzzy thresholding of remotely sensed images. In Pal, S.K., Ghosh, A., Kundu, M.K., eds.: Soft Computing for image processing, Physica-Verlag, Heidelberg, Germany (2000) 130–161
174. Pal, S.K., Shankar, B.U., Mitra, P.: Granular computing, rough entropy and object extraction. Pattern Recognition Letters **26** (2005) 2509–2517
175. Shankar, B.U., Murthy, C.A., Pal, S.K.: A new gray level based Hough transform for region extraction: An application to IRS images. Pattern Recognition Letters **19** (1998) 197–204
176. Angluin, D.: Queries and concept learning. Machine Learning **2** (1988) 319–342
177. Cohn, D., Atlas, L., Ladner, R.: Improving generalization with active learning. Machine Learning **15** (1994) 201–221
178. Pal, S.K., Mitra, P.: Pattern Recognition Algorithms for Data Mining: Scalability, Knowledge Discovery and Soft Granular Computing. Chapman & Hall/CRC, Boca Raton, Florida, USA (2004)
179. Campbell, C., Cristianini, N., Smola, A.: Query learning with large margin classifiers. In: Proc. 17th Int. Conf. on Machine Learning, Stanford, CA (2000) 111–118
180. Mitra, P., Murthy, C.A., Pal, S.K.: Data condensation in large databases by incremental learning with support vector machines. In: Proc. Int. Conf. on Pattern Recognition (ICPR2000), Barcelona, Spain (2000) 712–715
181. Tong, S., Koller, D.: Support vector machine active learning with application to text classification. J. Machine Learning Research **2** (2001) 45–66
182. Schohn, G., Cohn, D.: Less is more: Active learning with support vector machines. In: Proc. 17th Int. Conf. on Machine Learning, Stanford, CA (2000) 839–846
183. Vapnik, V.: Statistical Learning Theory. Wiley, New York (1998)
184. Burges, C.J.C.: A tutorial on support vector machines for pattern recognition. Data Mining and Knowledge Discovery **2** (1998) 1–47
185. Pal, S.K., Ghosh, A.: Fuzzy geometry in image analysis. Fuzzy Sets and Systems **48** (1992) 23–40
186. Pal, S.K., Ghosh, A.: Image segmentation using fuzzy correlation. Information Sciences **62** (1992) 223–250
187. Shannon, C.E.: A mathematical theory of communication. Bell System Technical Journal **27** (1948) 379–423
188. Bezdek, J.C., Pal, S.K., eds.: Fuzzy Models for Pattern Recognition: Methods that Search for Structures in Data. IEEE Press, New York (1992)
189. Pal, S.K., Majumder, D.D.: Fuzzy Mathematical Approach to Pattern Recognition. John Wiley, Halsted Press, New York (1986)
190. Murthy, C.A., Pal, S.K.: Bounds for membership function: Correlation based approach. Information Sciences **65** (1992) 143–171

191. Murthy, C.A., Pal, S.K.: Histogram thresholding by minimizing gray level fuzziness. Information Sciences **60** (1992) 107–135
192. Pal, S.K.: A note on the quantitative measure of image-enhancement through fuzziness. IEEE Trans. Pattern Analysis and Machine Intelligence **4** (1982) 204–208
193. Ghosh, A.: Use of fuzziness measures in layered networks for object extraction: A generalization. Fuzzy Sets and Systems **72** (1995) 331–348
194. Pal, N.R., Pal, S.K.: Entropy: A new definition and its applications. IEEE Trans. Systems, Man and Cybernetics **21** (1991) 1260–1270
195. Murthy, C.A., Pal, S.K., Majumder, D.D.: Correlation between two fuzzy membership functions. Fuzzy Sets and Systems **7** (1985) 23–38
196. Anderberg, M.R.: Cluster Analysis for Applications. Academic Press, New York (1973)
197. Pawlak, Z.: Rough Sets, Theoretical Aspects of Reasoning about Data. Kluwer Academic, Dordrecht (1991)
198. Skowron, A., Rauszer, C.: The discernibility matrices and functions in information systems. In Slowiński, R., ed.: Intelligent Decision Support, Handbook of Applications and Advances of the Rough Sets Theory, Kluwer Academic, Dordrecht (1992) 331–362
199. Komorouski, J., Pawlak, Z., Polkowski, L., Skowron, A.: Rough sets: A tutorial. In Pal, S.K., Skowron, A., eds.: Rough Fuzzy Hybridization: A New Trend In Decision-Making, Springer, Singapore (1999) 3–98
200. Wojcik, Z.: Rough approximation of shapes in pattern recognition. Computer Vision, Graphics and Image Processing **40** (1987) 228–249
201. Pal, S.K.: Fuzzy image processing and recognition: Uncertainties handling and applications. Int. J. Image and Graphics **1** (2001) 69–195
202. Hough, P.V.C.: A method and means for recognizing complex patterns. Technical report, (U. S. Patent 3069654) (1962)
203. Risse, T.: Hough transform for line recognition: Complexity of evidence accumulation and cluster detection. Computer Vision, Graphics and Image Processing **46** (1989) 327–345
204. Duda, R.O., Hart, P.E.: Use of the Hough transform to detect lines and curves in pictures. Communications ACM **15** (1972) 11–15
205. Lo, R., Tsai, W.: Gray-scale Hough transform for thick line detection in gray-scale images. Pattern Recognition **28** (1995) 647–661
206. Basak, J., Pal, S.K.: Theoretical quantification of shape distortion in fuzzy Hough transform. Fuzzy Sets and Systems **154** (2005) 227–250
207. Ballard, D.H.: Generalizing the Hough transform to detect arbitrary shapes. Pattern Recognition **13** (1981) 111–122
208. Duda, R.O., Hart, P.E.: *Pattern Classification and Scene Analysis*. Wiley, New York. (1973)
209. Beaubouef, T., Petry, F.E., Arora, G.: Information measure for rough and fuzzy sets and application to uncertainty in relational databases. In Pal, S.K., Skowron, A., eds.: Rough Fuzzy Hybridization: A new trend in decision-making, Springer, Singapore (1999) 200–214
210. Duntsch, I., Gediga, G.: Uncertainty measures of rough set prediction. Artificial Intelligence **106** (1998) 109–137

# Author Index

# Lecture Notes in Computer Science

For information about Vols. 1–4331

please contact your bookseller or Springer

Vol. 4380: S. Spaccapietra, P. Atzeni, F. Fages, M.-S. Hacid, M. Kifer, J. Mylopoulos, B. Pernici, P. Shvaiko, J. Trujillo, I. Zaihrayeu (Eds.), Journal on Data Semantics VIII. XV, 219 pages. 2007.

Vol. 4378: I. Virbitskaite, A. Voronkov (Eds.), Perspectives of Systems Informatics. XIV, 496 pages. 2007.

Vol. 4377: M. Abe (Ed.), Topics in Cryptology – CT-RSA 2007. XI, 403 pages. 2006.

Vol. 4376: E. Frachtenberg, U. Schwiegelshohn (Eds.), Job Scheduling Strategies for Parallel Processing. VII, 257 pages. 2007.

Vol. 4374: J.F. Peters, A. Skowron, I. Düntsch, J. Grzymała-Busse, E. Orłowska, L. Polkowski (Eds.), Transactions on Rough Sets VI, Part I. XII, 499 pages. 2007.

Vol. 4373: K. Langendoen, T. Voigt (Eds.), Wireless Sensor Networks. XIII, 358 pages. 2007.

Vol. 4372: M. Kaufmann, D. Wagner (Eds.), Graph Drawing. XIV, 454 pages. 2007.

Vol. 4371: K. Inoue, K. Satoh, F. Toni (Eds.), Computational Logic in Multi-Agent Systems. X, 315 pages. 2007. (Sublibrary LNAI).

Vol. 4370: P.P Lévy, B. Le Grand, F. Poulet, M. Soto, L. Darago, L. Toubiana, J.-F. Vibert (Eds.), Pixelization Paradigm. XV, 279 pages. 2007.

Vol. 4369: M. Umeda, A. Wolf, O. Bartenstein, U. Geske, D. Seipel, O. Takata (Eds.), Declarative Programming for Knowledge Management. X, 229 pages. 2006. (Sublibrary LNAI).

Vol. 4368: T. Erlebach, C. Kaklamanis (Eds.), Approximation and Online Algorithms. X, 345 pages. 2007.

Vol. 4367: K. De Bosschere, D. Kaeli, P. Stenström, D. Whalley, T. Ungerer (Eds.), High Performance Embedded Architectures and Compilers. XI, 307 pages. 2007.

Vol. 4366: K. Tuyls, R. Westra, Y. Saeys, A. Nowé (Eds.), Knowledge Discovery and Emergent Complexity in Bioinformatics. IX, 183 pages. 2007. (Sublibrary LNBI).

Vol. 4364: T. Kühne (Ed.), Models in Software Engineering. XI, 332 pages. 2007.

Vol. 4362: J. van Leeuwen, G.F. Italiano, W. van der Hoek, C. Meinel, H. Sack, F. Plášil (Eds.), SOFSEM 2007: Theory and Practice of Computer Science. XXI, 937 pages. 2007.

Vol. 4361: H.J. Hoogeboom, G. Păun, G. Rozenberg, A. Salomaa (Eds.), Membrane Computing. IX, 555 pages. 2006.

Vol. 4360: W. Dubitzky, A. Schuster, P.M.A. Sloot, M. Schroeder, M. Romberg (Eds.), Distributed, High-Performance and Grid Computing in Computational Biology. X, 192 pages. 2007. (Sublibrary LNBI).

Vol. 4358: R. Vidal, A. Heyden, Y. Ma (Eds.), Dynamical Vision. IX, 329 pages. 2007.

Vol. 4357: L. Buttyán, V. Gligor, D. Westhoff (Eds.), Security and Privacy in Ad-Hoc and Sensor Networks. X, 193 pages. 2006.

Vol. 4355: J. Julliand, O. Kouchnarenko (Eds.), B 2007: Formal Specification and Development in B. XIII, 293 pages. 2006.

Vol. 4354: M. Hanus (Ed.), Practical Aspects of Declarative Languages. X, 335 pages. 2006.

Vol. 4353: T. Schwentick, D. Suciu (Eds.), Database Theory – ICDT 2007. XI, 419 pages. 2006.

Vol. 4352: T.-J. Cham, J. Cai, C. Dorai, D. Rajan, T.-S. Chua, L.-T. Chia (Eds.), Advances in Multimedia Modeling, Part II. XVIII, 743 pages. 2006.

Vol. 4351: T.-J. Cham, J. Cai, C. Dorai, D. Rajan, T.-S. Chua, L.-T. Chia (Eds.), Advances in Multimedia Modeling, Part I. XIX, 797 pages. 2006.

Vol. 4349: B. Cook, A. Podelski (Eds.), Verification, Model Checking, and Abstract Interpretation. XI, 395 pages. 2007.

Vol. 4348: S.T. Taft, R.A. Duff, R.L. Brukardt, E. Ploedereder, P. Leroy (Eds.), Ada 2005 Reference Manual. XXII, 765 pages. 2006.

Vol. 4347: J. Lopez (Ed.), Critical Information Infrastructures Security. X, 286 pages. 2006.

Vol. 4346: L. Brim, B. Haverkort, M. Leucker, J. van de Pol (Eds.), Formal Methods: Applications and Technology. X, 363 pages. 2007.

Vol. 4345: N. Maglaveras, I. Chouvarda, V. Koutkias, R. Brause (Eds.), Biological and Medical Data Analysis. XIII, 496 pages. 2006. (Sublibrary LNBI).

Vol. 4344: V. Gruhn, F. Oquendo (Eds.), Software Architecture. X, 245 pages. 2006.

Vol. 4342: H. de Swart, E. Orłowska, G. Schmidt, M. Roubens (Eds.), Theory and Applications of Relational Structures as Knowledge Instruments II. X, 373 pages. 2006. (Sublibrary LNAI).

Vol. 4341: P.Q. Nguyen (Ed.), Progress in Cryptology - VIETCRYPT 2006. XI, 385 pages. 2006.

Vol. 4340: R. Prodan, T. Fahringer, Grid Computing. XXIII, 317 pages. 2007.

Vol. 4339: E. Ayguadé, G. Baumgartner, J. Ramanujam, P. Sadayappan (Eds.), Languages and Compilers for Parallel Computing. XI, 476 pages. 2006.

Vol. 4338: P. Kalra, S. Peleg (Eds.), Computer Vision, Graphics and Image Processing. XV, 965 pages. 2006.

Vol. 4337: S. Arun-Kumar, N. Garg (Eds.), FSTTCS 2006: Foundations of Software Technology and Theoretical Computer Science. XIII, 430 pages. 2006.

Vol. 4336: V.R. Basili, D. Rombach, K. Schneider, B. Kitchenham, D. Pfahl, R.W. Selby, Empirical Software Engineering Issues. XVII, 193 pages. 2007.

Vol. 4335: S.A. Brueckner, S. Hassas, M. Jelasity, D. Yamins (Eds.), Engineering Self-Organising Systems. XII, 212 pages. 2007. (Sublibrary LNAI).

Vol. 4334: B. Beckert, R. Hähnle, P.H. Schmitt (Eds.), Verification of Object-Oriented Software. XXIX, 658 pages. 2007. (Sublibrary LNAI).

Vol. 4333: U. Reimer, D. Karagiannis (Eds.), Practical Aspects of Knowledge Management. XII, 338 pages. 2006. (Sublibrary LNAI).

Vol. 4332: A. Bagchi, V. Atluri (Eds.), Information Systems Security. XV, 382 pages. 2006.